Number Theory and
Its Applications II

Herrn Professor Dr. Christopher Deninger zu seinem 60sten Geburtstag
gewidmet mit Respekt und freundlichen Grüssen

Number Theory and Its Applications II

Hailong Li
Weinan Normal University, China

Fuhuo Li
Sanmenxia SuDa Transportation Energy Saving Technology Co., Ltd., China

Nianliang Wang
Shangluo University, China

Shigeru Kanemitsu
Kyushu Institute of Technology, Japan

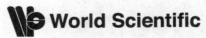 **World Scientific**

NEW JERSEY · LONDON · SINGAPORE · BEIJING · SHANGHAI · HONG KONG · TAIPEI · CHENNAI

Published by

World Scientific Publishing Co. Pte. Ltd.

5 Toh Tuck Link, Singapore 596224

USA office: 27 Warren Street, Suite 401-402, Hackensack, NJ 07601

UK office: 57 Shelton Street, Covent Garden, London WC2H 9HE

Library of Congress Cataloging-in-Publication Data

Names: Li, Hailong (Mathematician), author. | Li, Fuhuo, author. |
    Wang, Nianliang, author. | Kanemitsu, Shigeru, author.
Title: Number theory and its applications II / by Hailong Li (Weinan Normal University, China),
    Fuhuo Li (Sanmenxia SuDa Transportation Energy Saving Technology Co., Ltd., China),
    Nianliang Wang (Shangluo University, China),
    Shigeru Kanemitsu (Kyushu Institute of Technology, Japan).
Description: New Jersey : World Scientific, 2017. | Sequel to: Number Theory and Its Applications.
    Hackensack, N.J. : World Scientific, 2013. | Includes bibliographical references and index.
Identifiers: LCCN 2017053355 | ISBN 9789813231597 (hardcover : alk. paper)
Subjects: LCSH: Number theory--Textbooks. | Algebraic number theory--Textbooks.
Classification: LCC QA241 .L58594 2017 | DDC 512.7--dc23
LC record available at https://lccn.loc.gov/2017053355

British Library Cataloguing-in-Publication Data
A catalogue record for this book is available from the British Library.

For any available supplementary material, please visit
http://www.worldscientific.com/worldscibooks/10.1142/10753#t=suppl

Printed in Singapore

Preface

This Part II (hereafter abbreviated as **II**) of "Number theory and its applications" is not an almost independent book that complements Part I (**I**) [LiwK (2012)] in a similar way as **II** of "Vistas of special functions" [Chakraborty, Kanemitsu and Tsukada (2009)] does for Part **I** [Kanemitsu (2007)]. That is, each chapter of **II** corresponds more or less to the counterpart chapter of **I** and incorporates new material while keeping the record, and making free use, of material developed in **I**. This **II** aims as another important mission at becoming a prelude to the three more detailed coming books [Kitaoka], [Liu] and [LiFH]. Chapters 1 and 2 correspond to [Kitaoka] on algebraic number theory, Chapter 5 to [Liu] on Kuznetsov sum formula and Chapter 6 to [LiFH] on control theory and the use of operators.

In recent times more and more practical applications of number theory have been made [Burr (1991)], [Hejhal (1999)], etc. A glance at the contents of these books immediately give the range of applications which are graph-theoretical, chaotic and physical as has been the case in the last decades including an apparent and direct application to coding theory and pseudo-random number generations.

Applications of mathematics in general and number theory in particular to other disciplines are made possible through some underlying principles including classification, symmetry, nested or networked structure etc. in addition to their transparency and perfection.

We recall the first passage in the book [Davis and Hersh (1986)].

"In November 1619, René Descartes, a twenty-three-year-old Frenchman, dreamed of a world unified by mathematics, a world in which all intellectual matters could be dealt with rationality by logical computation. 18 years since then he wrote the most famous book 'Discourse'."

In [Kotre (1995), p. 25] the author refers to the "great Cartesian Theater," meaning apparently the notion of reason as described by "Cogito ergo sum."

We understand the preceding as the power of mathematics as liquidating block-head rigidity and perfection of computers.

Speaking of Descartes, his mathematically most important invention is the idea of coordinates, known as **Cartesian product**. This concept appears in various places and can unify the scattered notions quite well.

One example is the following. In [Conrey and Gosh (1993), p. 44], in relation to **homeostasis**, a chemo-dynamical stability of a cell in a variable environment, a description is made of the equilibrium after replication. The cell is brought out of a stable regime, to become dynamically speaking an *attractor*, and the stability after replication is regained in a different framework: the dynamical features of the system reappear in the *multiplicity* of nearly identical cells represented by the *Cartesian product* of many copies of the attractor, and ensures the information by perpetuation.

We illustrate the above conflict "logical computation vs. music of reason = math." by digital music. It is often said, especially by audience which has trained musical ears, that music played by digital devices sound rather **flat** and even boring compared with live performance. There is a good reason for that. In converting analogue signal into digital signal, the A/D transformer is used, which makes quantization and approximates the sampled signal at digital levels. This **rounding-off** of analogue signals give them a sort of **unpleasant perfection**. A good example is a karaoke estimator which would give bad marks for those professional singers who can get the audience moved, on the ground that they don't sing as in the scores. We shall come to this chopping-off problem in Chapter 6.

Not only in music but in flower arrangement, it is said that what is important is rather the blank space surrounding the arranged flowers. We recall a poem of R. Kinoshita:

Peony flowers
So stable and in full bloom
The solidness of the position
The flowers occupy.

Also "Praise of Shadows" by J. Tanizaki explores the Japanese sense of beauty—the subtle interplay of **shade and light** in several important aspects of Japanese life. That is, it puts stress on what is missing.

–"As is true with all artifacts, completed ones are less interesting; those which are unfinished and left at that are more appealing and give liveliness." from *An hermite's miscellany* by K. Urabe.

Here of course, "an hermite's" is a play-on-words of math. flavor. For the word "hermit" is naturally associated with the name Hermite and since in French, the 'h' is not pronounced, the indefinite article is to be 'an'. K. Urabe is more well-known as Kenko Yoshida.

–Hans describes the crystal of snow as "This is too regular. A living thing cannot be so symmetric. Things with such perfection smell of death. The Creator, it seems, has designed all that exists in his world in a fashion slightly deviated from perfect symmetry"–*Der Zauberberg* by Thomas Mann.

Thus digital art smells of death.

We say a few words about mathematical principles that appear in applications.

By way of classification many things can be made simpler, a typical example being the square integrable functions (in Lebesgue's sense) which are not ordinary functions but equivalence classes of functions modulo their values at measure 0 set.

Symmetry has been a main guiding principle of many cultural activities including art, architecture, design of some devices, etc. Their applications to engineering disciplines may be also made effectively through the use of operators in Hilbert-like spaces.

We expose rudiments of the theory of Hilbert spaces in Chapter 5 to that extent which enables the reader to follow the argument involving them. This is a further step forward from linear algebra whose aspects are exhibited in Chapter 1, **II** and matches that theory of N. P. Romanoff which, together with an independent work of A. Wintner, is the first instance of "Hilbert space and number theory."

Another mathematical structure appearing in many systems is a directed graph, or a **network** structure consisting of vertices (nodes) and edges (sides) which may assume on various meanings according to the system. For example, in [Conrey and Gosh (1993), p. 27] a cell is thought of as a chemical machine in constant interaction with the environment and vertices are marked by various chemical compounds and the edges represent chemical transformations. The direction of the flow in the graph indicates the predominant reactions which are taking in nutrients and oxygen from outside, transforming them into energy to fuel cellular machinery. They are organized as the above graph structure known as **metabolic pathways**.

A human brain is a complicated network in which connections are realized by neurons, i.e. edges are wireless signals. This suggests the most advanced EV2GG (Electric Vehicle to Green Grid) system which is a network of a special tree type such that edges are wireless signals and the vertices are the main control center at the root with many branch stations which are responsible for surveying and controlling the networked EVs (which are the vertices at the end) moving in their respective regions. In the smart grid project in progress the grid may depend on nuclear power plant but our proposal is the **Green Grid** in which electricity generation depends mostly on natural resources, mainly on solar generation including wind and hydro-generation. In the hitherto suggested system each EV is in grid-connected mode (or otherwise stand-alone mode) when it parks at a charging station and there ancillary services will be given and the driver can buy or sell the electricity. But for our future EV2GG system, as soon as an EV starts moving, it gets wirelessly connected to the GG and it will be under constant surveillance of GG which will give a fail-safe guard. This and related topics will be included in Chapter 6, **II**.

In electrical engineering there is a heavy use of impulse function, which is the Dirac delta function. This is also the case in control theory in Chapter 6, **I**. To make rigorous study of electric circuits, it is inevitable to treat the Dirac delta-function as a generalized function [Rosenfel'd and Yahinson (1966)].

Chapter 6, **I** was devoted to the basics of control theory as compared with the number-theoretic mean square averages, which is rather unique. Prior to this, there were two trends of applications of mathematics to control theory, i.e. application of functional analysis cf. e.g. [Helton (1982)], [Helton (1998)] and that of algebraic geometry [Falb (1990)], [Falb (1999)].

There we remarked that there is great similarity in control theory and number theory in their treatment of the signals in time domain (t) and frequency domain (ω) or an expanded frequency domain $(s = \sigma + j\omega)$ which is conducted by the Laplace transform in the case of control theory while in the theory of zeta-functions, this role is played by the Mellin transform, both of which are special cases of the Fourier transforms and convert the signals in time domain to those in the right half-plane (expanded frequency domain), cf. e.g. [Kanemitsu (2007), Chapter 7]. This remark is quite natural from our higher-stand point that *all phenomena are avatars of the niryana (core) and that all are waves and thus Fourier analysis comes in at the fundamental stage.* For example, in the study of linear systems, in place of a state space equation in time domain, one considers the transfer function (which is the ratio of the Laplace transforms of input and output signals) in the frequency domain. By the visualization principle, in its state space representation, it is to be embedded in the state equation. It is true in general that analysis of signals in frequency domain is far easier than that in time domain, and it is the governing procedure to study the signals in frequency domain and pull back the results to those in time domain. This idea applies to many problems in life, taking averages and drawing out a general tendency which one applies to analysis of general phenomena including anomalies or singularities.

domain	positive real axis	right-half plane	upper half-plane
name	time domain	expanded frequency domain	Lobachevski plane
variable	$t > 0$	$\sigma = \operatorname{Re} s > 0$	$\operatorname{Im} \tau > 0$
mapping	$t \leftrightarrow \frac{1}{t}$	$s \leftrightarrow 1 - s$	$\tau \leftrightarrow -\frac{1}{\tau}$
group	$\mathbb{R}_+^\times$	symmetry: $s \leftrightarrow 1 - s$	$\Gamma \subset \operatorname{SL}_2(\mathbb{R})$

Table 1. Domains and group structure

Here the variables are connected by

$$\sigma + j\omega = s = -i\tau. \tag{0.1}$$

More precisely, basic results in Chapter 1, **I**, "Elements of algebra" will be employed rather freely in Chapter 1, **II**, "Linear-algebraic approach to algebraic number theory." Chapter 1, **II** substitutes part of Chapter 2, **I** and some other parts are replaced by Chapter 2, **II** "Group-theoretic aspects of algebraic number theory" in which we state the basics of Galois theory and Artin L-functions. Then we add new material "pseudo-random number generators" based on intractable problems in zeta-functions. Chapter 5, **I**, "Around Dirichlet L-functions" is included as §12.2.

In Chapter 6, **II** we shall state basics of "number-theoretic" control theory with a new addition of "control theory in the unit disc" as a counterpart of Chapter 4, **II**. In these we are concerned with boundary behavior of functions analytic in the unit disc or in the upper half-plane and one can see a close interplay between various disciplines.

The authors would like to express their hearty thanks to Dr. J. Mehta for his help in processing the data into a complete form and to Ms. Kwong Lai Fun for her devoted help and constant encouragement throughout the period of writing of this Part II.

Contents

Chapter 1

Linear algebraic approach to algebraic number theory

Abstract: The purpose of this chapter is to abridge a big gap between elementary number theory and algebraic number theory with the emphasis on similarity of the material with that of linear algebra. This is also a prelude to the theory of Hilbert spaces in Chapter 5. In elementary number theory what is essential is the unique factorization property of the ring (or rather an integral domain) $\mathbb{Z}$ of (rational) integers. Algebraically speaking, this is because $\mathbb{Z}$ is first of all a Euclidean domain. Then there is a theorem (Theorem 1.5, **I**) to the effect that a Euclidean domain is a PID (Principal Ideal Domain) in which all the ideals are principal. Another theorem (Exercise 16, **I**) asserts that a PID is a UFD (Unique Factorization Domain) in which every element is expressed as a product of irreducible elements up to units and order of irreducible elements. One often refers to irreducible elements as prime elements since they are the same in UFD. Since $\mathbb{Z}$ is an integral domain, it has the quotient field $S^{-1}\mathbb{Z}$, where $S = \mathbb{Z} \setminus \{0\}$ is a multiplicatively closed subset. It is the rational number field $\mathbb{Q}$ of characteristic 0.

One of the driving forces has been an essential generalization of the concept of integers. In algebraic number theory, this is achieved by introducing algebraic integers. Any root of a monic polynomial with integer coefficients $\in \mathbb{Z}[x]$ is called an **algebraic integer**. Any integer $a \in \mathbb{Z}$ being the root of the polynomial $x - a \in \mathbb{Z}[x]$ is an algebraic integer, and is sometimes referred to as a rational integer.

We assemble some most general results (Theorems 1.11, 1.12) on f.g. (finitely generated) modules over a Dedekind domain, which is meant for submodules of the integer ring over the one in the underlying algebraic number field. In most of the books, mention is made only to the case of f.g. modules over a PID. In Chapters 1 and 2, sparing no pains, we state special cases of the theorems according to each context for the readers' convenience. Theorem 1.6 is a special case of the corresponding theorem in §2.1, which in turn is a special case of §1.4. All these are more or less generalizations of the well-known fact that every vector space has a basis and so any subspace, being a vector space itself, has a basis.

In Chapters 1 and 2 we change the notation $\mathcal{O}$ for the ring of algebraic integers to $\mathfrak{o}$.

1

1.1 Algebraic preliminaries

1.1.1 *Vector spaces*

We may also speak of a vector subspace generated by a subset S of a vector space V over F which consists of all linear combinations of elements of S with F-coefficients: $FS = \{\sum a_i s_i | a_i \in F, s_i \in S\}$, where all sums are finite. Indeed, whatever an algebraic system is, the principle is the same and the subsystem is defined as the intersection of all subsystems that contain a given set S, which is the smallest subsystem that contains S.

Example 1.1. Any additive abelian group V may be thought of as a $\mathbb{Z}$-module by defining the action by $m \in \mathbb{Z}$ as the sum of m a's for $m > 0$ and $(-m)$ $(-a)$'s for $m < 0$, with $0a = 0$. Hence a $\mathbb{Z}$-module is nothing other than an additive abelian group and we refer to a module. In the next subsection we consider finitely generated abelian groups ($\mathbb{Z}$-modules). In algebraic number theory, there are many modules appearing including the ring of integers (1.15) which is a free $\mathbb{Z}$-module and all ideals (which are also free). The module M in the proof of Theorem 1.5 is a finitely generated module.

Remark 1.1. The most often used fact from linear algebra is the theorem on homogeneous equations. Let A be a square matrix with entries in a ring R (which we always assume commutative). Consider the homogeneous system of equations

$$xA = o. \tag{1.1}$$

If the equation has a non-zero solution, then $\det A = 0$. This is often used in a context such as

$$\alpha(w_1, \cdots, w_n) = (w_1, \cdots, w_n)A,$$

where α is an element of an extension ring of R, A is a square matrix with entries in R, and $(w_1, \cdots, w_n)$ is a non-zero vector (often a basis). We refer to this fact as a **theorem on homogeneous equation**.

1.1.2 *Integral domains*

Definition 1.1. Cf. definition 1.3, **I**. Let R be a commutative ring. A subset I of R is called an **ideal** of R if

1. I is a submodule of R.
2. If $a \in I$, then $ra \in I$ holds for every $r \in R$.

Example 1.2.

1. For a commutative ring R, $\{0\}$ and R itself are ideals.
2. Every submodule of $\mathbb{Z}$ is an ideal of $\mathbb{Z}$.
3. Let R be a commutative ring and $a_1, \cdots, a_n \in R$. Then
$$(a_1, \cdots, a_n) = \{r_1 a_1 + \cdots + r_n a_n \mid r_1, \cdots, r_n \in R\}$$
is an ideal of R called the ideal generated by $\{a_1, \cdots, a_n\}$. When $n = 1$,
$$(a) = \{ra \mid r \in R\}$$
is called a **principal ideal**.

Definition 1.2. A commutative ring with unity is called an **integral domain** or sometimes just a domain if it has no zero-divisors. We often refer to them as rings.

By definition, an ideal $I \subset R$ is a prime ideal if and only if the residue class ring R/I is an integral domain and *a fortiori* a field is a domain.

Exercise 1. Let R be a commutative ring and let $= R \setminus \{0\}$. In analogy to the construction of fractions from integers, we introduce the relation between two elements $(a_1, s_1), (a_2, s_2) \in S \times R$
$$(a_1, s_1) \sim (a_2, s_2) \iff a_1 s_2 = a_2 s_1. \tag{1.2}$$
Show that this is an equivalence relation.

Definition 1.3. Let R, S be as in the above Exercise. We denote the equivalence class containing (a, s) by a/s. The set $S^{-1}R$ of all equivalence classes a/s $(a \in R, s \in S)$ is called the **quotient field** of R (or the field of fractions).

Theorem 1.1. *A Euclidean domain is a PID. A PID is a **unique factorization domain** abbreviated UFD, in which every element may be uniquely decomposed as the product of irreducible elements up to units and order of irreducible elements.*

Corollary 1.1. *The ring of rational integers and the polynomial ring over a field are Euclidean, so that they are PID and in turn UFD.*

By Corollary 1.1, every polynomial $f(x)$ of $\deg f(x) \geq 1$ is uniquely decomposed as
$$f(x) = a p_1(x)^{n_1} \cdots p_m(x)^{n_m}$$
where $a \in F^\times$, $p_1(x), \cdots, p_m(x)$ are monic irreducible mutually distinct polynomials, and $n_1, \cdots, n_m$ are natural numbers. Here a and the set $\{p_1(x), \cdots, p_m(x)\}$ and multiplicities n_k are uniquely determined by $f(x)$.

1.1.3   *Field extensions*

Let K, F be fields. If $K \supset F$, then K is called an **extension** of F, and F is called a **subfield** of K. Then K is a vector space over F with sum and product in K. If K is finite-dimensional as a vector space over F, K is called a **finite extension** of F, and the dimension of K as a vector space over F is called the **extension degree** of K over F and is denoted by

$$[K : F].$$

Proposition 1.1. *Let L, K, F be fields and suppose that K is a finite extension of F and L is a finite extension of K. Then L is a finite extension of F, and $[L : F] = [L : K][K : F]$.*

Proof. Let $\{v_1, \cdots, v_m\}$ be a basis of L as a vector space over K, and similarly $\{w_1, \cdots, w_n\}$ a basis of K over F. By definition, we have $m = [L : K]$ and $n = [K : F]$. We may show that $\{v_i w_j \mid 1 \leq i \leq m, 1 \leq j \leq n\}$ is a basis of L over F.   □

Let K be an extension of a field F and $a \in K$. If there is a non-zero polynomial $f(x)$ over F, i.e. $f(x) \in F[x]$ such that $f(a) = 0$, then a is called **algebraic** over F.

Let K be an extension of a field F and suppose that $a \in K$ is algebraic over F. Then

$$I = \{g(x) \in F[x] \mid g(a) = 0\} \ (\neq \{0\}) \tag{1.3}$$

is an ideal of $F[x]$, and so there is a polynomial $f(x)$ such that $I = (f(x))$ by Proposition 1.13, **I**. It is irreducible, because $g_1(a)g_2(a) = 0$ $(g_1(x), g_2(x) \in F[x])$ implies $g_1(a) = 0$ or $g_2(a) = 0$. We may assume that it is monic and then it is uniquely determined. The monic irreducible polynomial $f(x)$ is called the **minimal polynomial** of a over F denoted $\mathrm{Irr}(a, F, x)$.

The minimal polynomial $f(x)$ divides any polynomial $g(x) \in F[x]$ such that $g(a) = 0$, and it is characterized by the property that $f(x)$ is the monic polynomial in $F[x]$ of the least degree such that $f(a) = 0$.

We note that if $f(x) \in F[x]$ is a monic irreducible polynomial and $a \in K$ is a root of $f(x) = 0$, then $f(x)$ is the minimal polynomial of the algebraic element a over F. Hence $f(x)$ is also the minimal polynomial over F for all roots of $f(x) = 0$ in a field containing F.

If every element of an extension K of F is algebraic over F, then K is called an **algebraic** extension of F.

Proposition 1.2. *If K/F is a finite extension, then K is algebraic over*

F. I.e. for every element $\alpha \in K$ let

$$A(\alpha) = (a_{ij}), \tag{1.4}$$

which is to be defined by (1.6) in the proof.

  Then α is a root of the polynomial

$$p(x) = p_\alpha(x) = \det(x 1_n - A(\alpha)) \in F[x] \tag{1.5}$$

*called the **characteristic polynomial** of α with respect to K/F (cf. (2.7)). This is independent of the choice of the basis of K over F.*

Proof. Let $\{w_1, \cdots, w_n\}$ be a basis of K over F. For $\alpha \in K$, we define $a_{ij} \in F$ by

$$\alpha w_j = \sum_{i=1}^{n} a_{ij} w_i,$$

which is equivalent to

$$\alpha(w_1, \cdots, w_n) = (w_1, \cdots, w_n) A(\alpha). \tag{1.6}$$

  For this cf. Remark 1.2 below. By the theorem on homogeneous equation, we conclude that α is a root of the polynomial $p(x)$ which is a non-zero polynomial in $F[x]$ of degree n. Thus α is algebraic over F.

  To prove that $p(x)$ is independent of a choice of a basis, we let $\{w'_1, \cdots, w'_n\}$ be another basis of K over F. Then there is a regular matrix $B = (b_{ij})$ $(b_{ij} \in F)$ such that

$$(w'_1, \cdots, w'_n) = (w_1, \cdots, w_n) B,$$

and so we have

$$B^{-1} \alpha 1_n B(w'_1, \cdots, w'_n) = \alpha(w'_1, \cdots, w'_n) = \alpha(w_1, \cdots, w_n) B$$
$$= (w_1, \cdots, w_n) A(\alpha) B = (w'_1, \cdots, w'_n) B^{-1} A(\alpha) B,$$

whence the independence follows. $\qquad\qquad\square$

Remark 1.2. The correspondence $\alpha \mapsto A(\alpha)$ in (1.4) is called the **regular representation** of K/F. It depends on the choice of a basis. Viewing $A(\alpha)$ as a given matrix, the equation (1.5) is the eigen-equation for eigen-values of $A(\alpha)$, whence the name arises. The constant term and the coefficient of the x^{n-1} of the characteristic polynomial $p(x)$ which are elements in F and are independent of the choice of a basis, play a very eminent role in algebraic number theory, we give them special names in the following definition.

Definition 1.4. Cf. Definition 2.2. We put

$$N_{K/F}(\alpha) = \det A(\alpha), \ Tr_{K/F}(\alpha) = tr A(\alpha), \tag{1.7}$$

and call them the **norm** and the **trace** of α from K to F, respectively.

It is easy to see that

$$A(\alpha\beta) = A(\alpha)A(\beta), \ A(\alpha + \beta) = A(\alpha) + A(\beta) \quad (\alpha, \beta \in K). \tag{1.8}$$

Thus from (1.8) we have

Proposition 1.3. *Let K be a finite extension of F of degree n. Let $\alpha, \beta \in K$. Then we have*

$$N_{K/F}(\alpha\beta) = N_{K/F}(\alpha)N_{K/F}(\beta), \ Tr_{K/F}(\alpha + \beta) = Tr_{K/F}(\alpha) + Tr_{K/F}(\beta)$$

and

$$N_{K/F}(a) = a^n, \ Tr_{K/F}(a) = na \ \text{if} \ a \in F.$$

This is continued on to §1.1.4.

Proposition 1.4. *Let K be an extension of a field F, and suppose that $\alpha, \beta \in K$ are algebraic over F. Then $\alpha \pm \beta, \alpha\beta, \alpha/\beta$ ($\beta \neq 0$) are also algebraic.*

Proof. We may assume $\alpha \neq 0, \beta \neq 0$. Let $f(x) = x^m + a_{m-1}x^{m-1} + \cdots + a_0$, $g(x) = x^n + b_{n-1}x^{n-1} + \cdots + b_0 \in F[x]$ be the minimal polynomials of α, β over F, respectively. Hence we have

$$\alpha^m = -a_{m-1}\alpha^{m-1} - \cdots - a_0, \ \beta^n = -b_{n-1}\beta^{n-1} - \cdots - b_0.$$

and so the subspace U spanned by $\alpha^i\beta^j$ ($i \geq 0, j \geq 0$) over F in K is spanned by $\alpha^i\beta^j$ ($0 \leq i < m, 0 \leq j < n$). It is clear that $\gamma U \subset U$ for $\gamma = \alpha \pm \beta, \alpha\beta$. Hence for a vector

$$\boldsymbol{u} = (1, \alpha, \cdots, \alpha^{m-1}, \cdots, 1 \cdot \beta^{n-1}, \alpha\beta^{n-1}, \cdots, \alpha^{m-1}\beta^{n-1}),$$

we have $\boldsymbol{u}\gamma = \boldsymbol{u}(c_{ij})$ for $^{\exists}c_{ij} \in F$, although c_{ij} may be not unique. Therefore γ is a root of the characteristic polynomial of (c_{ij}), and so γ is algebraic over F. The algebraicity of β^{-1} follows from $\beta^{-n}g(\beta) = 1 + b_{n-1}(\beta^{-1}) + \cdots + b_0(\beta^{-1})^n = 0$. Applying the above to the product α/β of α an β^{-1}, α/β is algebraic over F. $\square$

Corollary 1.2. *Denote by $\overline{\mathbb{Q}}$ the set of all algebraic elements in $\mathbb{C}$. Then $\overline{\mathbb{Q}}$ is a field, called the **algebraic closure** of $\mathbb{Q}$.*

For a polynomial $f(x) = \sum_{i=0}^{n} a_i x^i$ with integer coefficients a_i, the greatest common divisor $(a_0, a_1, \cdots, a_n)$ is called the **content** of $f(x)$, and if it is 1, i.e. $a_0, \cdots, a_n$ are pairwise relatively prime, then $f(x)$ is called **primitive**.

Lemma 1.1. *If $g(x), h(x)$ are primitive, then $g(x)h(x)$ is also primitive.*

Proof. Obviously all the coefficients of $g(x)h(x)$ are integers. Suppose that $g(x)h(x)$ is not primitive and thence that there is a prime number p such that the content is divisible by p. Put $g(x) = \sum_{i=0}^{m} a_i x^i$, $h(x) = \sum_{j=0}^{m} b_j x^j$, and put $i_0 = \min\{i \mid a_i \not\equiv 0 \bmod p\}$, $j_0 = \min\{j \mid b_j \not\equiv 0 \bmod p\}$. Then the coefficient $c_{i_0+j_0}$ of $x^{i_0+j_0}$ of $g(x)h(x)$ is equal to

$$\sum_{i+j=i_0+j_0} a_i b_j = \sum_{i+j=i_0+j_0, 0 \leq i < i_0} a_i b_j + \sum_{i+j=i_0+j_0, i>i_0} a_i b_j + a_{i_0} b_{j_0}.$$

In the first term on the right-hand side, a_i is divisible by p, in the middle term b_j is divisible by p since $j < j_0$ and the last term is not divisible by p. Thus $c_{i_0+j_0}$ is not divisible by p which contradicts the assumption. $\square$

Lemma 1.2. *Let $f(x) = x^n + a_{n-1}x^{n-1} + \cdots + a_0$ $(^{\forall}a_i \in \mathbb{Z})$ be a product of polynomials $g(x) = \sum_{i=0}^{r} b_i x^i$, $h(x) = \sum_{i=0}^{s} c_i x^i$. We assume that all coefficients b_i, c_i are rational numbers and $b_r = 1$. Then we have $^{\forall}b_i$, $^{\forall}c_i \in \mathbb{Z}$.*

Proof. It is easy to see that $c_s = 1$. Take positive rational numbers α, β such that $\alpha g(x), \beta h(x)$ are primitive; then all the coefficients of $\alpha g(x)$ and $\beta h(x)$ are integers, in particular $\alpha = \alpha b_r, \beta = \beta c_s \in \mathbb{Z}$. Since $\alpha g(x) \cdot \beta h(x)$ is primitive by Lemma 1.1, and in the identity $\alpha \beta f(x) = \alpha g(x) \cdot \beta h(x)$, three polynomials $f(x), \alpha g(x), \beta h(x)$ are primitive, we have $\alpha \beta = 1$ and so $\alpha = \beta = 1$, whence the assertion follows. $\square$

Theorem 1.2 (Eisenstein). *Let p be a prime number and let $f(x) = a_n x^n + a_{n-1}x^{n-1} + \cdots + a_1 x + a_0$ be a polynomial with integer coefficients a_i such that*

$$p \nmid a_n, \ p \mid a_i \ (0 \leq i \leq n-1) \text{ and } p^2 \nmid a_0.$$

Then $f(x)$ is irreducible over $\mathbb{Q}$, i.e. is not a product of (non-constant) polynomials with rational coefficients.

Proof. We note that a polynomial with rational coefficients is a multiple of a primitive polynomial by a rational number. Suppose that $f(x)$ is reducible over $\mathbb{Q}$; then $f(x) = ag(x)h(x)$ where $a \in \mathbb{Q}$ and $g(x), h(x)$ are

primitive. Put $r = \deg g(x) \geq 1, s = \deg h(x) \geq 1$. By Lemma 1.1, $g(x)h(x)$
is primitive, and so a is an integer, by comparing contents of both sides of
$f(x) = ag(x)h(x)$. Put $g(x) = \sum_{i=0}^{r} b_i x^i$, $h(x) = \sum_{j=0}^{s} c_j x^j$ $(b_i, c_j \in \mathbb{Z})$.
By $ab_r c_s = a_n \not\equiv 0 \bmod p$, we get $a \not\equiv 0 \bmod p$. Hence taking into account
that $ab_0 c_0 = a_0 \equiv 0 \bmod p, \not\equiv 0 \bmod p^2$, we may assume without loss of
generality $b_0 \equiv 0 \bmod p$ and $c_0 \not\equiv 0 \bmod p$. Then putting

$$i_0 = \min\{i \mid b_i \not\equiv 0 \bmod p\},$$

the primitiveness of $g(x)$ implies $0 < i_0 \leq r < n$. Then $a_{i_0} = a\sum_{i+j=i_0} b_i c_j = a\{b_{i_0}c_0 + \sum_{i+j=i_0, i<i_0} b_i c_j\} \equiv ab_{i_0}c_0 \bmod p$, whence we
get $a_{i_0} \not\equiv 0 \bmod p$, which contradicts the assumption $p \mid a_i$ for $i < n$. □

Corollary 1.3. *For a prime number p, the polynomial $f(x) = x^{p-1} + x^{p-2} + \cdots + x + 1$ is irreducible over $\mathbb{Q}$.*

Proof. Since $f(x) = (x^p - 1)/(x - 1)$, $f(y+1) = ((y+1)^p - 1)/y = \sum_{k=0}^{p-1} a_k y^k$, where the coefficients a_k satisfy $a_k = \binom{p}{k+1} \equiv 0 \bmod p$ for
$0 \leq k \leq p-2$ and $a_0 = p \not\equiv 0 \bmod p^2$. Thus $f(y+1)$ is irreducible and so
$f(x)$ is irreducible over $\mathbb{Q}$. □

Example 1.3. The irreducible polynomials $x^2 + 1$ and $x^2 + x + 1$ yield the
Gaussian field $\mathbb{Q}(i)$ and the Eisensteinian field $\mathbb{Q}(\rho)$, respectively. These
are examples of cyclotomic fields as well as those of quadratic fields.

Theorem 1.3. (Existence of a primitive element) *A finite separable exten-
sion is a simple extension, i.e. if F is a subfield $\overline{\mathbb{Q}}$ and $K (\subset \overline{\mathbb{Q}})$ is a finite
extension of F, then there is an element $\gamma \in K$ such that $K = F(\gamma)$.*

If the characteristic of a field F is positive, an irreducible polynomial
of $F[x]$ may have multiple roots. However, in case that the characteristic
of F is 0, no irreducible polynomial $f(x)$ has any multiple roots in $\overline{F}[x]$,
in view of $(f(x), f'(x)) = 1$. Once we admit the existence of an algebraic
closure, the argument below are true for a field of characteristic 0.

1.1.4 *Norm and trace*

This subsection is a continuation from the passage preceding Proposition
1.3 and continues on to the last half of §2.2.

Proposition 1.5. *Let K be a finite extension of F and $a \in K$. Put*
$$t = [K : F], m = [F(a) : F] \text{ and } n = [K : F(a)].$$
Let $f(x) = x^m + a_{m-1}x^{m-1} + \cdots + a_0$ be the minimal polynomial of a over F. Then we have
$$N_{K/F}(a) = ((-1)^m a_0)^n, \ Tr_{K/F}(a) = (-a_{m-1})n.$$
Moreover, we have
$$\det(x1_t - A(a)) = f(x)^n,$$
where $A(a)$ is defined by (1.4).

Proof. Let $\{w_1, \cdots, w_n\}$ be a basis of K over $F(a)$; noting that $\{1, a, \cdots, a^{m-1}\}$ is a basis of $F(a)/F$ (cf. Theorem 1.3), we see
$$\{1 \cdot w_1, \cdots, a^{m-1}w_1, \cdots, 1 \cdot w_n, \cdots, a^{m-1}w_n\}$$
is a basis of K over F (cf. Proposition 1.1). Putting
$$A = \begin{pmatrix} 0 & 0 & \cdots & 0 & -a_0 \\ 1 & 0 & & 0 & -a_1 \\ & & \ddots & & \\ 0 & \cdots & 1 & 0 & -a_{m-2} \\ 0 & \cdots & 0 & 1 & -a_{m-1} \end{pmatrix},$$
we have $a(1, \cdots, a^{m-1}) = (1, \cdots, a^{m-1})A$, and so
$$a(1 \cdot w_1, \cdots, a^{m-1}w_1, \cdots, 1 \cdot w_n, \cdots, a^{m-1}w_n)$$
$$= (1 \cdot w_1, \cdots, a^{m-1}w_1, \cdots, 1 \cdot w_n, \cdots, a^{m-1}w_n) \begin{pmatrix} A & & 0 \\ & \ddots & \\ 0 & & A \end{pmatrix},$$
whence we see $N_{K/F}(a) = (\det A)^n = ((-1)^m a_0)^n$ and $Tr_{K/F}(a) = n(tr A) = n(-a_{m-1})$.

For the above choice of a basis of K over F, we have
$$\det(x1_t - A(a)) = \det(x1_m - A)^n,$$
and
$$\det(x1_m - A) = \begin{vmatrix} x & 0 & \cdots & 0 & a_0 \\ -1 & x & & 0 & a_1 \\ & & \ddots & & \\ 0 & \cdots & -1 & x & a_{m-2} \\ 0 & \cdots & 0 & -1 & x + a_{m-1} \end{vmatrix} = f(x),$$
expanding with respect to the mth column. $\qquad\square$

Corollary 1.4. *If $K = F(a)$ is a finite extension, the characteristic polynomial and the minimal polynomial of a with respect to K/F coincide.*

Proof. The assertion follows easily from the assumption $n = 1$.   □

Proposition 1.6. *Let $K (\subset \overline{\mathbb{Q}})$ be a finite extension of F and $a \in K$. Let $\{\sigma_1, \cdots, \sigma_k\}$ be all isomorphisms over F from K into $\overline{\mathbb{Q}}$. Then we have*

$$N_{K/F}(a) = \prod_{i=1}^{k} a^{\sigma_i} \text{ and } Tr_{K/F}(a) = \sum_{i=1}^{k} a^{\sigma_i}.$$

Proof. Let $f(x) = x^m + a_{m-1}x^{m-1} + \cdots + a_0$ be the minimal polynomial of a over F, and $f(x) = \prod_{i=1}^{m}(x - \alpha_i)$ in $\overline{\mathbb{Q}}[x]$. Then $a \mapsto \alpha_i$ induces an isomorphism η_i over F from $F(a)$ into $\overline{\mathbb{Q}}$. Denote an extension of η_i to an isomorphism from K into $\overline{\mathbb{Q}}$ by the same letter η_i (cf. Exercise 18, p. 29, I). If an isomorphism σ from K into $\overline{\mathbb{Q}}$ is identical to η_i on $F(a)$, then $\sigma\eta_i^{-1}$ fixes $F(a)$ element-wisely. Therefore, letting $\{\tau_1, \cdots, \tau_n\}$ $(n = [K : F(a)])$ be the set of isomorphisms over $F(a)$ from K into $\overline{\mathbb{Q}}$, we see that $\{\tau_j\eta_i\}$ is the set of isomorphisms from K into $\overline{\mathbb{Q}}$ over F. Thus we have

$$\prod_{i=1}^{k} a^{\sigma_i} = \prod_{i=1}^{m}\prod_{j=1}^{n} a^{\tau_j\eta_i} = (\prod_{i=1}^{m} a^{\eta_i})^n = (\prod_{i=1}^{m} \alpha_i)^n = ((-1)^m a_0)^n = N_{K/F}(a),$$

$$\sum_{i=1}^{k} a^{\sigma_i} = n\sum_{i=1}^{m} a^{\eta_i} = n\sum_{i=1}^{m} \alpha_i = n(-a_{m-1}) = Tr_{K/F}(a),$$

by the previous proposition.   □

Proposition 1.7. *Let $L \supset K \supset F$ be subfields of $\overline{\mathbb{Q}}$ and suppose $[L : F] < \infty$. For an element $a \in L$, we have*

$$N_{L/F}(a) = N_{K/F}(N_{L/K}(a)), \ Tr_{L/F}(a) = Tr_{K/F}(Tr_{L/K}(a)).$$

Proof. Let $\{\tau_1, \cdots, \tau_n\}$ be the set of all isomorphisms over K from L into $\overline{\mathbb{Q}}$ and let $\{\eta_1, \cdots, \eta_m\}$ be the set of all isomorphisms over F from K into $\overline{\mathbb{Q}}$ and extend them to isomorphisms from $\overline{\mathbb{Q}}$ on $\overline{\mathbb{Q}}$.

Let σ be an isomorphism over F from K into $\overline{\mathbb{Q}}$; then σ coincides with $^\exists\eta_i$ on K and so $a^\sigma = a^{\eta_i}$ for $^\forall a \in K$ and so $\sigma\eta_i^{-1}$ is the identity on K. Thus $\sigma\eta_i^{-1} = {}^\exists\tau_j$, that is $\{\tau_j\eta_i\}$ is the set of isomorphisms from L into $\overline{\mathbb{Q}}$ over F. Thus we have

$$N_{L/F}(a) = \prod_{i,j} a^{\tau_j\eta_i} = \prod_i(\prod_j a^{\tau_j})^{\eta_i} = \prod_i N_{L/K}(a)^{\eta_i} = N_{K/F}(N_{L/K}(a))$$

and similarly $Tr_{L/F}(a) = Tr_{K/F}(Tr_{L/K}(a))$.   □

1.2 Algebraic number fields

We denote by $\overline{\mathbb{Q}}$ the set of all algebraic elements $\alpha \in \mathbb{C}$ over $\mathbb{Q}$ (i.e. those for which there exists a non-zero polynomial $f(x) \in \mathbb{Q}[x]$ such that $f(\alpha) = 0$), which forms a field and is called the **algebraic closure** of $\mathbb{Q}$. Elements in $\overline{\mathbb{Q}}$ are called **algebraic numbers**.

If k is a finite extension over $\mathbb{Q}$ contained in the algebraic closure $\overline{\mathbb{Q}}$, then it is called an **algebraic number field** or simply a **number field**. By Corollary 1.5, p. 29, **I**, it is a finitely generated algebraic extension. All the integral elements over $\mathbb{Q}$ in k forms a Dedekind domain $\mathfrak{o}_k$ (also often denoted by $\mathcal{O}$ etc.) and is called the **ring of algebraic integers**. The notation originates from the term "order" ([Butzer and Stark (1986), Definition, p. 88]) (eine "Ordnung" in German). Then k is the quotient field of $\mathfrak{o}_k$. For more precise statements, cf. §1.3.

$\mathfrak{o}_k$ is a Dedekind ring and so the prime decomposition holds true in the sense of Theorem 1.10, (ii), i.e. every integral ideal $\mathfrak{a}$ may be expressed uniquely as a product of prime ideals $\mathfrak{a} = \mathfrak{p}_1^{e_1} \cdots \mathfrak{p}_1^{e_\nu}$, $0 \le e_i \in \mathbb{Z}$. Therefore we may speak of the gcd, lcm just as in the ring of rational integers, which is UFD, being Euclidean.

Let k be a number field. Since char $\mathbb{Q} = 0$, it follows from Theorem 1.20, p. 30, **I** that k is a simple extension $k = \mathbb{Q}(\theta)$. Hence we may apply Theorem 1.15, p. 28, **I**: If $f(X) = \mathrm{Irr}(\theta, X, \mathbb{Q})$ with $\deg f = n$, then $k = \mathbb{Q} \oplus \mathbb{Q}\theta \oplus \cdots \oplus \mathbb{Q}\theta^{n-1}$.

We may suppose θ has r_1 real conjugate roots $\theta^{(j)}$, $1 \le j \le r_1$ and $2r_2$ imaginary conjugate roots $\theta^{(j)}, \overline{\theta^{(j)}}$, $1 \le i \le r_2$. Here $n = r_1 + 2r_2$. Then there arise corresponding r_1 real conjugates $k^{(j)}$, $1 \le j \le r_1$ and $2r_2$ imaginary conjugates $k^{(j)}, \overline{k^{(j)}}$, $1 \le i \le r_2$. If they all coincide, then $k/\mathbb{Q}$ is a normal extension and so by Theorem 2.4, it is a Galois extension.

Lemma 1.3. *Let k be an algebraic number field. Define a bilinear form $B_k(x, y)$ on $k \times k$ by $\mathrm{Tr}(xy)$ for $x, y \in k$, where $\mathrm{Tr} = Tr_{F/\mathbb{Q}}$. If $B_k(x, y) = 0$ for all $y \in k$, then we have $x = 0$, and if $\omega_1, \cdots, \omega_n$ is a basis of k over $\mathbb{Q}$, then the matrix $(B_F(\omega_i, \omega_j))$ is regular.*

Proof. Suppose that $x \in k$ is not the zero and $B_F(x, y) = 0$ for all $y \in F$. Putting $y = 1/x$, we have the contradiction $[F : \mathbb{Q}] = Tr(1) = Tr(x \cdot 1/x) = 0$. If the rational matrix $(B_k(\omega_i, \omega_j))$ is not regular, there is a non-zero rational vector $\boldsymbol{x} = (x_1, \cdots, x_n) \in \mathbb{Q}^n$ such that

$$\boldsymbol{x}(B_F(\omega_i, \omega_j)) = (B_F(\sum x_i\omega_i, \omega_1), \cdots, B_F(\sum x_i\omega_i, \omega_n)) = (0, \cdots, 0),$$

whence $B_F(\sum x_i \omega_i, y) = 0$ for all $y \in F$. This and the former part imply the contradiction $\sum x_i \omega_i = 0$, i.e. $x_1 = \cdots = x_n = 0$. □

Lemma 1.4. *Let $\omega_1, \cdots, \omega_n$ be a basis of F over $\mathbb{Q}$. Then there is a basis $\omega_1', \cdots, \omega_n'$ of F over $\mathbb{Q}$ such that*

$$B_F(\omega_i, \omega_j') = \delta_{ij} = \begin{cases} 1 & \text{if } i = j, \\ 0 & \text{otherwise.} \end{cases}$$

Proof. Put $M = (B_F(\omega_i, \omega_j))$ and $(\omega_1', \cdots, \omega_n') = (\omega_1, \cdots, \omega_n)M^{-1}$. Writing $M^{-1} = (m_{ij})$, we have

$$B_F(\omega_i, \omega_j') = B_F(\omega_i, \sum_k \omega_k m_{kj}) = \sum_k B_F(\omega_i, \omega_k)m_{kj} = \delta_{ij}.$$

If $\sum_j a_j \omega_j' = 0$ for $a_j \in \mathbb{Q}$, then we have $a_i = B_F(\omega_i, \sum_j a_j \omega_j') = 0$. Thus $\omega_1', \cdots, \omega_n'$ are linearly independent over $\mathbb{Q}$. □

$\omega_1', \cdots, \omega_n'$ is called the **dual basis** of $\omega_1, \cdots, \omega_n$.

The Chinese remainder theorem (Theorem 1235, p. 32, **I**) may be generalized as

Lemma 1.5. *Let $\mathfrak{a}_1, \cdots, \mathfrak{a}_m$ be mutually relatively prime integral ideals. Then there are elements $a_i \in \mathfrak{o}_k$ $(1 \leq i \leq m)$ such that*

$$a_i \equiv \begin{cases} 1 \bmod \mathfrak{a}_i, \\ 0 \bmod \mathfrak{a}_j & \text{if } j \neq i. \end{cases} \tag{1.9}$$

Proof. If $j \neq i$, then $\mathfrak{a}_i + \mathfrak{a}_j = \mathfrak{o}_k$ implies that there are $b_i \in \mathfrak{a}_i$ and $b_j \in \mathfrak{a}_j$ such that $b_i + b_j = 1$. Therefore we have

$$b_j \equiv \begin{cases} 1 \bmod \mathfrak{a}_i, \\ 0 \bmod \mathfrak{a}_j. \end{cases}$$

Thus $a_i = \prod_{j \neq i} b_j$ satisfies the assertion of the lemma. □

Theorem 1.4. *Let $\mathfrak{a}_1, \cdots, \mathfrak{a}_m$ be mutually relatively prime integral ideals, and let $x_1, \cdots, x_m \in \mathfrak{o}_k$. Then there is an element $x \in \mathfrak{o}_k$ such that*

$$x \equiv x_i \bmod \mathfrak{a}_i \quad (i = 1, \cdots, m).$$

Proof. Let a_i be that in the previous lemma. Then $x = \sum x_i a_i$ satisfies the assertion. □

This theorem is called the (weak) approximation theorem. With this, Proposition 1.10, p. 23 **I** can be stated as

Proposition 1.8. *Let* $\mathfrak{a}$ *and* $\mathfrak{b}$ *be integral ideals of F. Then we have*

$$\mathfrak{o}_k/\mathfrak{b} \simeq \mathfrak{a}/\mathfrak{ab},$$

in particular

$$(\mathfrak{o}_k : \mathfrak{b}) = (\mathfrak{a} : \mathfrak{ab}).$$

Proof. Let $\mathfrak{a} = \prod_{i=1}^m \mathfrak{p}_i^{a_i}$ and $\mathfrak{b} = \prod_{i=1}^m \mathfrak{p}_i^{b_i}$ $(a_i, b_i \geq 0)$, where $\mathfrak{p}_i$ are distinct prime ideals. We take $\alpha \in \mathfrak{o}_k$ such that

$$\alpha \in \mathfrak{p}_i^{a_i} \setminus \mathfrak{p}_i^{a_i+1} \text{ for } 1 \leq i \leq m, \tag{1.10}$$

using Theorem 1.4. Then we shall show that

$$\mathfrak{ab} + (\alpha) = \mathfrak{a}. \tag{1.11}$$

Since $\mathfrak{ab} + (\alpha) \subset \mathfrak{p}_i^{a_i}$ for $^\forall i$, we have $\mathfrak{ab} + (\alpha) \subset \mathfrak{a}$. If $\mathfrak{ab} + (\alpha) \neq \mathfrak{a}$, then there is a prime ideal $\mathfrak{p}$ such that $\mathfrak{ab} + (\alpha) \subset \mathfrak{ap}$. We have $\prod \mathfrak{p}_i^{a_i+b_i} = \mathfrak{ab} \subset \mathfrak{ab} + (\alpha) \subset \mathfrak{ap} \subset \mathfrak{p}$, which yields $\mathfrak{p} \supset \mathfrak{p}_i$ for some i. Hence $\mathfrak{p} = \mathfrak{p}_i$ follows and then $\alpha \in \mathfrak{ap}_i \subset \mathfrak{p}_i^{a_i+1}$ holds, which contradicts (1.10), thereby showing (1.11).

Consider the translation

$$f : \mathfrak{o}_k \to \mathfrak{a}/\mathfrak{ab}; \quad f(a) = a\alpha, \tag{1.12}$$

which is well-defined and surjective by (1.11).

Note that Ker f consists of those a's, for which $a\alpha \in \mathfrak{ab}$, or $a \in \alpha^{-1}\mathfrak{ab}$. Since (1.10) implies $(\alpha) = \mathfrak{a} \prod \mathfrak{q}_i^{c_i}$, where prime ideals $\mathfrak{q}_i$ are different from any $\mathfrak{p}_j$ and $c_i \geq 0$, it follows that $a \in (\mathfrak{a} \prod \mathfrak{q}_i^{c_i})^{-1}\mathfrak{ab}$, and so $a \in \mathfrak{b} \prod \mathfrak{q}_i^{-c_i} \cap \mathfrak{o}_k = \mathfrak{b}$ by (1.20). Ker $f \supset \mathfrak{b}$ being clear, we have Ker $f = \mathfrak{b}$ and so Theorem 1.10, p. 17, **I** applies. $\qquad\square$

Remark 1.3. Proposition 1.8 is similar to Theorem 1.11, p. 30, **I**. Application of Proposition 1.8 yields multiplicativity of norms (cf. Remark 1.4). The Dirichlet series

$$\zeta_k(s) = \sum_{\mathfrak{a} \neq o} \frac{1}{N\mathfrak{a}^s} = \sum_{n=1}^\infty \frac{F(n)}{n^s}, \quad \sigma > 1 \tag{1.13}$$

is called the **Dedekind zeta-function** and $F(n) = \sum_{N\mathfrak{a}=n} 1$ is the number of ideals with norm n, called the Idealfunktion (ideal function). The Dedekind zeta-functions play equally important role in algebraic number theory as does the Riemann zeta-function in rational number field. There will appear several applications. Cf. Example 3.11.

We introduce the notion of the (absolute) **discriminant**. Let $w_1, \cdots, w_n$ be a $\mathbb{Z}$-basis of $\mathfrak{o}_k$. The (absolute) **discriminant**, denoted by D_k, is defined by

$$D_k = \det(Tr_{K/\mathbb{Q}}(w_i w_j)),$$

where $Tr = Tr_{K/\mathbb{Q}}$. It is a non-zero integer (cf. Lemmas 1.3 and 1.7). It is independent of the choice of a $\mathbb{Z}$-basis. For let $w'_1, \cdots, w'_n$ be another $\mathbb{Z}$-basis of $\mathfrak{o}_k$; then there is a matrix $U = (u_{ij})$ $(u_{ij} \in \mathbb{Z})$ such that

$$(w'_1, \cdots, w'_n) = (w_1, \cdots, w_n)U, \ \det U = \pm 1.$$

It is easy to see that

$$(Tr_{K/\mathbb{Q}}(w'_i w'_j)) = {}^t U(Tr_{K/\mathbb{Q}}(w_i w_j))U,$$

whence $\det(Tr_{K/\mathbb{Q}}(w'_i w'_j)) = \det(Tr_{K/\mathbb{Q}}(w_i w_j))$.

1.3   Algebraic integers

If an algebraic number α is a root of a monic polynomial with integer coefficients

$$x^n + a_{n-1}x^{n-1} + \cdots + a_1 x + a_0 \ (a_0, a_1, \cdots, a_{n-1} \in \mathbb{Z}), \qquad (1.14)$$

then α is called an **algebraic integer**. Every integer a in $\mathbb{Z}$, being a root of $x - a$, is an algebraic integer. We often call algebraic integers simply integers, and elements in $\mathbb{Z}$ rational integers.

Lemma 1.6. $\alpha \in \overline{\mathbb{Q}}$ *is an algebraic integer if and only if there is a finitely generated non-zero $\mathbb{Z}$-module M in $\overline{\mathbb{Q}}$ such that $\alpha M \subset M$.*

Proof. Suppose that α is an algebraic integer and is a root of (1.14). Then $\alpha^n = -a_{n-1}\alpha^{n-1} - \cdots - a_1\alpha - a_0$; we put

$$M = \mathbb{Z}[\alpha] = \{f(\alpha) \mid f(x) \in \mathbb{Z}[x]\}.$$

We see that $M = \mathbb{Z} + \mathbb{Z}\alpha + \cdots + \mathbb{Z}\alpha^{n-1}$, which is a $\mathbb{Z}$-module generated by finitely many elements $1, \alpha, \cdots, \alpha^{n-1}$, and clearly $\alpha M \subset M$. Conversely suppose $M = \mathbb{Z}\alpha_1 + \cdots + \mathbb{Z}\alpha_n$ $(\alpha_1, \cdots, \alpha_n \neq 0)$ and $\alpha M \subset M$; then

$$(\alpha\alpha_1, \cdots, \alpha\alpha_n) = (\alpha_1, \cdots, \alpha_n)A,$$

where every entry of A is in $\mathbb{Z}$. Since the equation $(x_1, \cdots, x_n)(\alpha 1_n - A) = 0$ has a non-trivial zero $(\alpha_1, \cdots, \alpha_n)$, it follows from the theorem on homogeneous equation that $\det(\alpha 1_n - A) = 0$. Hence α is a root of $\det(x 1_n - A) = 0$, which is a monic polynomial with rational integer coefficients. Hence α is algebraic. $\qquad \square$

Theorem 1.5. *Let $\alpha, \beta \in \overline{\mathbb{Q}}$ be algebraic integers. Then $\alpha \pm \beta, \alpha\beta$ are also algebraic integers.*

Proof. Let α, β be roots of $f(x) = x^n + a_{n-1}x^{n-1} + \cdots + a_0 \in \mathbb{Z}[x]$, $g(x) = x^m + b_{m-1}x^{m-1} + \cdots + b_0 \in \mathbb{Z}[x]$, respectively. Then

$$M = \sum_{0 \le i < n, 0 \le j < m} \mathbb{Z}\alpha^i\beta^j$$

is a finitely generated $\mathbb{Z}$-module and clearly $\alpha M \subset M$, $\beta M \subset M$. Hence we have $(\alpha \pm \beta)M \subset M$, $\alpha\beta M \subset M$, and so by Lemma 1.6, $\alpha \pm \beta, \alpha\beta$ are algebraic integers. $\qquad\square$

Proposition 1.9: *Let α be algebraic over $\mathbb{Q}$. α is an algebraic integer if and only if the minimal polynomial of α over $\mathbb{Q}$ is in $\mathbb{Z}[x]$. If, in particular α is an algebraic integer, then α^σ is also an algebraic integer for any isomorphism from $\overline{\mathbb{Q}}$ into $\overline{\mathbb{Q}}$.*

Proof. Since the minimal polynomial is monic, "only if"-part is obvious.

Suppose that α is an algebraic integer and let $f(x) (\in \mathbb{Z}[x])$ be a monic polynomial of which α is a root. Let $g(x)$ be the minimal polynomial of α over $\mathbb{Q}$. Then there is a polynomial $h(x) (\in \mathbb{Z}[x])$ such that $f(x) = g(x)h(x)$. By Lemma 1.2, we obtain $g(x) \in \mathbb{Z}[x]$.

The second assertion follows from $f(\alpha^\sigma) = (f(\alpha))^\sigma = 0$. $\qquad\square$

Corollary 1.5. *If a rational number a is an algebraic integer, a is a rational integer.*

Proof. Since the minimal polynomial of a over $\mathbb{Q}$ is $x - a$, the assertion follows immediately from the proposition. $\qquad\square$

Proposition 1.10. *Let α be an algebraic number; then there is a natural number a such that $a\alpha$ is an algebraic integer.*

Proof. Let $f(x) = x^n + a_{n-1}x^{n-1} + \cdots + a_0 \in \mathbb{Q}[x]$ be the minimal polynomial of α. We take a natural number a such that $aa_i \in \mathbb{Z}$ $(i = 0, \cdots, n-1)$. Then $F(x) = x^n + aa_{n-1}x^{n-1} + \cdots + a^k a_{n-k}x^{n-k} + \cdots + a^n a_0 \in \mathbb{Z}[x]$ and $F(a\alpha) = a^n f(\alpha) = 0$ is easy to see. Hence $a\alpha$ is an algebraic integer. $\qquad\square$

Proposition 1.11. *Suppose that α is a root of $x^n + \alpha_{n-1}x^{n-1} + \cdots + \alpha_0 = 0$, where $\alpha_{n-1}, \cdots, \alpha_0$ are algebraic integers. Then α is also an algebraic integer.*

Proof. Put

$$M = \left\{ \sum c_{\boldsymbol{a}} \alpha^a \alpha_0^{a_0} \cdots \alpha_{n-1}^{a_{n-1}} \mid c_{\boldsymbol{a}} \in \mathbb{Z}, \boldsymbol{a} = (a, a_0, \cdots, a_{n-1}) \in \mathbb{Z}^{n+1} \right.$$

with $0 \leq a \leq n-1, a_0, \cdots, a_{n-1} \geq 0 \Big\}.$

M is a module satisfying $\alpha M \subset M$. Moreover, it is finitely generated. In fact for the minimal polynomial $f_i(x) = x^{n_i} + c_{n_i-1}x^{n_i-1} + \cdots + c_0 \in \mathbb{Z}[x]$ of α_i we have $\alpha_i^{n_i} = -(c_{n_i-1}\alpha_i^{n_i-1} + \cdots + c_0)$, hence we may assume $0 \leq a_i \leq n_i - 1$ in the definition of M. Thus α is an algebraic integer by Lemma 1.6. □

Proposition 1.11 implies that if $\alpha \in F$ is an algebraic integer over $\mathfrak{o}_F$, then it is an algebraic integer, whence $\alpha \in F \cap \mathfrak{o}_F = \mathfrak{o}_f$, i.e. $\mathfrak{o}_F$ is integrally closed.

$$\mathfrak{o}_F = \{\alpha \in F \mid \alpha \text{ is an algebraic integer}\}, \qquad (1.15)$$

which is a ring by Theorem 1.5 called the **ring of algebraic integers** (in F).

Lemma 1.7. *If* $\alpha \in \mathfrak{o}_F$. *Then we have* $N_{F/\mathbb{Q}}\,\alpha \in \mathbb{Z}$ *and* $Tr_{F/\mathbb{Q}}\,\alpha \in \mathbb{Z}$.

Proof. The minimal polynomial of α is in $\mathbb{Z}[x]$ by Proposition 1.9. Hence Proposition 1.5 yields the assertion. □

Recall that a $\mathbb{Z}$-module is called a free $\mathbb{Z}$-module of rank n if it is isomorphic to $\mathbb{Z}^n$. Although the next theorem is the same as Lemma 2.2 which in turn is a corollary to Theorem 2.2, we again give a direct proof.

Theorem 1.6. *Let* N *be a free* $\mathbb{Z}$-module of rank n. *Then a subgroup* M *of* N *is a free* $\mathbb{Z}$-module of rank $m\,(\leq n)$.

Proof. Let $v_1, \cdots, v_n$ be a basis of N and put

$$N_k = \{a_1 v_1 + \cdots + a_k v_k \mid a_1, \cdots, a_k \in \mathbb{Z}\} \text{ and } M_k = M \cap N_k;$$
$$M_n = M, \ M_0 = \{0\}.$$

Since

$$I_k = \{a_k \in \mathbb{Z} \mid a_1 v_1 + \cdots + a_k v_k \in M\}$$

is a submodule of $\mathbb{Z}$, there is an integer $b_k\,(\geq 0)$ such that $I_k = b_k\mathbb{Z}$ (by Theorem 1.5, **I**). Take an element $u_k = b_{1,k}v_1 + \cdots + b_{k,k}v_k \in M$ $(b_{k,k} = b_k)$.

Clearly $u_k \in M_k$. If $b_k = 0$, then we choose 0 as u_k. Then for any element $v \in M_k$, there is an integer c such that

$$v - cu_k \in M_{k-1} \, (= M \cap N_{k-1}).$$

Let $v \in M$. Applying the above argument to v in the descending order, we see that there are integers $c_n, \cdots, c_1$ such that

$$v - c_n u_n - \cdots - c_k u_k \in M_{k-1} \, (k = n, n-1, \cdots, 1).$$

Thus we have $M = \{c_1 u_1 + \cdots + c_n u_n \mid c_1, \cdots, c_n \in \mathbb{Z}\}$.

Let $u_{i_1}, \cdots, u_{i_m} \, (1 \le i_1 < \cdots < i_m \le n)$ be all non-zero elements in $\{u_1, \cdots, u_n\}$. Then we have

$$M = \{c_{i_1} u_{i_1} + \cdots + c_{i_m} u_{i_m} \mid c_{i_1}, \cdots, c_{i_m} \in \mathbb{Z}\}.$$

Next, suppose that $c_{i_1} u_{i_1} + \cdots + c_{i_\ell} u_{i_\ell} = 0$ with $c_{i_\ell} \neq 0$; then the definition of N_k implies

$$c_{i_\ell} u_{i_\ell} \notin N_{i_\ell - 1}.$$

On the other hand, we have

$$c_{i_\ell} u_{i_\ell} = -(c_{i_1} u_{i_1} + \cdots + c_{i_{\ell-1}} u_{i_{\ell-1}}) \in N_{i_{\ell-1}} \subset N_{i_\ell - 1},$$

by $i_{\ell-1} \le i_\ell - 1$. This is a contradiction, and so $u_{i_1}, \cdots, u_{i_m} \, (1 \le i_1 < \cdots < i_m \le n)$ is a basis of M. $\qquad \square$

Remark 1.4. A slight modification of Theorem 1.6 shows that the group index $(\mathfrak{o} : \mathfrak{a})$ is finite $(\in \mathbb{N} \cup \{0\})$ for any integral ideal $\mathfrak{a} \subset \mathfrak{o}$. It is denoted by $N\mathfrak{a}$ and called the **norm** of $\mathfrak{a}$:

$$N\mathfrak{a} = (\mathfrak{o} : \mathfrak{a}). \qquad (1.16)$$

By Remark 1.4, every ideal has a finite norm. Thus algebraic number theory may be described in one aspect as a study on modules over a Dedekind domain **with finite norm property**. Cf. toward the end of §1.4.

Theorem 1.7. *Let F be an algebraic number field. Then $\mathfrak{o}_F$ is a free $\mathbb{Z}$-module of rank $[F : \mathbb{Q}]$, i.e. there are elements $\omega_1, \cdots, \omega_n \in F$ such that*

$$\mathfrak{o}_F = \mathbb{Z}\omega_1 + \cdots + \mathbb{Z}\omega_n \quad (n = [F : \mathbb{Q}]),$$

and $\omega_1, \cdots, \omega_n$ is a basis of F over $\mathbb{Q}$.

Proof. Take an element $\theta \in F$ such that $F = \mathbb{Q}(\theta)$ by Corollary 1.3. Moreover we may assume $\theta \in \mathfrak{o}_F$ by Proposition 1.10. Since $u_i = \theta^{i-1}$ $(i = 1, \cdots, n)$ is a basis of F over $\mathbb{Q}$, we can take a dual basis $u'_1, \cdots, u'_n$ by Lemma 1.4.

Let $w \in \mathfrak{o}_F$ and $w = \sum a_i u'_i$ $(a_i \in \mathbb{Q})$. Since $w, u_i \in \mathfrak{o}_F$, we have $a_i = B_F(w, u_i) \in \mathbb{Z}$ by Lemma 1.7. Thus $\mathfrak{o}_F$ is a submodule of $\mathbb{Z}u'_1 + \cdots + \mathbb{Z}u'_n$. Theorem 1.6 implies that $\mathfrak{o}_F$ has a basis $\omega_1, \cdots, \omega_m$ $(m \le n)$.

Since $\mathfrak{o}_F \subset \mathbb{Z}\omega_1 + \cdots + \mathbb{Z}\omega_m$, we have $F \subset \mathbb{Q}\omega_1 + \cdots + \mathbb{Q}\omega_m$, i.e. $n = \dim_{\mathbb{Q}} F \le m$. Thus we obtain $n = m$. $\square$

We call $\omega_1, \cdots, \omega_n$ a $\mathbb{Z}$-**basis** of $\mathfrak{o}_F$ and write

$$\mathfrak{o}_F = \mathbb{Z}[\omega_1, \cdots, \omega_n]^1.$$

Let us determine a $\mathbb{Z}$-basis of a quadratic field, i.e. a quadratic extension of $\mathbb{Q}$.

Proposition 1.12. *Let $m\,(\ne 0, 1)$ be a square-free integer, i.e. m is not divisible by p^2 for any prime number p. Put $F = \mathbb{Q}(\sqrt{m})$ and*

$$D = \begin{cases} m & \text{if } m \equiv 1 \bmod 4, \\ 4m & \text{if } m \equiv 2, 3 \bmod 4. \end{cases}$$

Then we have

$$\mathfrak{a}_F = \begin{cases} \mathbb{Z}[1, (1+\sqrt{m})/2] & \text{if } m \equiv 1 \bmod 4, \\ \mathbb{Z}[1, \sqrt{m}] & \text{if } m \equiv 2, 3 \bmod 4 \end{cases}$$

$$\mathbb{Z} = \mathbb{Z}[1, (D + \sqrt{D})/2].$$

Proof. Suppose $m \equiv 1 \bmod 4$. Let $\alpha = a + b(1 + \sqrt{m})/2$ $(a, b \in \mathbb{Q})$; then it implies $b^2 m = (b\sqrt{m})^2 = (2\alpha - 2a - b)^2 = 4\alpha^2 - 4(2a + b)\alpha + (2a + b)^2$, whence α is a root of

$$f(x) = x^2 - (2a + b)x + a^2 + ab + b^2(1 - m)/4.$$

Suppose $a, b \in \mathbb{Z}$; then $f(x) \in \mathbb{Z}[x]$ and so α is an algebraic integer.

Suppose, conversely α is an algebraic integer. If $b = 0$, then $\alpha = a \in \mathfrak{a}_F \cap \mathbb{Q} = \mathbb{Z}$ implies $a \in \mathbb{Z}$, hence $\alpha \in \mathbb{Z}[1, (1+\sqrt{m})/2]$. Suppose $b \ne 0$; then by $\alpha \notin \mathbb{Q}$, $f(x)$ is the minimal polynomial of α over $\mathbb{Q}$ and by Proposition 1.9 it is in $\mathbb{Z}[x]$. Thus we have $c := 2a + b \in \mathbb{Z}$, $d := a^2 + ab + b^2(1 - m)/4 \in \mathbb{Z}$. Substituting $a = (c - b)/2$, we have $4d = c^2 - b^2 m$, i.e. $b^2 m = c^2 - 4d \in \mathbb{Z}$.

---

[1] For a ring R, $R[a]$ means $\{f(a) \mid f(x) : \text{polynomial over } R\}$, and $R[a_1, \cdots, a_n]$ means $Ra_1 + \cdots + Ra_n$. When $n = 1$, they are distinguished by the context.

Since m is square-free, b is an integer. Then $b^2 m = c^2 - 4d$ implies $b^2 \equiv c^2 \bmod 4$, hence $b \equiv c \bmod 2$, which implies $2a = c - b$ is an even integer. Hence we have $a, b \in \mathbb{Z}$.

Suppose $m \equiv 2, 3 \bmod 4$. Let $\alpha = a + b\sqrt{m}$ $(a, b \in \mathbb{Q})$; then $b^2 m = (\alpha - a)^2$ yields that α is a root of

$$f(x) = x^2 - 2ax + a^2 - b^2 m.$$

Suppose $a, b \in \mathbb{Z}$; then $f(x) \in \mathbb{Z}[x]$, and hence α is an algebraic integer.

Suppose conversely $\alpha \in \mathfrak{o}_F$. If $b = 0$, we have $\alpha = a \in \mathfrak{a}_F \cap \mathbb{Q} = \mathbb{Z}$, whence $\alpha \in \mathbb{Z}[1, \sqrt{m}]$. Suppose $b \neq 0$; then $\alpha \notin \mathbb{Q}$, $f(x)$ is the minimal polynomial of α over $\mathbb{Q}$, whence it is in $\mathbb{Z}[x]$ by Proposition 1.9. Thus we have $c := 2a \in \mathbb{Z}$ and $d := a^2 - b^2 m \in \mathbb{Z}$. Therefore $4b^2 m = c^2 - 4d$ follows. Since m is square-free and $c^2 - 4d \in \mathbb{Z}$, $n := 2b$ is an integer. Hence $n^2 m = c^2 - 4d$ yields $c^2 \equiv 2n^2$ of $c^2 \equiv 3n^2 \bmod 4$ according to $m \equiv 2, 3 \bmod 4$. As $\ell^2 \equiv 0, 1 \bmod 4$ holds for any integer ℓ, the above yields $c \equiv n \equiv 0 \bmod 2$. Thus a, b are integers by $c = 2a$, $n = 2b$.

The last equality is easily shown. $\qquad\square$

When $m > 0$ in the proposition above, $\mathbb{Q}(\sqrt{m})$ is called a **real quadratic field**, otherwise a **imaginary quadratic field**.

Proposition 1.13. *Let F be an algebraic number field. If $\alpha \in \overline{\mathbb{Q}}$ is an algebraic integer, then all roots of the minimal polynomial $p(x)$ of α over F are also algebraic integers and all coefficients of $p(x)$ belong to $\mathfrak{o}_F$.*

Proof. Let $m(x)$ be the minimal polynomial of α over $\mathbb{Q}$. Since $m(\alpha) = p(\alpha) = 0$ and $m(x) \in F[x]$, $p(x)$ divides $m(x)$, hence the roots of $p(x)$ are roots of $m(x)$. Thus the roots of $p(x)$ are algebraic integers, and expressing coefficients of $p(x)$ by its roots, coefficients of $p(x)$ are algebraic integers. $\qquad\square$

1.3.1 Ideal class group

Let F be an algebraic number field of $[F : \mathbb{Q}] = n$. We define a **fractional ideal** (simply an **ideal**) $\mathfrak{a}$ of F by the condition that there is an element $\gamma \in F^\times$ such that $\gamma \mathfrak{a} = \{\gamma \alpha \mid \alpha \in \mathfrak{a}\}$ is a non-zero ideal of $\mathfrak{o}_F$ in the sense of Definition 1.1, i.e.

(i) $\alpha \pm \beta \in \mathfrak{a}$ for $\alpha, \beta \in \mathfrak{a}$,
(ii) $\omega\alpha \in \mathfrak{a}$ for $\omega \in \mathfrak{o}_F, \alpha \in \mathfrak{a}$,
(iii) $\{0\} \neq \gamma\mathfrak{a} \subset \mathfrak{o}_F$ for $^\exists \gamma \in F^\times$.

Hence a non-zero ideal of $\mathfrak{o}_F$ in the sense of the section 3.2 is also an ideal here, taking $\gamma = 1$. If an ideal $\mathfrak{a}$ is in $\mathfrak{o}_F$, we call $\mathfrak{a}$ an **integral ideal**. We note that we exclude $\{0\}$ from ideals hereafter.

Let $\mathfrak{a}$ be an ideal and $\gamma\mathfrak{a} \subset \mathfrak{o}_F$ ($\gamma \in F^{\times}$). Since $\mathfrak{o}_F$ is a free $\mathbb{Z}$-module of rank $[F : \mathbb{Q}]$ and $\mathfrak{a} \subset \gamma^{-1}\mathfrak{o}_F$, $\mathfrak{a}$ is also a free $\mathbb{Z}$-module by Theorem 2.1. Let $\alpha (\neq 0) \in \mathfrak{a}$; then $\alpha\mathfrak{o}_F \subset \mathfrak{a}$ holds by the property (ii), and so we have,

$$\alpha\mathfrak{o}_F \subset \mathfrak{a} \subset \gamma^{-1}\mathfrak{o}_F. \tag{1.17}$$

We may prove that $n = \mathrm{rank}_{\mathbb{Z}}\, \alpha\mathfrak{o}_F \leq \mathrm{rank}_{\mathbb{Z}}\, \mathfrak{a} \leq \mathrm{rank}_{\mathbb{Z}}\, \gamma^{-1}\mathfrak{o}_F = n$, hence $\mathrm{rank}_{\mathbb{Z}}\, \mathfrak{a} = \mathrm{rank}_{\mathbb{Z}}\, \mathfrak{o}_F = [F : \mathbb{Q}]$.

Proposition 1.14. *Let $\mathfrak{a}$ be an ideal of F. Then there is a natural number a such that $a\mathfrak{a} \subset \mathfrak{o}_F$.*

Proof. Let $\omega_1, \cdots, \omega_n$ be a $\mathbb{Z}$-basis of $\mathfrak{o}\mathfrak{o}_F$ and for γ in (1.17), we take a natural number a such that $a\gamma^{-1}\omega_i \in \mathfrak{o}_F$ ($i = 1, \cdots, n$). Then we have $a\mathfrak{a} \subset a\gamma^{-1}\mathfrak{o}_F \subset \mathfrak{o}_F$. $\qquad\qquad\square$

Let $\mathfrak{a}, \mathfrak{b}$ be ideals of F. The product is defined as follows:

$$\mathfrak{a}\mathfrak{b} = \{\sum \alpha_i\beta_i \mid \alpha_i \in \mathfrak{a}, \beta_i \in \mathfrak{b}\},$$

where the sum is a finite sum.

The product $\mathfrak{a}\mathfrak{b}$ is also an ideal. Indeed, the conditions (i) and (ii) are easily checked. About (iii), take $\gamma_1, \gamma_2 \in F^{\times}$ such that $\gamma_1\mathfrak{a}, \gamma_2\mathfrak{b} \subset \mathfrak{o}_F$, and then $(\gamma_1\gamma_2)\mathfrak{a}\mathfrak{b} \subset \mathfrak{o}_F$ is clear.

It is also easy to see that $\mathfrak{a}\mathfrak{b} = \mathfrak{b}\mathfrak{a}$, $(\mathfrak{a}\mathfrak{b})\mathfrak{c} = \mathfrak{a}(\mathfrak{b}\mathfrak{c})$ and $\mathfrak{a}\mathfrak{o}_F = \mathfrak{a}$ for ideals $\mathfrak{a}, \mathfrak{b}, \mathfrak{c}$.

We define a principal ideal (α) for $\alpha \in F^{\times}$ by

$$(\alpha) = \{\alpha\omega \mid \omega \in \mathfrak{o}_F\}.$$

We denote by I_F the set of all ideals of F and by P_F the set of all principal ideals of F. If an integral ideal $\mathfrak{a}$ is a prime ideal of $\mathfrak{o}_F$, it is called simply a **prime ideal** of F. The rest of this subsection is devoted to prove

Theorem 1.8. *I_F is a group with respect to the product of ideals above and P_F is a subgroup of I_F of finite index h_F. $\mathfrak{a}$ is uniquely written as a product of prime ideals.*

The finite group I_F/P_F is called the **ideal class group** of F and the order h_F of the group is called the **class number**. The study of the structure of I_F/P_F is a basic problem in algebraic number theory and there are many interesting results.

Lemma 1.8. *Let $\mathfrak{a}$ be an integral ideal of F. Then there are only a finite number of integral ideals $\mathfrak{b}$ such that $\mathfrak{a} \subset \mathfrak{b}$. In particular, for a sequence of integral ideals $\mathfrak{a}_1 \subset \mathfrak{a}_2 \subset \cdots$, there is a natural number N such that $\mathfrak{a}_N = \mathfrak{a}_{n+1} = \cdots$.*

Proof. Consider the residue class ring $\mathfrak{o}_F/\mathfrak{a}$, which is a finite set by Remark 1.4. Hence the possibility of $\mathfrak{b}/\mathfrak{a}$ as a subset of $\mathfrak{o}\mathfrak{o}_F/\mathfrak{a}$ is finite. Thus the possibility of $\mathfrak{b}$ is finite. $\qquad\square$

Lemma 1.9. *Let $\mathfrak{p}$ be a prime ideal of F. Then $\mathfrak{p}$ is a maximal ideal of $\mathfrak{o}_F$ and $\mathfrak{o}_F/\mathfrak{p}$ is a finite field.*

Lemma 1.10. *Let $\mathfrak{a}$ be an ideal. There is an ideal $\mathfrak{a}'$ such that $\mathfrak{a}\mathfrak{a}' = \mathfrak{o}_F$. $\mathfrak{a}'$ is equal to*

$$\mathfrak{a}^{-1} := \{\alpha \in F \mid \alpha\mathfrak{a} \subset \mathfrak{o}_F\}. \tag{1.18}$$

Proof. By Proposition 1.14, there exist a natural number t and $\gamma \in F^\times$ such that $\mathfrak{a}^t = (\gamma)$. Hence we have $\mathfrak{a} \cdot \gamma^{-1}\mathfrak{a}^{t-1} = \mathfrak{o}_F$, hence $\mathfrak{a}' = \gamma^{-1}\mathfrak{a}^{t-1}$.

Let us see that the set $\mathfrak{a}^{-1}$ defined by the right-hand side of (1.18) is an ideal. It is easy to see that $\alpha \pm \beta \in \mathfrak{a}^{-1}$ if $\alpha, \beta \in \mathfrak{a}^{-1}$ and $\mathfrak{o}_F\mathfrak{a}^{-1} \subset \mathfrak{a}^{-1}$. For any non-zero element $\gamma \in \mathfrak{a}$, $\gamma\mathfrak{a}^{-1} \subset \mathfrak{o}_F$ follows from the definition of $\mathfrak{a}^{-1}$. Thus $\mathfrak{a}^{-1}$ has been proved to be an ideal.

$\mathfrak{a}\mathfrak{a}' = \mathfrak{o}_F$ yields $\mathfrak{a}' \subset \mathfrak{a}^{-1}$.

If, conversely $\alpha \in \mathfrak{a}^{-1}$, then $\mathfrak{b} := \alpha\mathfrak{a} \subset \mathfrak{o}_F$ implies $\alpha \in (\alpha) = \alpha\mathfrak{a}\mathfrak{a}' = \mathfrak{b}\mathfrak{a}' \subset \mathfrak{o}_F\mathfrak{a}' \subset \mathfrak{a}'$, therefore $\mathfrak{a}^{-1} \subset \mathfrak{a}'$. Thus we have proved $\mathfrak{a}' = \mathfrak{a}^{-1}$. $\qquad\square$

Lemma 1.11. *The set I_F of all ideals of F is an abelian group, where the identity is $\mathfrak{o}_F$ and the inverse element of $\mathfrak{a}$ is given by $\mathfrak{a}^{-1}$.*

Proof. This follows from the definition of the product of ideals and the previous lemma. $\qquad\square$

1.4 Dedekind domains

Let I be an ideal of a commutative ring R, and suppose $I \neq R$, as above. If any ideal J satisfying $I \subset J \subset R$ is I or R, then I is called a **maximal ideal**.

Proposition 1.15. *Let I be an ideal of R. Then I is maximal if and only if the residue class ring R/I is a field.*

Corollary 1.6. *A maximal ideal is a prime ideal.*

For the residue class ring R/I is a field whence a domain.

Theorem 1.9. *Let $R = \mathbb{Z}$ or $R = F[x]$ for a field F. If $I\,(\neq 0)$ is a prime ideal of R, then I is maximal.*

Definition 1.5. An integral domain is called a **Dedekind domain** if it is integrally closed, noetherian and such that prime ideals are maximal. A domain is said to be integrally closed if all integers in its quotient field again belong to it.

It is shown in Proposition 1.11 that the ring $\mathfrak{o}_F$ (1.15) of algebraic integers is integrally closed. It being noetherian is proved in Lemma 1.8. And maximality of a prime ideal is proved in Lemma 1.9. Thus the ring of algebraic integers is a Dedekind domain. Since unique factorization into irreducible elements does not hold in general, one considers unique factorization of fractional ideals into prime ideals. By Theorem 1.10 a Dedekind domain satisfies unique factorization of fractional ideals and *a fortiori* so does the ring of integers. In Theorem 1.8 we prove that the ring of integers satisfies unique factorization of fractional ideals.

One of the main purposes of algebraic number theory is the study on modules over the ring of algebraic integers. It is known that a Dedekind domain is a unique factorization domain if and only if the ideal class group is trivial. Hence, more or less one studies modules (or lattices) over Dedekind domains with class number finite (Theorem 1.8).

1.5 Modules over Dedekind domains

This section is a more general description of algebraic structure to be stated in §2.1 and they can be read together side by side with benefits.

Definition 1.6. Let M be an Abelian group and let R be a ring with 1. We say that M is a left R-module if R acts on M (cf. Definition 1.19, p. 38, **I**) and the action satisfies additivity: if for any $a \in R, x \in M$, we have $ax \in M$ and
(i) $a, b \in R, \quad x \in M, \quad (ab)(x) = a(bx)$
(ii) $1x = x$

(iii) $a \in R,\quad x, y \in M,\quad a(x + y) = ax + ay$

(iv) $a, b \in R,\quad x \in M,\quad (a + b)x = ax + by.$

If the action is given from the right and it satisfies the counterpart conditions, M is called a right R-module. In what follows we do not distinguish the left and right, and refer to a left R- module as an R-module. Modules over a ring is a generalization of the notion of a vector space over a field and the action corresponds to scalar multiplication. If M is a set and R is a group and conditions (i)-(iii) are satisfied, then R is called a group with an **operator domain** M (or an M-group). Those elements $x \in M$ are called **torsion** elements if for some non-zero r $in R$, $rx = 0$. The set of all torsion elements form a subgroup of M, called the **torsion subgroup**. An R-module M is said to have a **basis** $\{e_i\} \subset M$ if every element $x \in M$ can be expressed unique as $x = \sum_i r_i e_i$, where r_i are 0 except for finitely many i. If M has a basis, it is called a **free** R-module. A free $\mathbb{Z}$-module is called a **free Abelian group**.

Let R be an integral domain and K its quotient field. An R-submodule $\mathfrak{a}$ of K is called a **fractional ideal** if there is a non-zero element $r \in R$ so that $r\mathfrak{a} \subset R$ (if $r \in K$, we may clear the denominator by multiplying an element of R). In contrast to fractional ideals, ordinary ideals of R are called **integral ideals**.

Proposition 1.16. *Two fractional ideals $\mathfrak{a}, \mathfrak{b}$ are isomorphic as R-modules if and only there exists an $r \in K$ such that $\mathfrak{a} = r\mathfrak{b}$.*

Proof. Plainly $\mathfrak{a} \simeq r\mathfrak{a}$. It suffices therefore to prove the converse for integral ideals. For a homomorphism $f : \mathfrak{a} \to \mathfrak{b}$, $a^{-1}f(a)$ ($0 \neq a \in \mathfrak{a}$) is constant. Indeed, for another $0 \neq a' \in \mathfrak{a}$, $af(a') = f(a'a) = f(aa') = a'f(a')$. Hence putting $a^{-1}f(a) = r$, we have $f(a) = ra$, which gives an isomorphism, completing the proof. $\square$

For a fractional ideal $\mathfrak{a}$, let $\mathfrak{a}^{-1} = \{x \in K | x\mathfrak{a} \subset R\}$ be the **inverse fractional ideal** of $\mathfrak{a}$. $\mathfrak{a}^{-1}\mathfrak{a} \subset R$ holds.

Exercise 2. Prove that $\mathfrak{a}^{-1}$ is a fractional ideal.

Solution. We may easily check that $\mathfrak{a}'$ is an Abelian group. It therefore suffices to find an element to clear the denominators. Choose a non-zero element $r \in R$ such that $r\mathfrak{a} \subset R$. Then for any $0 \neq a \in \mathfrak{a}$, $ra \in R$ and $ra \in \mathfrak{a}$ since $\mathfrak{a}$ is an R-module. Hence $ra \in R \cap \mathfrak{a}$. Hence for any $x \in \mathfrak{a}^{-1}$, $rax \in R$ by definition. Hence $ra\mathfrak{a}^{-1} \subset R$.

Definition 1.7. If $\mathfrak{a}^{-1}\mathfrak{a} = R$ holds, $\mathfrak{a}$ is called **invertible**. A domain is called a **Dedekind domain** if all its fractional ideals are invertible. An element of an extension domain S of a domain R is called **integral** over R if it is a root of a monic polynomial $\in R[X]$. If all integral elements in the quotient field K of R are only those in R, then R is said to be **integrally closed**. A ring is called **Noetherian** if one of the following equivalent conditions holds true.

(i) (maximal condition) In any set of ideals in R there exists a maximal element.

(ii) (ascending chain condition) Any infinite chain of ideals

$$\mathfrak{a}_1 \subset \mathfrak{a}_2 \subset \cdots$$

terminates, i.e. from some number N on, they are all equal.

(iii) Any ideal of R are finitely generated.

Theorem 1.10. *A Dedekind domain is characterized by one of the following two equivalent conditions.*

(i) It is an integrally closed, Noetherian domain in which every prime ideal is maximal.

(ii) Any fractional ideal may be expressed uniquely as a product of prime ideals $\mathfrak{p}_1^{e_1} \cdots \mathfrak{p}_1^{e_\nu}$, $e_i \in \mathbb{Z}$.

In view of Proposition 1.16, it is natural to classify modules according to the isomorphism. Each equivalence class of isomorphic ideals is called an **ideal class**. They form a group called the **ideal class group**. Its order is called the **class number**.

Exercise 3. Prove the following assertions.

(i) In a Dedekind domain, the set I of all fractional ideals form a group and the set P of all principal ideals forms its subgroup.

(ii) The ideal class group is isomorphic to the factor group I/P. A PID is a Dedekind domain whose ideal class group is the unit group.

Example 1.4. Consider the imaginary quadratic field $\mathbb{Q}(\sqrt{-5}) = \{a + b\sqrt{-5} | a, b \in \mathbb{Q}\}$. Its ring of integers is given by $\mathbb{Z}(\sqrt{-5}) = \{a + b\sqrt{-5} | a, b \in \mathbb{Z}\}$. We illustrate by the decomposition $6 = 2 \cdot 3 = (1 - \sqrt{-5})(1 + \sqrt{-5})$ and show that $\mathbb{Z}(\sqrt{-5})$ is not a PID. Indeed, since $N2 = 4$, $N3 = 9$, $N(1 \pm \sqrt{-5}) = 6$, these cannot be associates. Thus we are to consider a substitute for ordinary prime decomposition. And it is the decomposition into prime ideals given by Theorem 1.10.

In the times of E. Kummer, it was thought that all rings of integers are UFD and Kummer thought that he proved Fermat's last theorem. Kummer's investigation depended on the UF property of rings and does not apply to all rings. To avoid this difficulty, he introduced the idea of **ideal numbers**, which eventually got developed into modern theory of ideals by Dedekind (L. Kronecker, E. Noether). Algebraic number theory has been developed in order to assure the *prime decomposition*.

We quote the following fundamental theorem on a finitely generated module over a Dedekind domain (cf. [Narkiewicz (2004), Theorems 1.10 and 1.11]).

Theorem 1.11. *Let R be a Dedekind domain and let M be a finitely generated R-module. Then M is isomorphic to*

$$R^k \oplus \mathfrak{i} \oplus A,$$

where A is the torsion subgroup of M, $k \in \mathbb{N}$ and $\mathfrak{i}$ is some fractional ideal of R. The expression is unique up to orders of entries.

In most of books, the above theorem is stated for the case of R being a *PID*. Since we consider only algebraic number fields over $\mathbb{Q}$, the ring of integers is a free $\mathbb{Z}$-module, and (1.19) applies (cf. also Lemma 4.3, **I**). When we consider relative extensions, we are to appeal to Theorem 1.11. Indeed, what is more interesting is the case of relative extensions. Indeed, the celebrated class field theory is a theory of relative Abelian extensions. For Abelian extensions over $\mathbb{Q}$ we may apply the Kronecker-Weber Theorem (Theorem 2.10) and we may make detailed analysis. Similarly, class field theory enables us to make a detailed study on relative Abelian extensions.

Theorem 1.12. *Let R be a PID and let M be a finitely generated R-module. Then M is isomorphic to*

$$R/e_1 R \oplus \cdots \oplus R/e_l R$$

where e_i may be chosen so that $e_i | e_{i+1}$ $(1 \le i \le l-1)$ and in choosing so, they are uniquely determined up to associates. Or more concretely,

$$\mathbb{Z}/a_1\mathbb{Z} \oplus \cdots \oplus \mathbb{Z}/a_l\mathbb{Z} \oplus \underbrace{\mathbb{Z} \oplus \cdots \oplus \mathbb{Z}}_{r \text{ times}}, \tag{1.19}$$

where $1 < a_1$, $a_i | a_{i+1}$ $(1 \le i \le l-1)$ and the product $a = a_1 \cdots a_l$ and r are uniquely determined.

$e_1, \cdots, e_l$ or $a_1, \cdots, a_l$ are called **elementary divisors**. r is called the
R-rank of G.

Exercise 4. (i) If $\mathfrak{a} = \prod_{i=1}^{m} \mathfrak{p}_i^{a_i}$ and $\mathfrak{b} = \prod_{i=1}^{m} \mathfrak{p}_i^{b_i}$, $0 \le a_i, b_i \in \mathbb{Z}$, then

$$\mathfrak{a} + \mathfrak{b} = \prod_{i=1}^{m} \mathfrak{p}_i^{\min\{a_i, b_i\}}, \quad \mathfrak{a} \cap \mathfrak{b} = \prod_{i=1}^{m} \mathfrak{p}_i^{\max\{a_i, b_i\}}$$

(cf. Remark 1.2).

(ii) Every fractional ideal $\mathfrak{a}$ may be given in the form $\mathfrak{a} = \prod_{i=1}^{m} \mathfrak{p}_i^{a_i} \prod_{i=1}^{n} \mathfrak{p}_i^{-b_i}$, $0 \le a_i, b_i \in \mathbb{Z}$. Prove that

$$\mathfrak{a} \cap \mathfrak{o}_k = \prod_{i=1}^{m} \mathfrak{p}_i^{a_i}. \tag{1.20}$$

Chapter 2

Group-theoretic aspects of algebraic number theory

Abstract: In this chapter we state group-theoretic aspects of algebraic numbers, which together with the linear-algebraic approach in Chapter 1 hopefully gives an easily accessible introduction to algebraic number theory. By this we present basic results without proof and we illustrate them by starting with the structure theorem on f.g. Abelian group applied to establishing the Dirichlet unit theorem, and we state rudiments of Galois theory since this is the main role played by group theory. We state some facts about cyclotomic fields since they are a good example of the theory and could work for an introduction to class field theory.

Those which can be read more light-heartedly than very serious introductions such as [Lang (1970)], [Weil (1976)] are [Goldstein (1971)], [Kitaoka], [Narkiewicz (2004)] etc. with [Butzer and Stark (1986)] as an intermediate. There are many more good books on algebraic number theory and we cannot exhaust all of them.

Then we state some elements of zeta-functions associated with more advanced objects including elliptic curves and group representations. Then we consider the theory of Ashnell-Goldfeld about a plausible PRN generator depending on an intractable problem in zeta-function theory. The style is very sketchy and the interested reader should read suitable literature on the subject. There are numerous good books on elliptic curves.

2.1 Finitely generated abelian group

In this section, we give several results on the structure of a finitely generated abelian group, which are essentially used in algebraic number theory. Although the main structure theorem, Theorem 2.2, is a special case of Theorem 1.12 which in turn is a special case of Theorem 1.11 on f.g. modules over a Dedekind domain, we repeat similar proofs here for the readers' merit. It will be beneficial to compare the proofs here and in §1.5. A highlight is the Dirichlet unit theorem, Theorem 2.3, which is a group-theoretic object.

Denote by $\mathbb{Z}^n$ the direct sum (Cartesian product) of n copies of $\mathbb{Z}$, that is

$$\mathbb{Z}^n = \{(a_1, \cdots, a_n) \mid a_1, \cdots, a_n \in \mathbb{Z}\}.$$

A module isomorphic to $\mathbb{Z}^n$ is called a **free $\mathbb{Z}$-module** of rank n. For a free $\mathbb{Z}$-module G of rank n, there are elements $v_1, \cdots, v_n \in G$, called a $\mathbb{Z}$-**basis** of G such that

$$G = \{a_1 v_1 + \cdots + a_n v_n \mid a_1, \cdots, a_n \in \mathbb{Z}\} \tag{2.1}$$

and

$$a_1 v_1 + \cdots + a_n v_n = 0 \text{ if and only if } a_1 = \cdots = a_n = 0. \tag{2.2}$$

The last equality follows by taking the isomorphic images $v_1, \cdots, v_n$ of $(1, 0, \cdots, 0), \cdots, (0, \cdots, 0, 1)$.

Lemma 2.1. *Let G be a module. G is a free $\mathbb{Z}$-module of rank n if and only if there are elements $v_1, \cdots, v_n \in G$ satisfying (2.1) and (2.2).*

Proof. We have only to show the "if" part. Suppose the existence of elements $v_1, \cdots, v_n \in G$ satisfying (2.1) and (2.2). Define the mapping φ from $\mathbb{Z}^n$ to G by

$$\varphi((a_1, \cdots, a_n)) = \sum_{i=1}^{n} a_i v_i.$$

Then φ is clearly a homomorphism. Surjectivity follows from the condition (2.1), and the condition (2.2) gives the injectivity, hence φ is the isomorphism. $\qquad\square$

In analogy to the existence of a basis in a vector subspace, we have

Lemma 2.2. *Let N be a free $\mathbb{Z}$-module of rank n. Then a subgroup M of N is a free module of rank $m \, (\leq n)$.*

We can say more about a basis.

Theorem 2.1. *Let N be a free module of rank n and let $M \, (\neq 0)$ be a submodule of N. Then there exist a basis $v_1, \cdots, v_n$ of N and natural numbers $a_1, \cdots, a_m \, (m \leq n)$ such that*

$$M = \{b_1(a_1 v_1) + \cdots + b_m(a_m v_m) \mid b_1, \cdots, b_m \in \mathbb{Z}\}$$

and

$$a_1 \mid \cdots \mid a_m.$$

In particular, $a_1 v_1, \cdots, a_m v_m$ is a basis of M.

Proof. We use the induction on the rank of N. If it is 1, the assertion is nothing but Corollary 1.1. Let f be a homomorphism from M to $\mathbb{Z}$. Since $f(M)$ is a submodule of $\mathbb{Z}$, there is an integer $a_f \, (\geq 0)$ such that

$$f(M) = a_f \mathbb{Z}.$$

Let $u_1, \cdots, u_n$ be a basis of N. By the assumption $M \neq 0$, there is a vector $v = c_1 u_1 + \cdots + c_n u_n \in M$ with $c_j \neq 0$ for some j. Defining a homomorphism f by $f(\sum_i b_i u_i) = b_j$, we have $f(v) = c_j \neq 0$ and so $f(M) \neq 0$, i.e. $a_f > 0$. Thus we can choose a homomorphism g from M to $\mathbb{Z}$ so that

$$a_g = \min\{a_f \mid a_f > 0\},$$

where f stands for a homomorphism from M to $\mathbb{Z}$, and put $a_1 = a_g$; then

$$g(M) = a_1 \mathbb{Z}$$

and so there is an element $w \in M$ such that

$$g(w) = a_1.$$

Let us show that $f(w)$ is a multiple of a_1 for every homomorphism f from M to $\mathbb{Z}$. Suppose that $f(w)$ is not divisible by a_1; then putting $c = (a_1, f(w))$, $0 < c < a_1$ holds. Writing $c = c_1 a_1 + c_2 f(w)$ $(c_1, c_2 \in \mathbb{Z})$, we see that $(c_1 g + c_2 f)(M) \ni (c_1 g + c_2 f)(w) = c \, (< a_1)$. This contradicts the minimality of a_1. Thus $f(w)$ is a multiple of a_1 for every homomorphism f from M to $\mathbb{Z}$.

Define a homomorphism f_j by

$$f_j\left(\sum b_j u_j\right) = b_j.$$

By what we have proved, $f_j(w)$ is a multiple of a_1 for $j = 1, \cdots, n$. Thus $w = a_1 v_1$ holds for some $v_1 \in N$ and $g(v_1) = 1$ by $g(w) = a_1$. Put

$$N_1 = \ker g, \quad M_1 = M \cap N_1.$$

Let us see

$$N = \mathbb{Z} v_1 \oplus N_1 \quad \text{and} \quad M = \mathbb{Z} a_1 v_1 \oplus M_1. \tag{2.3}$$

Take $v \in N$; then $g(v - g(v)v_1) = g(v) - g(v) = 0$ is easy and so $v - g(v)v_1 \in N_1$, which implies $v = g(v)v_1 + (v - g(v)v_1) \in \mathbb{Z} v_1 + N_1$, and so $N = \mathbb{Z} v_1 + N_1$. If $a v_1 \in \mathbb{Z} v_1 \cap N_1$, then $a = g(a v_1) = 0$ yields $\mathbb{Z} v_1 \cap N_1 = 0$. Thus $N = \mathbb{Z} v_1 \oplus N_1$ holds.

Next, take $u \in M$. Since $g(u)$ is divisible by a_1, $g(u)/a_1$ is an integer and $g(u - \frac{g(u)}{a_1} w) = g(u) - g(u) = 0$ holds, which gives $u = \frac{g(u)}{a_1} w + (u - \frac{g(u)}{a_1} w) \in \mathbb{Z} w + M_1 (\subset M)$. Thus we have proved $M = \mathbb{Z} w + M_1$. $\mathbb{Z} w \cap M_1 \subset \mathbb{Z} v_1 \cap N_1 = 0$ concludes $M = \mathbb{Z} w \oplus M_1$. Therefore (2.3) has been shown.

Now we can show

$$f(M_1) \subset a_1 \mathbb{Z}$$

for every homomorphism f from M_1 to $\mathbb{Z}$. Define an integer c by $f(M_1) = c\mathbb{Z}$. To see $c \equiv 0 \bmod a_1$, we assume that $c \not\equiv 0 \bmod a_1$ and hence $d = (c, a_1) < a_1$. Writing $d = c_1 c + c_2 a_1$ $(c_1, c_2 \in \mathbb{Z})$, we define a homomorphism f_1 from M to $\mathbb{Z}$ by $f_1(aa_1 v_1 + z) = aa_1 + f(z)$ $(z \in M_1)$, by using (2.3). For an element $z_0 \in M_1$ with $f(z_0) = c$, we have $f_1(c_2 a_1 v_1 + c_1 z_0) = c_2 a_1 + c_1 c = d < a_1$, which contradicts the minimality of a_1.

Since M, N_1 are free $\mathbb{Z}$-modules by Lemma 2.2 and the rank of N_1 is $n-1$, by the induction assumption there exists a basis $v_2, \cdots, v_n$ of N_1 and integers $a_2 \mid \cdots \mid a_m$ such that
$$M_1 = \{b_2 a_2 v_2 + \cdots + b_m a_m v_m \mid b_2, \cdots, b_m \in \mathbb{Z}\}.$$
Applying the above to the homomorphism f from M to $\mathbb{Z}$ defined by
$$f(b_1 a_1 v_1 + \cdots + b_m a_m v_m) = b_2 a_2,$$
we have $a_2 \mathbb{Z} = f(M) \subset a_1 \mathbb{Z}$ and hence $a_1 \mid a_2$. $\qquad\square$

Exercise 5. Let M, N be those in the theorem and let p be a prime number not dividing a_m. Show that if $u \in M$ satisfies $u \in pN$, then $u \in pM$.

Remark 2.1. $a_1, \cdots, a_m$ are called **elementary divisors**. It is known that they are uniquely determined (see the following theorem).

The following theorem is known as **Fundamental Theorem of Finitely Generated Abelian Groups**.

Theorem 2.2. *Let G be a finitely generated abelian group, that is there are finitely many elements $g_1, \cdots, g_n \in G$ such that*
$$G = \{g_1^{b_1} \cdots g_n^{b_n} \mid b_1, \cdots, b_n \in \mathbb{Z}\}.$$
Then G is isomorphic to
$$\mathbb{Z}/(a_1) \oplus \cdots \oplus \mathbb{Z}/(a_\ell) \oplus \underbrace{\mathbb{Z} \oplus \cdots \oplus \mathbb{Z}}_{r},$$
where $1 < a_1 \leq \cdots \leq a_\ell$ and a_{i+1} is a multiple of a_i $(i = 1, \cdots, \ell-1)$. The product a of $a_1, \cdots, a_\ell$, and r are uniquely determined by G.

As a direct consequence of this, we have

Theorem 2.3 (Dirichlet unit theorem). *Let K be an algebraic number field and define non-negative integers r_1, r_2 such that $r_1 + 2r_2 = [K : \mathbb{Q}]$ and isomorphisms σ_i from K into $\overline{\mathbb{Q}}$ so that $K^{\sigma_i} \subset \mathbb{R}$ if and only if $1 \leq i \leq r_1$ and $\sigma_{i+r_2} = \sigma_i J$ for $r_1 + 1 \leq i \leq r_2$, where J is the complex conjugation. Then there exist $u_1, \cdots, u_{r_1+r_2-1} \in \mathfrak{o}_K^\times$ such that $u \in \mathfrak{o}_K^\times$ is written as*
$$u = \zeta u_1^{a_1} \cdots u_{r_1+r_2-1}^{a_{r_1+r_2-1}} \quad (\zeta \in W_K, a_1, \cdots, a_{r_1+r_2-1} \in \mathbb{Z}),$$
and $\zeta, a_1, \cdots, a_{r_1+r_2-1}$ are uniquely determined by u.

$r = r_1 + r_2 - 1$ is called the rank of the unit group $o_K^\times$.

Let K be an algebraic number field. Put

$$o_K^\times = \{u \in o_K \mid u^{-1} \in o_K\}.$$

$o_K^\times$ is a group with respect to the product, and an element of $o_K^\times$ is called a **unit** of K.

$\alpha \in o_K$ is a unit if and only if $N_{K/\mathbb{Q}}(\alpha) = \pm 1$.

$o_\mathbb{Q}^\times = \{\pm 1\}$ is clear.

Proposition 2.1. *Let K be an imaginary quadratic field, i.e. $K = \mathbb{Q}(\sqrt{m})$ for a negative square-free integer m. Then $o_K^\times$ is a finite group consisting of roots of unity.*

Proof. We take a $\mathbb{Z}$-basis of o_K (cf. Proposition 1.12 above) so that

$$o_F = \begin{cases} \mathbb{Z}[1, (1 + \sqrt{m})/2] & \text{if } m \equiv 1 \bmod 4, \\ \mathbb{Z}[1, \sqrt{m}] & \text{if } m \equiv 2, 3 \bmod 4. \end{cases}$$

Suppose $m \equiv 1 \bmod 4$, and put $u = a + b(1 + \sqrt{m})/2 \in o_K^\times$ $(a, b \in \mathbb{Z})$; then $\pm 1 = N_{K/\mathbb{Q}}(u) = u\bar{u} = (a + \frac{b}{2})^2 + \frac{b^2}{4}|m|$ yields $(2a + b)^2 + b^2|m| = 4$. If $u \neq \pm 1$ (whence $b \neq 0$), we have $m = -3$ and $b = \pm 1, 2a + b = \pm 1$. Thus $o_K - \times$ is a finite group.

Suppose $m \equiv 2$ or $3 \bmod 4$ and put $u = a + b\sqrt{m} \in o_K^\times$. $N_{K/\mathbb{Q}}(u) = u\bar{u} = a^2 + b^2|m| = \pm 1$ has solutions $a = \pm 1, b = 0$ or $a = 0, b = \pm 1$ only when $m = -1$. Thus $o_K^\times$ is a finite group. $\square$

The proof shows that $o_K^\times = \{\pm 1\}$ for an imaginary quadratic field $K \neq \mathbb{Q}(i), \mathbb{Q}(\rho)$ in Example 1.3.

Exercise 6. Show that $o_K^\times = \{\pm 1, \pm i\}$ for $K = \mathbb{Q}(i) = \mathbb{Q}(\sqrt{-1})$, and $o_K^\times = \{\pm 1, \pm \rho, \pm \rho^2\}$ for $K = \mathbb{Q}(\rho) = \mathbb{Q}(\sqrt{-3})$.

Proposition 2.1 is in conformity with Theorem 2.3 with $r = 0$.

2.2   Galois extensions

Continuing the theory of field extensions in Chapter 1, **I** we shall state rudiments of Galois theory, which describes the structure of a polynomial through the group of permutations of its roots (acting on its splitting field).

Definition 2.1. If G is a subgroup of the group Aut L of all automorphisms (ring isomorphisms) of a field L, then

$$L^G = \{a \in L | a^\sigma = a, \quad \forall \sigma \in G\} \tag{2.4}$$

forms a subfield of L, called the **fixed field** of G. An algebraic extension L/K is called a **Galois extension** if for a subgroup G of $\mathrm{Gal}(L/K)$ of all K-automorphisms of L (ring isomorphisms form L onto L whose restriction to K is the identity),

$$K = L^G.$$

Or we may also say that L/K is Galois if

$$L^\sigma \subset L \text{ for } \forall \sigma \in \mathrm{Gal}(L/K); \quad L^{\mathrm{Gal}(L/K)} = L.$$

This notion is often used in the following context. If an element θ of a Galois extension L of K is fixed by $\forall \sigma \in \mathrm{Gal}(L/K) : \theta^\sigma = \theta$, then $\theta \in K$. E.g. if $f \in L[X]$ satisfies $f^\sigma = f$ for $\forall \sigma \in \mathrm{Gal}(L/K)$, then $f \in K[X]$.

Theorem 2.4. *An algebraic extension is a Galois extension if and only if it is a normal, separable extension.*

The following fundamental theorem in Galois theory reduces the study of intermediate fields of a Galois extension to that of subgroups of the Galois group. In particular, it is often used in concluding that if $M = L^G$, then $M = K$.

Theorem 2.5. (Fundamental theorem of Galois theory) *Let L/K be a finite Galois extension with Galois group G. Then the extension degree = the order of the Galois group,*

$$[L : K] = |G|, \tag{2.5}$$

and intermediate fields M of L/K and subgroups H of G are one-to-one correspondence under

$$H = \mathrm{Gal}(L/M), \quad M = L^H, \tag{2.6}$$

and $[L : M] = |H|, [M : K] = (G : H)$.

By Theorem 1.21, p. 30, **I**, we have

Corollary 2.1. *If L/K is a separable extension, then there exists a Galois extension L'/K containing L. If $[L : K] < \infty$, then $[L' : K]$ can be taken finite.*

The following is a supplement as well as a continuation of §1.1.4. Let L/K be a separable extension of degree n. Take a Galois extension L'/K containing L according to Corollary 2.1. By Theorem 1.19, p. 30, **I**, there are n distinct K-isomorphisms $\{\sigma_1, \cdots, \sigma_n\}$. For $\alpha \in L$, define its characteristic polynomial F_α (cf. (1.5)) by

$$p_\alpha(X) = (X - \alpha^{\sigma_1}) \cdots (X - \alpha^{\sigma_n}). \tag{2.7}$$

Exercise 7. Prove that $p_\alpha \in K[X]$.

Solution. For any $\tau \in \mathrm{Gal}(L/K)$, $\{\tau\sigma_1, \cdots, \tau\sigma_n\} = \{\sigma_1, \cdots, \sigma_n\}$. Hence $p_\alpha^\tau = p_\alpha$ and so by Theorem 2.5, $p_\alpha \in K[X]$.

Hence, in particular, all symmetric expressions in $\alpha^{\sigma_1}, \cdots, \alpha^{\sigma_n}$ are the elements of K.

Exercise 8. Let $f_\alpha(X) = \mathrm{Irr}(\alpha, K, X)$ be the minimal polynomial of α. Prove that $p_\alpha(X) = f_\alpha(X)^d$, where $d = [L : K(\alpha)]$.

Solution. Since all the roots of $p_\alpha(X)$ are conjugates of α (cf. Corollary 1.6, p. 30, **I**), it follows that irreducible factors of $p_\alpha(X)$ in $K[X]$ are only $f_\alpha(X)$. Hence p_α must be a power of f_α. Since $[K(\alpha) : K] = \deg \mathrm{Irr}(\alpha, K, X)$ and $\deg f_\alpha = [L : K] = n$, it follows that $d[K(\alpha) : K] = n = [L : K]$, whence $d = [L : K(\alpha)]$.

We rephrase Definition 1.4 as follows.

Definition 2.2.

$$\mathrm{Tr}_{L/K}\alpha = \alpha^{\sigma_1} + \cdots + \alpha^{\sigma_n} \in K \tag{2.8}$$

is called the **trace** of α (from L to K) while

$$\mathrm{N}_{L/K}\alpha = \alpha^{\sigma_1} \cdots \alpha^{\sigma_n} \in K \tag{2.9}$$

is called **norm** of α (from L to K). They appear as coefficients of p_α.

Example 2.1. Let L, K, L' be as above and $[L : K] = n$. Then if $\alpha \in K$, then $\mathrm{Tr}_{L/K}\alpha = n\alpha$ and $\mathrm{N}_{L/K}\alpha = \alpha^n$.

2.3   Galois extension of number fields and Artin L-functions

We state a description of the general case of a Galois extension of number fields and specify it to the special case with the ground field being the rational number field as in [Ashnel and Goldfeld (1997)]. The best reference for Artin L-functions is [Martinet (1977)] in which both definitions are exhibited, the first definition for unramfied primes in [Artin (1923)] and the general definition in [Artin (1930)]. Other good references are [Heilbronn (1967)], [Landau (1918), pp. 232-239], [Suetuna (1950), pp. 227-238] which contain the first definition (for unramfied primes) and the universal form for the logarithmic derivative (2.15).

Proposition 2.2. *Suppose that K/k is a finite Galois extension with Galois group $G = \mathrm{Gal}(K/k)$. Let $\mathfrak{P}$ and $\mathfrak{Q}$ be prime ideals of k. Then we have*

$$\mathfrak{P} \cap k = \mathfrak{Q} \cap F \iff \mathfrak{P} = \mathfrak{Q}^\sigma \ for \ ^\exists \sigma \in Gal(K/k).$$

Proposition 2.3. *Suppose K/k is as above and let $\mathfrak{O}_K$ denote the ring of integers in K. Given a prime ideal $\mathfrak{p}$ of k, we have*

$$\mathfrak{p} = \mathfrak{p}\mathfrak{O}_K = (\mathfrak{P}_1 \cdots \mathfrak{P}_g)^e, \ and \ [K:F] = efg \qquad (2.10)$$

where $\mathfrak{P}_i$ are distinct prime ideals of K lying above $\mathfrak{p}$: denoted $\mathfrak{P}_i \mid \mathfrak{p}$. Moreover $f = [\mathfrak{O}_K/\mathfrak{P}_i : \mathfrak{o}_k/\mathfrak{p}]$ is independent of i.

e is called the **ramification index** and f the **residue class degree**. For any $\mathfrak{P} \mid \mathfrak{p}$, we have $N\mathfrak{P} = N\mathfrak{p}^f$.

Let $Z_\mathfrak{P}$ denote the **decomposition group** (Zerlegungsgruppe) of $\mathfrak{P}$ which is a subgroup of G consisting of those automorphisms that fixes $\mathfrak{P}$:

$$Z_\mathfrak{P} = \{\sigma \in G | \sigma\mathfrak{P} = \mathfrak{P}\} \qquad (2.11)$$

and $T_\mathfrak{P}$ denote the **inertia group** (Trägheitsgruppe) of $\mathfrak{P}$ which is a normal subgroup of $G_\mathfrak{P}$ consisting of those automorphisms that induces the trivial automorphism on $(\mathfrak{O}_K/\mathfrak{P})/(\mathfrak{o}_k/\mathfrak{p})$.

We have a **Frobenius automorphism** $\sigma = \left[\frac{K/k}{\mathfrak{P}}\right]$ characterized by

$$\sigma\alpha \equiv \alpha^{N\mathfrak{P}} \pmod{\mathfrak{P}} \qquad (2.12)$$

for $\alpha \in \mathfrak{o}_k$ which is unique up to modulo $T_\mathfrak{P}$.

Let χ be a character of a representation of G (in a finite dimensional vector space over an algebraically closed field, e.g. in $\mathrm{GL}(n,\mathbb{C})$) and let T be a subset of G. We define

$$\chi(T) = \sum_{t \in T} \chi(t). \qquad (2.13)$$

For $m \in \mathbb{N}$ we define

$$\chi(\mathfrak{p}^m) = \frac{1}{e}\chi(\sigma^m T_\mathfrak{P}) = \frac{1}{e}\sum_{t \in T_\mathfrak{P}} \sigma^m t \qquad (2.14)$$

where σ is the Frobenius automorphism defined by (2.12). One can easily check that the notation in (2.14) involving only $\mathfrak{p}$ is justified.

We define the **Artin L-function** in its universal form by

$$\log L(s, \chi, K/k) = \sum_{\mathfrak{p}, m} \frac{\chi(\mathfrak{p}^m)}{mN\mathfrak{p}^{ms}} \tag{2.15}$$

or

$$L(s, \chi, K/k) = \prod_{\mathfrak{p}, m} \exp\left(\frac{\chi(\mathfrak{p}^m)}{mN\mathfrak{p}^{ms}}\right) \tag{2.16}$$

which are absolutely convergent for $\sigma > 1$. It can be proved that Artin L-functions satisfy the functional equation and have Euler products.

Artin conjectured all the Artin L-functions not associated with the principal character are *entire*.

We define $\mathcal{S}_{Artin}$ **class** to be all Artin L-functions introduced above (it seems that integrality is also assumed in Ashnell-Goldfeld [Ashnel and Goldfeld (1997)]). This will be used in §2.6 below.

Finally we introduce Chebotarev's density theorem (cf. e.g. [Lagarias (1997)]), which will be used in the final stage of construction of a PNG. Let K/F be a Galois extension of algebraic number fields, and let C be a subset of $Gal(K/F)$ such that $\sigma C \sigma^{-1} = C$ for $^\forall \sigma \in Gal(K/F)$. For a positive number x, we denote by $\pi_C(x, K/F)$ the number of unramified prime ideals $\mathfrak{p}$ of F such that $N_F(\mathfrak{p}) \leq x$ and there exists an unramified prime ideal $\mathfrak{P}$ of K lying above $\mathfrak{P}$ satisfying $\sigma_{K/F}(\mathfrak{P}) \in C$. Then we have

Theorem 2.6. (Chebotarev)

$$\pi_C(x, K/F) = \frac{\#C}{\#Gal(K/F)} x/\log x + o(x/\log x) \quad if \quad x \to \infty. \tag{2.17}$$

The more appropriate main term is $\mathrm{li}(x) = \int_2^x dt/\log t$ and the improvement of the error term $o(x/\log x)$ is important as in the PNT. Cf. §3.5.

2.4 Examples of number fields

2.4.1 *The quadratic field with the golden section unit*

In this subsection we illustrate the theory of §1.2 by the real quadratic field

$$k = \mathbb{Q}(\sqrt{5}) = \{\alpha = a + b\sqrt{5} \,|\, a, b \in \mathbb{Q}\}. \tag{2.18}$$

We determine those α's which are the roots of the equation

$$x^2 + Ax + B = 0, \ A, B \in \mathbb{Z} \tag{2.19}$$

and prove that these numbers (**algebraic integers** in k) form a ring (indeed, it is an integral domain but customarily called a ring) called the **ring of integers** in k and denoted by $\mathfrak{o} = \mathfrak{o}_k$.

Proof. If $\alpha = a + b\sqrt{5}$ is a root of (2.19), then the conjugate root is $\alpha' = a - b\sqrt{5}$, whose **norm** $N = N_{k/\mathbb{Q}}$ and **trace** $\mathrm{Tr} = \mathrm{Tr}_{k/\mathbb{Q}}$ being

$$N_{k/\mathbb{Q}}\,\alpha = a^2 - 5b^2, \quad \mathrm{Tr}_{k/\mathbb{Q}}\,\alpha = 2a.$$

Since $2a = \mathrm{Tr}\,\alpha = A$, we see that a is a half-integer: $a = \frac{a_1}{2}$, say, $a_1 \in \mathbb{Z}$. Hence the denominator of a^2 is at most 4 and so, in view of the fact that $a^2 - 5b^2 \in \mathbb{Z}$, that of $5b^2$ must be at most 4, or $b = \frac{b_1}{2}$, $b_1 \in \mathbb{Z}$. Hence it follows that $a^2 - 5b^2 = \frac{a_1{}^2 - 5b_1{}^2}{4}$ and $a_1{}^2 - 5b_1{}^2 \equiv a_1{}^2 - b_1{}^2 \bmod 4$, whence $a_1 . b_1$ must be of the same parity.

Then it is easily checked that α is of the form $a + b\tau$, $a, b \in \mathbb{Z}$, where

$$\tau = \frac{1 + \sqrt{5}}{2}, \tag{2.20}$$

sometimes denoted ϕ is the **golden ratio** and b is even or odd according as $a_1 \equiv b_1 \equiv 0 \bmod 2$ or $a_1 \equiv b_1 \equiv 1 \bmod 2$, whence we conclude that

$$\mathfrak{o}_k = \mathbb{Z} \oplus \tau\mathbb{Z}. \tag{2.21}$$

$\square$

Note that the proof used to derive (2.21) applies to any quadratic fields and gives the integral basis (cf. Proposition 1.12): If $k = \mathbb{Q}(\sqrt{d}) = \{\alpha = a + b\sqrt{d} | a, b \in \mathbb{Q}\}$, where d is a square-free integer, then the ring of integers is of the form $\mathfrak{O} = \mathbb{Z} \oplus \omega\mathbb{Z}$, where $\omega = \frac{1+\sqrt{d}}{2}$ in the form (2.20) if $d \equiv 1 \pmod 4$, while it is in the form $1 + \sqrt{d}$ if $d \equiv 2$ or $3 \pmod 4$. The discriminant D_k is d or $4d$, respectively.

The invertible elements of $\mathfrak{o}$ are called **units** which constitute a group $\mathfrak{o}^\times$ (**unit group**) and denoted by $U = U_k$. The following is known and easily proved (cf. e.g. [Hartman and Wintner (1938)]).

- If $\xi = x + y\tau$, $x, y \in \mathbb{Z}$, then $N\xi = x^2 + xy - y^2$, an **indefinite quadratic form**.

- $\xi \in U$ if and only if $|N\xi| = 1$.

- The only elements of U of finite order are ± 1, which constitute the **torsion subgroup** $< -1 >$, with $< \alpha >$ designating the subgroup generated by α.

- A generator (> 1) of the group $U/\{\pm 1\}$ is called a **fundamental unit**.

In our case, τ is a fundamental unit and

Theorem 2.7. (Dirichlet unit theorem) $U = \{\pm 1\} \times < \tau >$.

Thus the field $\mathbb{Q}(\sqrt{5})$ is sometimes called the **field of the golden section unit**.

Theorem 2.8. $\mathbb{Q}(\sqrt{5})$ *is a Euclidean field and hence is a UFD, or what amounts to the same thing, the* **class number** $h = h_k$, *the order of the* **ideal class group** $C = I/P$ *is 1.*

For more detailed information, cf. Theorem 2.12 below.

For curiosity as well as relevance to self-adjoint operators in Hilbert space in Chapter 5, we state a topic from atomic orbital theory involving the golden ratio.

In [Chakraborty, Kanemitsu and Tsukada (2009), §1.4], we applied Chebyshëv polynomials to the theory of molecular orbitals of hydrocarbon and compute their energy levels. First we make a brief statement on quantum mechanics.

In quantum mechanics, one assumes that the totality of all states of a system forms a Hilbert space V over $\mathbb{C}$ and that all (quantum) mechanical quantities are expressed as hermitian operators or self-adjoint operators $A : V \to V$ (cf. Definition 5.10). For a hermitian operator A, the (normalized) eigenvectors $\boldsymbol{v}$ belonging to its eigenvalue $\lambda \, (\in \mathbb{R})$ is viewed as the quantum state whose mechanical quantity is equal to λ. The hermitian operator H expressing the energy of a system is called the **Hamiltonian** and expressed as

$$H = -\frac{\hbar}{2m}\Delta^e + V(\boldsymbol{r}), \qquad (2.22)$$

where m is the mass, $V(\boldsymbol{r})$ the potential energy, of the particle and $\Delta^e = \frac{\partial^2}{\partial^2 x} + \frac{\partial^2}{\partial^2 y}\frac{\partial^2}{\partial^2 z}$ is the Euclidean Laplacian (cf. (5.150), Chapter 5) and $\hbar = \frac{h}{2\pi} = 1.0545887 \times 10^{-34} J \cdot s$ is called the **Planck constant** The quantum state of the system $\boldsymbol{v} = \boldsymbol{v}(t) = \boldsymbol{v}(t, \boldsymbol{r})$ varies with time variable t according to the **Schrödinger equation**

$$i\hbar\frac{\mathrm{d}}{\mathrm{d}t}\boldsymbol{v}(t) = H\boldsymbol{v}(t).$$

If $Hv(t) = Ev(t)$, E being real and called the **energy levels** of the system, the solution is given by $v(t) = e^{-\frac{iEt}{\hbar}} v(0)$ and is called the stationary state on the ground that its expectation does not change with time. The energy level means the values of the energy which the stationary state can assume.

We may study the difference between energy levels of molecular orbitals of a chain-figured polyene (1, 3-butadiene) and a ring-shaped one (1, 3-cyclobutadiene). Since the latter is straightforward, we dwell on the former.

We recall the rudiments of the method. We apply the simple LCAO (Linear Combination of Atomic Orbitals) method with the overlapping integrals $S_{ij} = 0$, whereby we also incorporate the simple Hückel method with Coulomb integral of the carbon atom in the $2p$ orbit be α, and the resonance integral between neighboring C-C atoms in the $2p$ orbit be β. We assume that the Coulomb integral of the atomic orbital ϕ_i is $\alpha + h_{ii}\beta$ ($1 \leq i \leq 6$), and the resonance integral between ϕ_i and ϕ_j is $h_{ij}\beta$. Further, for simplicity, we assume that only the resonance integrals of neighboring atoms are non-zero, and normalize so that $h_{ii} = 1$. Then the **secular determinant** of 1, 3-butadiene is

$$\begin{vmatrix} \alpha - E & \beta & 0 & 0 \\ \beta & \alpha - E & \beta & 0 \\ 0 & \beta & \alpha - E & \beta \\ 0 & 0 & \beta & \alpha - E \end{vmatrix} = 0, \qquad (2.23)$$

where α is the Coulomb integral and β the resonance integral. This reduces to

$$\begin{vmatrix} \lambda & 1 & 0 & 0 \\ 1 & \lambda & 1 & 0 \\ 0 & 1 & \lambda & 1 \\ 0 & 0 & 1 & \lambda \end{vmatrix} = 0 \qquad (2.24)$$

after setting $\lambda = \frac{\alpha - E}{\beta}$.

In order to determine the atomic orbitals of 1, 3-butadiene, we are to solve the system of equations

$$\begin{cases} c_1\lambda + c_2 & = 0 \\ c_1 + c_2\lambda + c_3 & = 0 \\ c_2 + c_3\lambda + c_4 = 0 \\ c_3 + c_4\lambda = 0 \end{cases} \qquad (2.25)$$

Let τ denote any one of ϕ, ϕ^{-1} and $-\phi^{-1}, -\phi$, the former being the roots of the equation $\tau^2 - \sqrt{5}\tau + 1 = 0$ and the latter $\tau^2 + \sqrt{5}\tau + 1 = 0$. The common property is that

$$\tau^2 - 1 - \tau^2(\tau^2 - 2) = \tau^4 - 3\tau^2 + 1 = 0, \qquad (2.26)$$

whence

$$\tau(\tau^2 - 2) = \tau - \frac{1}{\tau}. \qquad (2.27)$$

Since $\phi - \phi^{-1} = 1$ we may immediately fill the first column. And at the same time, the second column by multiplying the value of τ.

The coefficient matrix for the eigenspace $E_A(\tau)$ is

$$A_\tau = \begin{pmatrix} \tau & 1 & 0 & 0 \\ 1 & \tau & 1 & 0 \\ 0 & 1 & \tau & 1 \\ 0 & 0 & 1 & \tau \end{pmatrix} \sim \begin{pmatrix} 1 & 0 & 0 & \tau(2 - \tau^2) \\ 0 & 1 & 0 & 1 - \tau^2 \\ 0 & 0 & 1 & \tau \\ 0 & 0 & 0 & 0 \end{pmatrix} \qquad (2.28)$$

by fundamental row operations, where in the last step we use (2.26). Hence putting $c_4 = c$, we may obtain

$$c_1 = \tau(2 - \tau^2)c, \quad c_2 = (\tau^2 - 1)c, \quad c_3 = -c\tau, \qquad (2.29)$$

i.e. the first row of the table. We need to normalize the coefficients. Since the length of the vector is $2c^2(1 + \tau^2) = 2c^2\sqrt{5}|\tau|$, there are two cases: for $\tau = \phi, -\phi$, $c = \frac{1}{\sqrt{5+\sqrt{5}}} = \frac{1}{\sqrt{7.236}} = 0.37$ and for $\tau = \phi^{-1}, -\phi^{-1}$, $c = \frac{1}{\sqrt{5-\sqrt{5}}} = \frac{1}{\sqrt{2.764}} = 0.60$.

τ	$\tau(2-\tau^2)$	τ^2-1	c_1	c_2	c_3	c_4	c
ϕ	-1	ϕ	$-c$	$c\phi$	$-c\phi$	c	-0.37
ϕ^{-1}	1	$-\phi^{-1}$	c	$-c\phi^{-1}$	$-c\phi^{-1}$	c	0.60
$-\phi^{-1}$	-1	$-\phi^{-1}$	$-c$	$-c\phi^{-1}$	$c\phi^{-1}$	c	-0.60
$-\phi$	1	ϕ	c	$c\phi$	$c\phi$	c	$0,37$

Table 2.1. Values of coefficients

By this table, the molecular orbital of $1,3$-butadiene C_4H_6 can be immediately read off:

$$E_4 = \alpha - \phi\beta \qquad \psi_4 = 0.37\varphi_1 - 0.60\varphi_2 + 0.60\varphi_3 - 0.37\varphi_4 \qquad (2.30)$$
$$E_4 = \alpha - \phi^{-1}\beta \qquad \psi_4 = 0.60\varphi_1 - 0.37\varphi_2 - 0.37\varphi_3 + 0.60\varphi_4$$
$$E_4 = \alpha + \phi^{-1}\beta \qquad \psi_4 = 0.60\varphi_1 + 0.37\varphi_2 - 0.37\varphi_3 - 0.60\varphi_4$$
$$E_4 = \alpha + \phi\beta \qquad \psi_4 = 0.37\varphi_1 + 0.60\varphi_2 + 0.60\varphi_3 + 0.37\varphi_4$$

2.4.2 *Cyclotomic fields*

Let $\eta = \eta_k = e^{2\pi i \frac{1}{k}}$ denote the first primitive k-th root of 1 in $\mathbb{C}$ and let μ_k denote the group of all k-th roots of 1, which is generated by η, We make an important remark that *the set of all k-th roots of 1 and the set of all primitive d-th roots of 1 for all $d|k$ are the same.* This fact is often described as

$$X^k - 1 = \prod_{d|k} \Phi_d(X), \tag{2.31}$$

where $\Phi_d(X) = \prod_{(m,k)=1}(X - \eta^m)$ is the d-th **cyclotomic polynomial** whose roots are all the primitive d-th roots of 1. (2.31) implies in particular (1.17), p. 10, **I**, which is the inversion of (3.11).

We start by stating other equivalent formulations of $\varphi(q)$.

Recalling the cyclic group μ_q of q-th roots of 1 in Example 1.1, **I**, we have

$$\varphi(q) = \text{the number of \textbf{primitive } } q\text{-th roots of 1} \tag{2.32}$$
$$\text{(i.e. the generators) of } \mu_q.$$

Secondly, let ζ be a primitive q-th root of 1 and let $\mathbb{Q}(\zeta)$ be the q-th **cyclotomic field**. Since

$$\text{Gal}(\mathbb{Q}(\zeta)/\mathbb{Q}) \cong (\mathbb{Z}/q\mathbb{Z})^\times, \tag{2.33}$$

it follows that

$$\varphi(q) = |\text{Gal}(\mathbb{Q}(\zeta)/\mathbb{Q})| \tag{2.34}$$

and from Galois theory

$$\varphi(q) = [\mathbb{Q}(\zeta) : \mathbb{Q}], \tag{2.35}$$

the extension degree of $\mathbb{Q}(\zeta)$ over $\mathbb{Q}$.

Now we consider some interesting extensions of $\mathbb{Q}(\sqrt{5})$, which turn out to be of degree 4. First

Proposition 2.4. *The splitting field B of the polynomial $f(X) = X^4 - 8X^2 + 36$ is the biquadratic field $\mathbb{Q}(i, \sqrt{5})$, which is a Galois extension of $\mathbb{Q}$ with Galois group V_4, the Klein four group, i.e. the direct product of two cyclic groups of order 2. All the subfields of B (save for B and $\mathbb{Q}$) are $\mathbb{Q}(\sqrt{5})$, $\mathbb{Q}_4 = \mathbb{Q}(\sqrt{-4})$ and $k = \mathbb{Q}(\sqrt{-5})$.*

Proof. The quadratic equation $X^2 - 8X + 36 = 0$ has the roots $4 \pm 2\sqrt{5}i$. Since $\sqrt{4 \pm 2\sqrt{5}i} = i \pm \sqrt{5}$, the roots of $f(X)$ are $\pm(i \pm \sqrt{5})$. Since these are contained in B and

$$\frac{1}{2}\left(i + \sqrt{5} - (i - \sqrt{5})\right) = \sqrt{5}, \frac{1}{2}(i + \sqrt{5} + i - \sqrt{5}) = i,$$

we have $B = \mathbb{Q}(i, \sqrt{5})$. Since B is obtained as the adjoint of i to $k = \mathbb{Q}(\sqrt{5})$ and $X^2 + 1$ is irreducible over k, B is of degree 4 over $\mathbb{Q}$. As the splitting field of the irreducible polynomial f, B is a Galois extension whose Galois group $G = \mathrm{Gal}(K/\mathbb{Q})$ is not cyclic. Since the groups of order 4 are Abelian, it must be isomorphic to V_4 by Exercise 11, p. 19, **I**. It is given as $< \sigma > \times < \tau >$, where

$$\sigma(i) = -i, \sigma(\sqrt{5}) = \sqrt{5}, \tau(i) = i, \tau(\sqrt{5}) = -\sqrt{5},$$

and $\sigma^2 = 1, \tau^2 = 1, \sigma\tau = \tau\sigma$: $G = \{1, \sigma, \tau, \sigma\tau\}$. All the three subgroups of G and their fixed fields are given by

- $\{1, \sigma\}$ whose fixed field is $\mathbb{Q}(\sqrt{5})$,

- $\{1, \tau\}$ whose fixed field is $\mathbb{Q}(\sqrt{i})$,

- $\{1, \sigma\tau\}$ whose fixed field is $\mathbb{Q}(\sqrt{-5})$.

□

We use the following results from algebraic number theory to determine the class number $h = h_k$ and the Hilbert class field of $k = \mathbb{Q}(\sqrt{-5})$.

Theorem 2.9. *Let k be an arbitrary ground field (a finite extension of $\mathbb{Q}$) and let $\tilde{k}$ be its **Hilbert class field** (or **absolute class field**) of k, i.e. the maximal unramified Abelian extension of k. Then*

$$\mathrm{Gal}(\tilde{k}/k) \approx C = I/P. \tag{2.36}$$

By $\mathbb{Q}_m$ we mean the mth **cyclotomic field** $\mathbb{Q}(\zeta_m)$ with ζ_m designating the primitive mth root of 1 and $\mathbb{Q}(\zeta_m)_+ = \mathbb{Q}(\zeta_m + \zeta_m^{-1})$ its **maximal real subfield**, i. e. the real subfield k_+ with extension degree $[\mathbb{Q}(\zeta_m) : k_+] = 2$.

Theorem 2.10. (Kronecker-Weber theorem) *If $k/\mathbb{Q}$ is Abelian, then there exists an integer $2 < m \in \mathbb{N}$ such that*

$$k \subset \mathbb{Q}_m,$$

m being called an admissible modulus.

Corollary 2.2. *For an Abelian extension $k/\mathbb{Q}$, there exists the least admissible modulus $f = f_k$, called the **conductor**, such that $k \subset \mathbb{Q}_m$.*

Proposition 2.5. *If $k = \mathbb{Q}_m$ or $k = \mathbb{Q}_m^+ = \mathbb{Q}(\zeta_m + \zeta_m^{-1})$, then $f_k = m$. If $k = \mathbb{Q}(\sqrt{d})$ with d square-free, then*

$$f_k = \begin{cases} |d| & d \equiv 1 \bmod 4, \\ |4d| & d \equiv 2,3 \bmod 4. \end{cases}$$

It follows that for $k = \mathbb{Q}(\sqrt{5})$, $f_k = 5$ and $k = \mathbb{Q}_5^+$, the maximal real subfield of the 5th cyclotomic field. Since the Galois group $\mathrm{Gal}(\mathbb{Q}_5/\mathbb{Q})$ is isomorphic to $(\mathbb{Z}/5\mathbb{Z})^\times$ which is a cyclic group of order $\varphi(5) = 4$ and that of B is not cyclic, $k = \mathbb{Q}(\sqrt{5})$ and B are not the same. Cf. Exercise 9 below.

Theorem 2.11. (Conductor-ramification theorem) *Suppose $k/\mathbb{Q}$ be Abelian. Then the rational prime p ramifies in k if and only if $p|f_k$.*

The following formula is due to Dirichlet (cf. [Borevič and Šafarevič (1964), p. 346]) and proved in a general form by Funakura [Funakura (1990)]. A more general theorem is given as Theorem 5.11, p. 145 **I**.

Proposition 2.6. *Let $k = \mathbb{Q}(\sqrt{d})$, $d < -2$ and suppose the conductor f_k of k is odd. Then*

$$h_k = \frac{1}{2 - \left(\frac{d}{2}\right)} \sum_{\substack{0 < a < f_k/2 \\ (a, f_k) = 1}} \left(\frac{d}{a}\right), \tag{2.37}$$

where $\left(\frac{d}{a}\right)$ is the Kronekcer symbol.

Example 2.2. For $k = \mathbb{Q}(\sqrt{-5})$ we have $h_k = 2$. Hence $\mathfrak{o}_k$ is not a UFD. An example is $6 = 2 \cdot 3 = (1 + \sqrt{5})(1 - \sqrt{5})$. We also have $\tilde{k} = \mathbb{Q}(i, \sqrt{5})$.

Proof. Proposition 2.6 reads in this case

$$h_k = \frac{1}{2 - \chi_{-4}(2)} \sum_{a = 1, 3, 7, 9} \chi_{-4}(a) \left(\frac{a}{5}\right), \tag{2.38}$$

where χ_{-4} is the real primitive odd character to the modulus 4: $\chi_{-4}(-1) = -1$. Hence

$$h_k = \frac{1}{2}\left(1 + \chi_{-4}(-1)\left(\frac{-2}{5}\right) + \chi_{-4}(-1)\left(\frac{2}{5}\right) + \chi_{-4}(1)\left(\frac{2}{5}\right)\right) = 2 \tag{2.39}$$

because $\left(\frac{-1}{5}\right) = (-1)^{\frac{5-1}{2}} = 1$ and $\left(\frac{2}{5}\right) = (-1)^{\frac{5^2-1}{8}} = -1$.

Hence by Theorem 2.9, $|\mathrm{Gal}(\tilde{k}/k)| = 2$.

By Galois theory, this implies $[\tilde{k} : k] = 2$. Since $\mathbb{Q}(i, \sqrt{5})/k$ is unramified, we have $\mathbb{Q}(i, \sqrt{5}) \subset \tilde{k}$, whence we conclude the second assertion. $\square$

Exercise 9. Let p be an odd prime and let $\zeta_p = e^{\frac{2\pi i}{p}}$ be the first primitive pth root of 1. Prove that the cyclotomic field $\mathbb{Q}(\zeta_p)$ contains a unique quadratic subfield given either by $\mathbb{Q}(\sqrt{p})$ for $p \equiv 1 \,(\mathrm{mod}\,4)$ or $\mathbb{Q}(\sqrt{-4p})$ for $p \equiv -1 \,(\mathrm{mod}\,4)$.

Solution. We use some well-known facts on cyclotomic fields. The Galois group of $\mathbb{Q}(\zeta_p)/\mathbb{Q}$ is isomorphic to $(\mathbb{Z}/p\mathbb{Z})^\times$ which is a cyclic group of order $\varphi(p) = p - 1$.

Now all the subgroups of a cyclic group are cyclic and they are determined uniquely by their orders which are the divisors of $p - 1$. Hence for each divisor d of $p - 1$, there is a unique (up to isomorphism) subgroup with order d, and *a fortiori* there is a unique subgroup of order $\frac{\varphi(p)}{2} = \frac{p-1}{2}$ to which there corresponds, by Galois theory, a unique fixed field of degree 2. Since the possible discriminants in our case are p or $-4p$ according as $p \equiv 1 \,(\mathrm{mod}\,4)$ or $p \equiv -1 \,(\mathrm{mod}\,4)$ (cf. e.g. [Davenport (2000), p. 43]), the assertion follows.

Exercise 10. Solve the 5th cyclotomic equation $X^4 + X^3 + X^2 + X^2 + X + 1 = 0$ by two different ways.

To sum up, we have obtained

Theorem 2.12. *The quadratic field of the golden section unit* $k = \mathbb{Q}(\sqrt{5})$ *is*

- *one of known real quadratic Euclidean fields.*

- *the maximal real subfield (as well as the unique real quadratic subfield) of the 5th cyclotomic field* $\mathbb{Q}_5 = \mathbb{Q}(\zeta_5)$.

- *the unique real quadratic subfield of the bicyclic biquadratic field B which is the Hilbert class field of k.*

Toward the end of this subsection, we shall make use of (2.31) to prove a special case of the celebrated **Dirichlet prime number theorem** to the effect that every arithmetic progression a mod q, $(a, q) = 1$ contains infinitely many primes.

Theorem 2.13. (R. A. Smith [Smith (1981)]) *The congruence*

$$X^q \equiv 1 \pmod{p} \tag{2.40}$$

is solvable with an integer of order q mod p for infinitely many primes p, and a fortiori, there exist infinitely many primes p such that $p \equiv 1 \pmod{q}$.

For a proof we need a lemma.

Lemma 2.3. (T. Nagel) *If $f \in \mathbb{Z}[X]$ is not a constant, then the congruence*

$$f(x) \equiv 1 \pmod{p} \tag{2.41}$$

is solvable for infinitely many primes p.

Proof. We prove this in the form:
$f(x)$ $(x \in \mathbb{Z})$ has infinitely many prime divisors.

Suppose $f(X) = a_0 X^m + \cdots + a_{m-1} + a_m$, $m \in \mathbb{N}$. If $a_m = 0$, then all the primes are prime divisors of $f(x)$. Hence we may suppose $a_m \neq 0$. Suppose the contrary to the assertion, i.e. that $f(x)$ has only finitely many prime divisors $p_1, \cdots, p_r$, say. Let $b = p_1 \cdots p_r a_m$. Then clearly,

$$f(bY) = a_m g(Y),$$

where $g(Y) = A_m Y^m + \cdots + A_1 Y + 1$, and where all A_i's are divisible by $p_1 \cdots p_r$.

Every prime divisor of $g(y)$ is a prime divisor of $f(x)$ and none of $p_1, \cdots, p_r$ can be a prime divisor of $g(y)$ in view of its constant term 1. Hence $g(y) = \pm 1$ for all $y \in \mathbb{Z}$. However, this equation $g(y) = \pm 1$ has at most $2m$ roots, and so we may find an integer y such that $g(y)$ is divisible by a prime p_{r+1} other than $g(y) = \pm 1$. Since p_{r+1} is a prime divisor of $f(x)$, we have a contradiction, completing the proof. $\square$

Proof of Theorem 2.13 Recall the p-adic valuation in Example 2.3, p. 58 **I**: For $x \in \mathbb{Z}$, $x = p^{v_p(x)} y$ with $y \in \mathbb{Z}$, $(y, p) = 1$. Substituting the value of $x \in \mathbb{Z}$, $x \neq \pm 1$ in (2.31) and taking the p-adic valuation of both sides, we obtain

$$v_p(x^n - 1) = \sum_{d \mid n} v_p(\Phi_p(x)), \tag{2.42}$$

which is inverted to

$$v_p(\Phi_n(x)) = \sum_{d|n} \mu\left(\frac{n}{d}\right) v_p(x^n - 1) \tag{2.43}$$

by the Möbius inversion (3.4). By Lemma 2.3, there exist infinitely many primes p such that

$$\Phi_n(x) \equiv 1 \pmod{p} \tag{2.44}$$

is solvable in $\mathbb{Z}$. Fix such a prime p and choose an $x > n$ satisfying (2.44). Recall the notion of the order of an element of a group and let $f = o(x)$, i.e. the order of x in $(\mathbb{Z}/p\mathbb{Z})^\times$. Then clearly $f \mid n$. It therefore suffices to prove that $f = n$.

If $f \nmid d$, then $v_p(x^d - 1) = 0$. Hence (2.43) becomes

$$v_p(\Phi_n(x)) = \sum_{\substack{d|n \\ f|d}} \mu\left(\frac{n}{d}\right) v_p(x^n - 1) = \sum_{d|m} \mu\left(\frac{m}{d}\right) v_p(x^d f - 1) \tag{2.45}$$

on writing df for d, where $m = \frac{n}{f}$, which is relatively prime to p.

Now by Exercise below, each summand $v_p(x^d f - 1)$ amounts to $v_p(x^f - 1)$, which can therefore be factored out, whereby we obtain

$$v_p(\Phi_n(x)) = v_p(x^f - 1) \sum_{d|m} \mu\left(\frac{m}{d}\right). \tag{2.46}$$

By (3.7), the right-hand side of (2.46) is 0 if $m > 1$. But since $v_p(\Phi_n(x)) \geq 1$, it follows that $m = 1$, amounting to $f = n$. This complete the proof.

Exercise 11. Prove for $(d, p) = 1$ that $v_p(x^d f - 1) = v_p(x^f - 1)$.

2.4.3 *The dihedral group as a Galois group*

Generalizing the situation in Proposition 2.4 slightly, consider the bi-quadratic field $\mathbb{Q}(i, \sqrt{p})$ for an odd prime p, which is a Galois extension of $\mathbb{Q}$ with Galois group V_4 and its 3 subfields of $\mathbb{Q}(\sqrt{p})$, $\mathbb{Q}_4 = \mathbb{Q}(\sqrt{-4})$ and $\mathbb{Q}(\sqrt{-p})$. P. Chowla proves the following theorem.

Theorem 2.14. *Suppose $p \equiv 5 \bmod 8$ and let h and ε be the class number of $\mathbb{Q}(\sqrt{p})$ and let H be the class number of $\mathbb{Q}(\sqrt{-p})$. Further let $\left\{\frac{v}{2}\right\}$ denote the least (positive) solution of the congruence $2\left\{\frac{v}{2}\right\} \equiv v \bmod p$. Then*

$$\varepsilon^h = (-1)^m \prod_{v < \frac{p}{2}} 2 \cos \frac{2\pi}{p} \left\{\frac{v}{2}\right\}, \tag{2.47}$$

where v runs through all quadratic residues $< \frac{p}{2}$ and

$$m = \frac{1}{2} \left\{\frac{p-1}{4} + \frac{1}{2}H\right\}. \tag{2.48}$$

In this subsection we shall prove (following [Chakraborty, Kanemitsu and Kuzumaki (2013)]) that Theorem 2.14 does not give any relation between the class number of the real and imaginary quadratic fields $\mathbb{Q}(\sqrt{p})$, $\mathbb{Q}(\sqrt{-p})$ but is a restatement of the classical theorem of Dirichlet ([Borevič and Šafarevič (1964), Theorem 2, p. 362]).

Theorem 2.15. (Dirichlet) *If k is a real quadratic field with discriminant d, class number h and the fundamental unit $\varepsilon > 1$, then*

$$\varepsilon^h = \frac{\prod \sin \frac{\pi n}{d}}{\prod \sin \frac{\pi v}{d}}, \tag{2.49}$$

where v and n run through those integers in $(0, \frac{p}{2})$ such that $\left(\frac{v}{d}\right) = 1$ and $\left(\frac{n}{d}\right) = -1$.

Corollary 2.3. *If $p \equiv 5 \bmod 8$, then Chowla's theorem (Theorem 2.14) is a reformulation of Dirichlet's theorem (Theorem 2.15) in conjunction with another formula of Dirichlet*

$$2S_{1/4}(\chi_p) = H, \tag{2.50}$$

where $S_{1/4}(\chi_p) = \sum_{0<a<p/4} \left(\frac{a}{p}\right)$ with $\chi_p(a) = \left(\frac{a}{p}\right)$ the Legendre symbol.

Proof. Since $\left(\frac{2}{p}\right) = -1$, the non-residues are given in the form $2v$. Hence (2.49) may be written as

$$\varepsilon^h = \prod \frac{\sin \frac{2\pi v}{p}}{\sin \frac{\pi v}{p}} = \prod_{0<v<p/2} 2\cos \frac{\pi v}{p}. \tag{2.51}$$

Noting that $\{\frac{v}{2}\}$ also means the least positive residue $\bmod\, p$ of $2^{-1}v$, where 2^{-1} is the inverse of $2 \bmod p$:

$$\left\{\frac{v}{2}\right\} = 2^{-1}v - p\left[\frac{2^{-1}r}{p}\right],$$

we may express $\{\frac{v}{2}\}$ more concretely as

$$\left\{\frac{v}{2}\right\} = \begin{cases} \frac{v}{2} & v \text{ even}, \\ \frac{v+p}{2} & v \text{ odd}. \end{cases} \tag{2.52}$$

Hence if we replace v in (2.51) by $\{\frac{v}{2}\}$, then those terms with odd v are negative and there are m' of them, where m' indicates the number of odd residues in $(0, \frac{p}{2})$. Hence it follows that

$$\varepsilon^h = (-1)^{m'} \prod_{\substack{0<v<p/2 \\ 2\nmid v}} 2\cos 2\pi \left\{\frac{v}{2}\right\}. \tag{2.53}$$

The remaining part remains the same as in [Chowla (1968)] and we may prove by elementary argument that

$$m' = \frac{1}{2}\left\{\frac{p-1}{4} + \frac{1}{4}S_{1/4}\right\},$$

which is m in view of (2.51), completing the proof. $\qquad\square$

2.5 Ciphers vs. PRNG

Ciphers and codes are two sides of a coin. Codes are mainly used for sending data not only spacially but also temporarily. In order to be able to recipher the sent data for extremely long distance, like from Mars to the Earth, which are naturally disturbed by noise or to restore the saved data after preserving important data temporary as well as the elapse of time, which could be degraded, the (error-correcting) codes are essential.

On the other hand, ciphers have been used so that the secret messages would not be eavesdropped. Cf. [Kanemitsu (2010)], [Kanemitsu (2011)].

In our computer era, security of communication is all the more crucial since they are mostly sent electronically and are more feasible for eavesdropping as long as we rely on the public key crypto-system in which one uses a plausible one-way function, where a **one way function** is one whose values can be computed in polynomial time (abbreviated as P) while its inverse is not. The existing candidate one-way functions depend in one way or another on seemingly intractable problems in number theory, e.g. factorization of large integers, computation of the discrete logarithms etc. A typical one is the RSA (Rivest-Shamir-Adleman) crypto-system which uses the product of two large primes different by the order of a few digits and the product is made public. Cf. Exercise 17. Thus the crypt hinges on the difficulty of factorizing the publicized integer. But if a plausible quantum computers are realized, such crypts will become vulnerable.

The general class of questions for which some algorithm can provide an answer in **polynomial time** is called "class P" or just "P". For some questions, there is no known way to find an answer quickly, but it is possible to verify the (provided) answer quickly. The class of questions for which an answer can be verified in polynomial time is called NP, **nondeterministic polynomial time**. It is generally believed that $P \neq NP$ which is one of the most important unsolved problems in theoretical computers. There is no guarantee that one-way function exists even if $P \neq NP$. In view of this, it seems worthwhile pursuing for a candidate one-way function.

On the other hand, random number generation has been the main issue since the dawn of cryptography so as to secure secret messages. With development of computers, the situation has been changing in that instead of using genuine random number systems, one uses a system generated by computers, which are by definition deterministic and never genuinely random but just pseudo-random. However, since there is a limitation of the ability of the existing computers, anything that cannot be computed in **polynomial time** by them may be regarded as uncomputable.

In this regard, Goldreich et al's work [Goldreich, Krawczyk and Luby (1971)] is epoch-making, which is to the effect that if there exists a *one-to-one* one-way function, then there exists a pseudorandom number generators, PNG, i.e. they have established a link between one way functions and PRG.

Motivated by this, Ashnell and Goldfeld [Ashnel and Goldfeld (1997)] made use of the **abundant subclass** of the **Selberg class** of zeta-functions and constructed a conjectural PRN generator under the assumption that the Fourier projection for the subclass of zeta-functions associated with elliptic curves and the subclass of Artin L-functions is a one-way function. The first result in this direction was about the class of Dirichlet L-functions by Damgard [Damgard (1988)].

This direction of research will be of crucial importance not only in the intended application to the construction of the PRG but also on the possibility of attacking the most formidable remaining problem in whole mathematics, the generalized Riemann hypothesis (GRH) cf. Remark 2.2. This is so because Ashnell and Goldfeld remark that if GRH is true for the subclasses S_*, where $*$ refers to the Kronecker, Elliptic and Artin subclasses, then there is *at most one* candidate one way function. I.e. there may be no one-way function or may exist more than one, and the latter case implies the invalidity of the hypothesis. Thus, this direction is like hunting for two rabbits but surely will get one.

2.6　The abundant class of Selberg zeta-functions

We study a plausible construction of a PNG based on an intractable problem in number theory, especially the zeta-functions by generalizing [Ashnel and Goldfeld (1997)].

We adopt the following definition of a pseudo-random number generator due to [Blum and Micali (1994)] and [Yao (1992)].

Definition 2.3. A pseudo-random number generator PNG is a deterministic polynomial time algorithm which expands short bit sequences (seeds) into a longer ones such that the resulting ensemble is polynomial time indistinguishable from a target probability distribution.

Ashnell and Goldfeld [Ashnel and Goldfeld (1997)] presented an algorithm for cryptographically secure PNG that is based on the candidate one-way function for the class $\mathcal{S}_{Elliptic}$ and $\mathcal{S}_{Artin}$.

Definition 2.4. For an integer $n \geq 0$, we define its **bit size** $d_2(n)$ by

$$d_2(n) = \begin{cases} [\log_2 n] + 1 & \text{if } n > 0 \\ 1 & \text{if } n > 0 \end{cases}. \tag{2.54}$$

For any vector $\boldsymbol{n} = (n_1, \cdots, n_r) \in \mathbb{N} \cup \{0\}^r$ we define its **norm** by

$$\|\boldsymbol{n}\| = \sum_{i=1}^r d_2(n_i). \tag{2.55}$$

For positive integers r, s a function

$$f : \mathbb{N} \cup \{0\}^r \to \mathbb{N} \cup \{0\}^s$$

is called a **one-way function** if the following three conditions hold.

(i) there exists an integer $k > 0$ such that

$$\|\boldsymbol{n}\|^{1/k} \leq \|f(\boldsymbol{n})\| \leq \|\boldsymbol{n}\|^k \tag{2.56}$$

for $\boldsymbol{n} = (n_1, \cdots, n_r) \in \mathbb{N} \cup \{0\}^r$;

(ii) $f(\boldsymbol{n})$ can be computed in polynomial time in $\|\boldsymbol{n}\|$;

(iii) given $\boldsymbol{m} \in \mathbb{N} \cup \{0\}^s$ there does not exist a polynomial time algorithm which either computes a vector $\boldsymbol{n} \in \mathbb{N} \cup \{0\}^r$ such that $f(\boldsymbol{n}) = \boldsymbol{m}$ or indicates that no such value exists.

Condition (i) asserts that $f(\boldsymbol{n})$ is neither longer or shorter than $\boldsymbol{n}$.

In [Ashnel and Goldfeld (1997)] a new candidate one-way function is provided based on a seemingly intractable problem in the theory of zeta-functions. The candidate arises from the subclass of the Selberg class zeta-functions, the **feasible Selberg class**.

Definition 2.5. Every zeta-function in the **Selberg class** is given by the Dirichlet series

$$Z(s) = \sum_{m=1}^{\infty} \frac{a(n)}{n^s} \qquad (2.57)$$

absolutely convergent for $\sigma \gg 1$ (with $s = \sigma + it$). It is a meromorphic function satisfying the following conditions

(i)

$$a(n) = O(n^{\alpha}) \qquad (2.58)$$

under which the half-plane of absolute convergence is $\sigma > \alpha + 1$.

(ii) It satisfies the **functional equation**: there exist $A, r, \alpha_i > 0$, $\beta_i \in \mathbb{C}$ $i = 1, \cdots, N$, $w \in \mathbb{C}$ with $|w| = 1$ and a polynomial $P(s)$ such that $Z(s)$ satisfies the zeta-symmetry

$$\Lambda(s) := A^s P(s) \Delta(s) Z(s) = w \cdot \overline{\Lambda(r - \bar{s})}, \qquad (2.59)$$

where $\Delta(s)$ indicates the gamma factor

$$\Delta(s) Z(s) = \prod_{i=1}^{N} \Gamma(\alpha_i s + \beta_i) \qquad (2.60)$$

(iii) It admit the **Euler product**:

$$\log Z(s) = \sum_{n=1}^{\infty} \frac{b(n)}{n^s}, \qquad (2.61)$$

where $b(n) = 0$ unless $n = p^r$, a positive prime power.

The constant A in (3.244) is called the **conductor** of $Z(s)$.

This slightly differs from the original definition of Selberg. For the modified Selberg class, cf. §3.8.11 and especially (3.245).

Definition 2.6. A Selberg class zeta-function belongs to the class of **feasible Selberg class** S if in addition,
(iv) given a prime power p^r, there exists an algorithm to compute $b(p^r)$ in polynomial time.

Definition 2.7. The subclass S' of **abundant Selberg class** of the feasible Selberg class S if in addition,
(v) for every $\varepsilon > 0$, the number of distinct zeta-functions in the subfamily for which the conductor lies in an interval of length $B > 0$ is greater than $B^{1-\varepsilon}$ as $B \to \infty$.

Definition 2.8. An intractable problem for abundant subfamilies states that given a sequence

$$(a_1, \cdots, a_k) \tag{2.62}$$

of complex numbers, how difficult is it to determine whether or not there exists an abundant Selberg class zeta-function

$$Z(s) = \sum_{m=1}^{\infty} \frac{a(n)}{n^s} \in \mathcal{S}' \tag{2.63}$$

such that $a(n) = a_n$ for $n = 1, \cdots, k$ and the conductor of $Z(s)$ lies in a given interval. If such a function exists, how difficult is it to construct one such zeta-function.

Definition 2.9. For sufficiently large $B > 0$ let $\mathcal{S}^B$ denote the finite sub-class of the abundant class such that the conductor A lies in the interval $[B, 2B]$. For each such B, let k, m be positive integers satisfying

$$k \geq (\log B)^{\mu}, m \leq (\log B)^{\nu} \tag{2.64}$$

with $\mu, \nu > 0$ fixed and independent of B. Let

$$F : \mathcal{S}^B \to \mathbb{C}^k \tag{2.65}$$

denote the **Fourier projection function** defined by

$$F(Z(s)) = (a(m), \cdots, a(m+k)) \tag{2.66}$$

for $Z(s) \in \mathcal{S}^B$.

Conjecture (Ashnel and Goldfeld). For the classes $\mathcal{S}_{\text{Kronecker}}$, $\mathcal{S}_{\text{Elliptic}}$, $\mathcal{S}_{\text{Artin}}$, the associated Fourier projection is a one-way function.

Remark 2.2. If we assume the GRH (Generalized Riemann Hypothesis) for subclasses $\mathcal{S}_{\text{Kronecker}}$, $\mathcal{S}_{\text{Ellipic}}$ to be defined below and take $\mu > 2$, $\nu = 1$ in (2.64) as in (2.130), then F is a *one-to-one* function [Goldfeld and Hoffstein (1993)]. Then Goldreich et al [Goldfeld and Hoffstein (1993)] implies the existence of a PNG. But as remarked above, on the GRH, there is at most one such a function, thereby presenting a delicate problem lying at a threshold of the truth or falsity of such hypotheses.

For the RH etc, cf. §4.1.5.

Theorem 2.16. (Ashnell-Goldfeld [Ashnel and Goldfeld (1997), Theorem 5]) *Suppose $a, b \in \mathbb{Z}$ determines an elliptic curve $E : y - 2 = f(x)$, where $f(x) = x^3 + ax + b$. Let K be the field obtained by adjoining the roots of $f(x)$ to $\mathbb{Q}$ and let $d = [K : \mathbb{Q}]$. Let $c_E(n)$ be th coefficients of the L-function associated with E in (2.122). If $d = 3$, then the density of primes for which $c_E(p)$ is even is $\frac{1}{3}$, while for $d = 6$, it is $\frac{2}{3}$ and for $d = 1, 2$, $c_E(p)$ is even except for finitely many primes.*

For the proof of the probability distribution Chabotarev density theorem Theorem 2.6 is used.

Theorem 2.17. (Ashnell-Goldfeld [Ashnel and Goldfeld (1997), Theorem 7]) *Let E be an elliptic curve defined over $\mathbb{Q}$. Let K be the field obtained by adjoining the 2-torsion points of E to $\mathbb{Q}$ and let $d = [K : \mathbb{Q}]$. Then there exists an entire Artin L-function*

$$L_K(s) = \sum_{n=1}^{\infty} \frac{b(n)}{n^s} \in \mathcal{S}_{Artin} \qquad (2.67)$$

such that

$$b(p) \equiv c_E(p) \pmod 2. \qquad (2.68)$$

The Ashnell-Goldfeld scheme is as follows. Suppose for the field K in Theorem 2.17 that $d = [K : \mathbb{Q}] = 6$. Given a sequence of bits with nth bit being obtained by computing $c_E(p_n) \pmod 2$ for some elliptic curve as above, whereby p_n indicates the nth prime. By the conjecture for $\mathcal{S}_{elliptic}$, Theorem 2.16 assures that the bit sequence $c_E(p_n) \pmod 2$ is pseudo-random with probability distribution $\left(\frac{1}{3}, \frac{2}{3}\right)$. Theorem 2.17 implies that it is not possible in polynomial time to determine E from this sequence.

The aim of the subsequent part is to give basics of the L-functions associated with a Dirichlet character and with an elliptic curve over $\mathbb{Q}$ and then incorporate subsections on the Selberg subclasses $\mathcal{S}_{Kronecker}$ and $\mathcal{S}_{elliptic}$.

2.6.1 Dirichlet characters

Cf. the next subsection and also §3.5.

Definition 2.10. We write χ mod q for a group character of the multiplicative group $(\mathbb{Z}/q\mathbb{Z})^{\times}$.

We define an arithmetical function $\tilde{\chi}$ from χ mod q as follows. For $(a, q) = 1$, we let $\tilde{\chi}(a) = \chi(\bar{a})$, with $\bar{a}$ indicating the reduced residue class containing a. Then we call its 0-extension (i.e. $\tilde{\chi}(a) = 0$ if $(a, q) > 1$) a Dirichlet character $\tilde{\chi}$ mod q, and we use the same symbol χ to indicate this arithmetical function and refer to it as a **Dirichlet character** χ mod q. They inherit the **multiplicative** property $\chi(ab) = \chi(a)\chi(b)$ of residue class characters and the **principal Dirichlet character**, denoted by χ_0, is a periodic function of period q having the values $\chi_0(a) = 1$ or $\chi_0(a) = 0$ according as $(a, q) = 1$ or $(a, q) > 1$. A divisor q_1 of q is called a **defining** (or

admissible) modulus for χ if $(a, q) = 1$, $a \equiv 1 \pmod{q_1}$ imply $\chi(a) = 1$, which amounts to saying that $\chi(a) = \chi(b)$ if $a \equiv b \pmod{q_1}$, $(a, q) = (b, q) = 1$. This last statement is the same as χ being periodic of period q_1.

If $q_1, q_2 | q$ are defining moduli of $\tilde{\chi}$, then so is (q_1, q_2) and the principle implies that the least defining modulus $f = f_\chi$ (called the **conductor** of $\tilde{\chi}$, 'f' being from the German word "Fuerer" meaning a tram conductor) divides all other defining moduli, and in particular, (q_1, q_2). If the conductor f is q, then χ as well as $\tilde{\chi}$ is called a **primitive** character to the modulus f, or mod f (read: modulo f), otherwise **imprimitive**, i.e. if there is a proper divisor f of q such that f is a defining modulus of χ as well as $\tilde{\chi}$, or in other words, χ is **induced** from ψ through the canonical surjection $(\mathbb{Z}/q\mathbb{Z})^\times \to (\mathbb{Z}/f\mathbb{Z})^\times$. In this case, χ is identical with a primitive character ψ to the modulus f, where ψ is defined as follows. If $(n, q) = 1$, then $\psi(n) = \chi(n)$, and if $(n, q) > 1$ but $(n, f) = 1$, then we choose an integer a satisfying $(n + af, q) = 1$ and define $\psi(n) = \chi(n + af)$. We need to prove that this ψ is one of $\varphi(f)$ Dirichlet characters. For this we use the following characterization of Dirichlet characters, i.e. an *arithmetical function* X *is a Dirichlet character* mod q *if and only if it satisfies the following conditions: multiplicative, periodic of period* q, $X(n) = 0$ *for* $(n, q) > 1$ *but not always* 0.

Let

$$G_n(\chi) = \sum_{a=1}^{q} \chi(a) e^{2\pi i \frac{n}{q} a} \tag{2.69}$$

denote the (generalized) **Gauss sum** and let $\tau(\chi)$ be the normalized Gauss sum:

$$\tau(\chi) = G_1(\chi) = \sum_{a=1}^{q} \chi(a) e^{2\pi i \frac{a}{q}}. \tag{2.70}$$

For an extensive theory of these Gauss sums, cf. e.g. [LiwK (2012), Chapter 5].

Theorem 2.18. (Apostol) *The separability of the Gauss sum*

$$G_n(\chi) = \bar{\chi}(n)\tau(\chi) \tag{2.71}$$

characterizes the primitivity of χ.

Proof. If $(n, q) = 1$, then $\chi(a) = \chi(an)\bar{\chi}(n)$ and so

$$G_n(\chi) = \bar{\chi}(n) \sum_{a=1}^{q} \chi(an) e^{2\pi i \frac{na}{q}} = \bar{\chi}(n)\tau(\chi) \tag{2.72}$$

since an runs through all residue classes mod q. If now $(n, q) = 1$ and χ is primitive, then we can show that both sides of (2.72) are 0. Hence (2.71) holds for χ primitive.

Conversely, assume (2.71). Then by Lemma 2.5, $\tau(\chi) = G_1(\chi) \neq 0$ if and only $R = 1$, i.e. if and only if $\frac{q}{f}$ is square-free and $\left(\frac{q}{f}, f\right) = 1$, and then

$$\tau(\chi) = \mu\left(\frac{q}{f}\right) \psi\left(\frac{q}{f}\right) \tau(\psi) \tag{2.73}$$

and

$$G_n(\chi) = g(n)\tau(\chi), \tag{2.74}$$

where $g(n)$ is given by (2.82).

Now under (2.71), for $G_n(\chi) \neq 0$, we must have $\tau(\chi) \neq 0$, which is the case if and only if $R = 1$ and $g(n) = \bar{\chi}(n)$ for $\forall n \in \mathbb{N}$, i.e. $\tilde{q} = 1$, or $q = f$, i.e. χ is primitive. □

We note that (2.73) is true for all characters. Indeed, if $\frac{q}{f}$ is not square-free or $\left(\frac{q}{f}, f\right) > 1$, then the right-hand side of (2.73) is 0 and so $\tau(\chi) = 0$. We also prove the following well-known fact.

$$|\tau(\chi)| = \sqrt{q}. \tag{2.75}$$

Proof Multiplying (2.69) and its complex conjugate, we obtain

$$|G_n(\chi)|^2 = \sum_{a,b=1}^{q} \chi(a)\bar{\chi}(b)e^{2\pi i n \frac{a-b}{q}}. \tag{2.76}$$

Summing this over $n = 1, \cdots, q$ and noting the sum of geometric sequence, we deduce that

$$\sum_{n=1}^{q} |G_n(\chi)|^2 = q \sum_{a=1}^{q} |\chi(a)|^2 = q\varphi(q) \tag{2.77}$$

by orthogonality.

We recall the lemmas used above.

Lemma 2.4. ([LiwK (2012), Lemma 5.5, p. 142]) *Let χ be a character modulo q induced by the primitive character ψ with conductor f and we write for $n \in \mathbb{N}, n_0 = \frac{n}{(q,n)}, q_0 = \frac{q}{(q,n)}$. Then we have*

$$G_n(\chi) = \begin{cases} 0, & \text{if } f \nmid q_0, \\ \frac{\varphi(q)}{\varphi(q_0)}\mu\left(\frac{q_0}{f}\right) \psi\left(\frac{q_0}{f}\right) \overline{\psi}(n_0)\tau(\psi) & \text{if } f \mid q_0. \end{cases} \tag{2.78}$$

Specifying Lemma 2.4 with $(n, q) = 1$ we have

$$G_n(\chi) = \bar{\chi}(n)\tau(\chi). \tag{2.79}$$

Substituting this in (2.77), we find that the left-hand side is $|\tau(\chi)|^2\varphi(q)$, whence (2.75) follows.

Joris's evaluation of the Gauss sum is similar to Lemma 2.4. We denote the primitive character inducing χ to the modulus q by ψ with modulus f and introduce the notation

$$\tilde{q} = \prod_{\substack{p|q \\ p\nmid f}} p, \quad R = \frac{q}{f\tilde{q}}. \tag{2.80}$$

Lemma 2.5. ([LiwK (2012), Lemma 5.6, p. 143])

$$G_n(\chi) = \begin{cases} 0, & if\ R \nmid n, \\ \mu(\tilde{q})\varphi(\tilde{q})\tau(\psi)Rg\left(\frac{n}{R}\right) & if\ R \mid n, \end{cases} \tag{2.81}$$

where

$$g(n) = \mu((n, \tilde{q}))\varphi((n, \tilde{q}))\bar{\psi}(n). \tag{2.82}$$

2.6.2 *Dirichlet L-functions*

The **Dirichlet L-function** is defined by (sometimes L-series) $L(s, \chi)$

$$L(s, \chi) = \sum_{n=1}^{\infty} \frac{\chi(n)}{n^s} = \prod_p \left(1 - \frac{\chi(p)}{p^s}\right)^{-1}, \quad \sigma = \operatorname{Re} s > 1. \tag{2.83}$$

As most of the zeta- and L-functions, the Dirichlet L-function satisfies the functional equation, which takes the following concise form for χ a *primitive* character:

$$L(1 - s, \chi) = q^{s-1}(2\pi)^{-s}\Gamma(s)\left(e^{-\frac{\pi i}{2}s} + \chi(-1)e^{\frac{\pi i}{2}s}\right)\tau(\chi)L(s, \bar{\chi}), \tag{2.84}$$

which is usually stated in the form (see e.g. [Davenport (2000), Chapter 9])

$$\xi(1 - s, \chi) = \frac{i^a\sqrt{q}}{\tau(\chi)}\xi(s, \bar{\chi}), \tag{2.85}$$

where

$$\xi(s, \chi) = \left(\frac{\pi}{q}\right)^{-\frac{s+a}{2}}\Gamma\left(\frac{s+a}{2}\right)L(s, \chi), \tag{2.86}$$

and where $\mathfrak{a}$ is 0 or 1 according as $\chi(-1) = 1$ or $\chi(-1) = -1$ ($\chi(-1) = (-1)^{\mathfrak{a}}$):

$$\mathfrak{a} = \mathfrak{a}(\chi) = \frac{1 - \chi(-1)}{2} = \begin{cases} 0 & \chi \text{ even} \\ 1 & \chi \text{ odd} \end{cases}. \qquad (2.87)$$

We may prove that the two expressions (2.84) and (2.85) are equivalent by using two familiar properties of the gamma function (4.81) and (4.82).

Two most famous independent proofs are known of the functional equation (2.85), one depending on the theta transformation formula (cf. Example 2.3 below) and the other on the Hurwitz formula through the very convenient expression

$$L(s, \chi) = q^{-s} \sum_{a=1}^{q-i} \chi(a) \zeta\left(s, \frac{a}{q}\right), \qquad (2.88)$$

where $\zeta(s, \alpha)$ indicates the **Hurwitz zeta-function** defined for $0 < \alpha \leq 1$ by

$$\zeta(s, \alpha) = \sum_{n=1}^{\infty} \frac{1}{(n + \alpha)^s}, \quad \sigma > 1, \qquad (2.89)$$

whose special case amounts to the additive form definition of the Riemann zeta-function (3.31). Both of them are analytic in the right half-plane in the first instance and can be meromorphically continued to the whole complex plane. The functional equation for the Riemann zeta-function $\zeta(s)$ (as a consequence of (2.85)) with the even character χ_0^* reads

$$\pi^{-\frac{s}{2}} \Gamma\left(\frac{s}{2}\right) \zeta(s) = \pi^{-\frac{1-s}{2}} \Gamma\left(\frac{1-s}{2}\right) \zeta(1 - s) \qquad (2.90)$$

(cf. [Kanemitsu (2007)]).

We shall give a variant of the theta-transformation proof of (2.85) through the Lipschitz summation formula below, which is of interest in its own right. Indeed, on the way, we also achieve the proof depending on the Hurwitz zeta-function. For $\sigma > 1$, (2.88) is stated as [Kanemitsu (2007), (8.17), p. 171]. Note that for $0 < \sigma \leq 1$, the series (2.83) for $L(s, \chi)$ (χ non-trivial) is uniformly convergent in s, so that it is analytic in $\sigma > 0$. In the remaining range, the functional equation (cf. (2.91) below) for the Dirichlet L-function gives its analytic continuation and (2.88) is valid for all $s \neq 1$.

In contrast to (2.84), the formula

$$L(1 - s, \chi) = q^{s-1}(2\pi)^{-s} \Gamma(s) \left(e^{-\frac{\pi i}{2} s} + \chi(-1) e^{\frac{\pi i}{2} s}\right) \ell(s, \bar{\chi}) \qquad (2.91)$$

is valid for *all* χ, where

$$\ell(s,\chi) = \sum_{n=1}^{\infty} G_n(\chi)n^{-s} = \sum_{a=1}^{q-1} \chi(a)\ell_s\left(\frac{a}{q}\right) \tag{2.92}$$

is the ℓ-function considered by many authors including Joris [Joris (1977)], Neukirch [Neuenschwander (1978)], Kubert-Lang [Kubert (1981)] et al. Here $\ell_s(x) = \sum_{n=1}^{\infty} \frac{e^{2\pi ixn}}{n^s}$ is the **polylogarithm function**, a special case $L(x,s,1)$ of the **Lipschitz-Lerch transcendent** which we introduce by ([Srivatava and Choi (2001), (11), p. 122])

$$L(s,x,a) = \sum_{n=0}^{\infty} \frac{e^{2\pi ixn}}{(n+a)^s} \quad \sigma > 0, \ \alpha \in \mathbb{C}-\{0,-1,-2,\cdots\}, \ \mu \in \mathbb{R}\backslash\mathbb{Z}. \tag{2.93}$$

Then the Hurwitz formula (the functional equation for the Hurwitz zeta-function; see Corollary 2.4 below) is a consequence of the functional equation for the Lipschitz-Lerch transcendent stated in Theorem 2.20 below:

$$L(1-s,x,a)$$
$$= \frac{\Gamma(s)}{(2\pi)^s}\left\{e^{\pi i(\frac{1}{2}s-2ax)}L(s,-a,x) + e^{-\pi i(\frac{1}{2}s-2a(1-x))}L(s,a,1-x)\right\} \tag{2.94}$$

valid for $0 < x < 1$ ([Srivatava and Choi (2001), (10), p. 122]).

In what follows we shall deduce (2.94) from the Lipschitz summation formula, which in turn is a consequence of the Poisson summation formula. First we recall the latter as stated in Rademacher [Rademacher (1973)] (for a character analogue cf. (2.105) below).

Lemma 2.6. (Poisson summation formula) *Suppose* $f \in C^2(\mathbb{R})$ *and that integrals*

$$\int_{-\infty}^{\infty} f(x)\,dx \quad and \quad \int_{-\infty}^{\infty} |f''(x)|\,dx \tag{2.95}$$

exist. Let

$$a_n = \hat{f}(n) = \int_{-\infty}^{\infty} f(x)e^{-2\pi inx}dx \tag{2.96}$$

be the Fourier transform of f. *Then we have the* **Poisson summation formula**

$$S(u) := \sum_{n=-\infty}^{\infty} f(n+u) = \sum_{n=-\infty}^{\infty} a_n e^{2\pi inu} \tag{2.97}$$

uniformly in u *in any finite interval, and in particular*

$$S(0) := \sum_{n=-\infty}^{\infty} f(n) = \sum_{n=-\infty}^{\infty} \hat{f}(n). \tag{2.98}$$

Theorem 2.19. (Lipschitz summation formula) *For the complex variables* $z = x+iy$, $x > 0$, $s = \sigma + it$, $\sigma > 1$ *and the real parameter* $0 < \alpha \le 1$, *we have the Lipschitz summation formula*

$$\frac{(2\pi)^s}{\Gamma(s)} \sum_{m=0}^{\infty} (m+\alpha)^{s-1} e^{-2\pi z(m+\alpha)} = \sum_{n=-\infty}^{\infty} \frac{e^{2\pi i n\alpha}}{(z+ni)^s}. \tag{2.99}$$

Under the condition $0 < \alpha < 1$ *this formula holds in the wider half-plane* $\sigma > 0$.

Proof. The left-hand side series is absolutely convergent for all s in view of the convergence factor $e^{-2\pi x(m+\alpha)}$.

For $\sigma > 1$, the right-hand series is also absolutely convergent because the n-th term of a Majorant series is $O\left(\frac{1}{|(n+1)^\sigma|}\right)$.

The Poisson summation formula applied to $f(u) = \dfrac{e^{2\pi i u\alpha}}{(z+ui)^s}$ gives

$$S(0) = \sum_{n=-\infty}^{\infty} f(n) = \sum_{m=-\infty}^{\infty} A_m,$$

where

$$A_m = \int_{-\infty}^{\infty} \frac{e^{2\pi i u\alpha}}{(z+ui)^s} e^{-2\pi i m u} \, du.$$

Putting $z + ui = w$, we obtain

$$A_m = \frac{1}{i} \int_{z-i\infty}^{z+i\infty} \frac{e^{2\pi(w-z)(\alpha-m)}}{w^s} \, dw = \frac{e^{2\pi z(m-\alpha)}}{i} \int_{z-i\infty}^{z+i\infty} \frac{e^{-2\pi w(m-\alpha)}}{w^s} \, dw,$$

or more conveniently

$$A_{-m} = \frac{e^{-2\pi z(m+\alpha)}}{i} I_m,$$

say, where

$$\int_{z-i\infty}^{z+i\infty} \frac{e^{2\pi w(m+\alpha)}}{w^s} \, dw. \tag{2.100}$$

First we treat the terms with $m + \alpha > 0$. Putting $2\pi w(m+\alpha) = \zeta$, we obtain

$$A_{-m} = \frac{e^{-2\pi z(m+\alpha)}(2\pi)^s(m+\alpha)^{s-1}}{2\pi i} \int_{a-i\infty}^{a+i\infty} \frac{e^\zeta}{\zeta^s} \, d\zeta,$$

where we put $a = 2\pi(m+\alpha)\operatorname{Re}(z)$. By the **Laplace expression** for the gamma function

$$\frac{1}{\Gamma(s)} = \int_{(c)} \frac{e^z}{z^s} \, dz, \tag{2.101}$$

where (c) means the vertical line $z = c + iy$, $-\infty < y < \infty$ with $c > 0$, we have

$$A_{-m} = \frac{e^{-2\pi z(m+\alpha)}(2\pi)^s(m+\alpha)^{s-1}}{\Gamma(s)}.$$

In the remaining case $m + \alpha \leq 0$, the integrals (2.100) are 0.
Thus we deduce Formula (2.99). □

For $\sigma > 0$, we may prove the Laplace expression (2.101) for the gamma function. by recalling the definition of the gamma function as the Mellin transform.

$$\Gamma(s) = \int_0^\infty e^{-t}t^{s-1}\, dt$$

for $\sigma > 0$. For elements of Mellin and Laplace transforms, cf. [Kanemitsu (2007), Chapter 7].
Make the change of variable $t = zu$ with $\operatorname{Re} z > 0$ to obtain

$$\frac{\Gamma(s)}{z^s} = \int_0^\infty e^{-zt}t^{s-1}\, dt. \tag{2.102}$$

More precisely, the procedure is as follows. First we assume $z > 0$ and then we check that both sides are analytic in $\operatorname{Re} z > 0$ and so they must coincide. Indeed, the left-hand side is analytic in the cut plane $-\pi < \arg z \leq \pi$.
Now we think of (2.102) as the Laplace transform of the function t^{s-1}. Then by the inversion formula

$$t^{s-1} = \frac{1}{2\pi i}\int_{(c)} e^{zt}\frac{\Gamma(s)}{z^s}\, dz, \quad t > 0$$

or

$$\frac{1}{\Gamma(s)} = \frac{1}{2\pi i}\int_{(c)} e^{zt}\frac{1}{(zt)^s}\, d(zt),$$

which is (2.101).

Another proof is possible by applying the Hankel expression.

Theorem 2.19 as it stands refers to the summation of the divergent series $\sum_{m=0}^\infty (m+\alpha)^{s-1}$ by means of the convergence factor $e^{-2\pi z(m+\alpha)}$, $x > 0$, thus the Lipschitz summability arises. It leads, however, to a more essential proposition which is so crucial in the whole spectrum of number theory:

Theorem 2.20. *For $0 < \alpha < 1, 0 < \mu < 1$ we have the Lerch functional equation ((2.94))*

$$e^{2\pi i \mu \alpha} L(1-s, \mu, \alpha) \tag{2.103}$$

$$= \frac{\Gamma(s)}{(2\pi)^s} \left(e^{\frac{\pi i}{2}s} L(s, 1-\alpha, \mu) + e^{-\frac{\pi i}{2}s + 2\pi i \alpha} L(s, 1-\alpha, \mu) \right).$$

We now deduce (2.103) from (2.99). The right-hand side of (2.99) may be written for $\sigma > 1$ as

$$e^{\frac{\pi i}{2}s} \sum_{n=0}^{\infty} \frac{e^{2\pi i (1-\alpha)n}}{(n+\mu)^s} + e^{-\frac{\pi i}{2}s} \sum_{n=0}^{\infty} \frac{e^{2\pi i \alpha(n+1)}}{(n+1-\mu)^s},$$

which is the second factor of the right-hand side of (2.103).

On the other hand, the left-hand side of (2.99) is

$$\frac{(2\pi)^s}{\Gamma(s)} e^{2\pi i \mu \alpha} L(1-s, \mu, \alpha)$$

for $\sigma < 1$.

Now by Dirichlet's test, the series (2.93) is uniformly convergent for $\sigma > 0$ and represents an analytic function there. It follows that Formula (2.103) valid in the common domain $0 < \sigma < 1$ must hold true for all s but for singularities of $\Gamma(s)$. This completes the proof of (2.103).

Corollary 2.4. *For $0 < \alpha < 1$ the Hurwitz formula holds:*

$$\zeta(1-s, \alpha) = \frac{\Gamma(s)}{(2\pi)^s} \left(e^{\frac{\pi i}{2}s} \ell_s(1-\alpha) + e^{-\frac{\pi i}{2}s} \ell_s(\alpha) \right). \tag{2.104}$$

We note that (2.104) and (4.79) are a relation between two bases of the Kubert space (a complex vector space of dimension 2 consisting of real analytic functions which satisfy the Kubert identity), cf. [Milnor (1983), Theorem 1 et seq, p. 287].

We are now in a position to deduce the functional equation (2.84) from Corollary 2.4.

Proof of (2.84) Substituting (2.104) in (2.88), we deduce that

$$L(1-s, \chi) = q^{s-1} \frac{\Gamma(s)}{(2\pi)^s} \left(e^{\frac{\pi i}{2}s} \sum_{a=1}^{q-1} \chi(a) \ell_s\left(1 - \frac{a}{q}\right) + e^{-\frac{\pi i}{2}s} \sum_{a=1}^{q-1} \chi(a) \ell_s\left(\frac{a}{q}\right) \right),$$

which amounts to (2.84), completing the proof.

The following is stated as Exercise 72, p. 132 **I**, which we state here as

Proposition 2.7. *If the Dirichlet L-function $L(s, \chi)$ satisfies the functional equation (2.84), then χ is primitive.*

First proof [Berndt (1975), Theorem 2.3] states that if f is of bounded variation on $[a, b]$, $-\infty < a < b < \infty$, then the **character analogue of the Poisson summation formula**

$$\frac{1}{2} \sum_{a \leq n \leq x}{}' \chi(n)(f(n+0) + f(n-0)) \tag{2.105}$$

$$= \frac{4\tau(\chi)}{q} \sum_{n=1}^{\infty} \int_{a}^{b} f(u)\bar{\chi}(n)\frac{e^{2\pi i\frac{u}{q}n} + \chi(-1)e^{-2\pi i\frac{u}{q}n}}{2i^{\mathfrak{a}}} \, du.$$

holds, where $\mathfrak{a}$ is defined in (2.87).

If we choose $f(u) = e^{2\pi i \frac{m}{q} n}$ in (2.105), then we obtain (2.84), whence by Apostol's theorem, χ is primitive.

Second proof (Joris)   Formula (2.91) is true for all χ with $\ell(s, \chi)$ being the second member in (2.92). Comparing (2.84) and (2.91) yields (2.71) again.

For the third proof see Exercise 72, p. 132 **I**.

As another application of the Poisson summation formula we deduce the theta transformation formula which is in the long run equivalent to the functional equation for the associated zeta-function (i.e. the Riemann zeta-function). The theta transformation formula (2.109) also in the form (4.111) will be used to deduce both the Gauss quadratic reciprocity and differentiability of Riemann's function in §4.2.1 *at a stretch*. The following preliminaries are from **I** without proof.

Theorem 2.21. *Suppose that $f(x, t)$ is integrable in x on $[a, b]$ for all t near the point α and that the limit function $\lim_{t \to \alpha} f(x, t) = f(x, \alpha)$ is integrable on $[a, b]$. Further suppose that the limit is uniform in $x \in [a, b]$. Then we may take the limit under the integral sign:*

$$\lim_{t \to \alpha} \int_{a}^{b} f(x, t) \, dx = \int_{a}^{b} \lim_{t \to \alpha} f(x, t) \, dx. \tag{2.106}$$

Corollary 2.5. *Suppose that $f(x, t)$ is integrable in x on $[a, b]$ for all t in a certain domain T, that $f(x, t)$ is differentiable in $t \in T$ for every $x \in [a, b]$ and that $f_t = \frac{\partial f}{\partial t}$ is continuous in both the variables x and t. Then we may differentiate under the integral sign:*

$$\frac{d}{dt} \int_{a}^{b} f(x, t) \, dx = \int_{a}^{b} \frac{\partial}{\partial t} f(x, t) \, dx. \tag{2.107}$$

Example 2.3. (theta transformation formula) Let $\theta(t)$ denote the Jacobi elliptic theta function

$$\theta(t) = \sum_{n=-\infty}^{\infty} e^{-\pi n^2 t}, \quad t > 0. \tag{2.108}$$

Then it satisfies the **theta transformation formula**

$$\theta\left(\frac{1}{t}\right) = t^{1/2}\theta(t). \tag{2.109}$$

Proof In Lemma 2.6 we choose $f(x) = e^{-\pi n^2 x}$ and evaluate the Fourier coefficients.

$$\hat{f}(n) = \int_{-\infty}^{\infty} e^{-\pi x^2 t} e^{-2\pi i n x} \mathrm{d}x. \tag{2.110}$$

The integrand may be expressed as $e^{-\pi n^2/t} e^{-\pi(x\sqrt{t}+in/\sqrt{t})^2}$. Hence it amounts to establishing

$$\int_{-\infty}^{\infty} e^{-\pi(x+iu)^2} \mathrm{d}x = 1 \tag{2.111}$$

for any $u \in \mathbb{R}$, which is done in Exercise 71, **I**.

Under (2.111), we have $\hat{f}(n) = t^{1/2} e^{-\pi n^2/t}$. Hence (2.98) leads to (2.109), completing the proof.

2.6.3   The $\mathcal{S}_{Kronecker}$ class

We consider those Dirichlet L-functions associated with quadratic fields, which are in turn associated with Kronecker characters $\chi_d(n) = \left(\frac{n}{d}\right)$, where d is the discriminant of the field. Or in equivalent theory of binary quadratic forms $ax^2 + bxy + cy^2$, it is called a fundamental discriminant. It is any product of relatively prime factors of the form

$$-4, 4, -8, (-1)^{\frac{1}{2}(p-1)}p, \tag{2.112}$$

where p indicates an odd prime. These exhaust all real primitive characters modulo $|d|$.

We define the $\mathcal{S}_{Kronecker}$ class to be the one of all such L-functions associated to Kronecker characters.

$$L_d(s) = L\left(s, \left(\frac{d}{\cdot}\right)\right) = \sum \frac{\left(\frac{d}{n}\right)}{n^s} = \prod_p \frac{1}{1 - \frac{\left(\frac{d}{n}\right)}{p^s}} \tag{2.113}$$

for $\sigma > 1$.

By (2.84) and (2.75) we confirm that the Dirichlet L-function with Kronecker character satisfies the functional equation in Definition 2.5 and the conductor is $\sqrt{q}$.

Then Ashnel-Goldfeld Conjecture in this case reads:

For sufficiently large $B > 0$ let $\mathcal{S}_{Kronecker}^{B}$ denote the finite subclass of the abundant class such that the absolute values of the discriminant $|d|$ lies in the interval $[B, 2B]$. For each such B, we let

$$b = [(\log B)^{\mu}], \qquad (2.114)$$

the integral part, with $\mu > 2$. Given a list of signs (2.133) with each $a_j = \pm 1$, the Fourier projection function defined by (2.66)

$$F : \mathcal{S}^B_{Kronecker} \to \{\pm 1\}^k \qquad (2.115)$$

with

$$F(L_d(s)) = (a_1, \cdots, a_b) \qquad (2.116)$$

for $Z(s) \in \mathcal{S}^B_{Kronecker}$ is a one-way function.

The intractable problem corresponding to Definition 2.8 reads as follows. Given a sequence

$$(a_1, \cdots, a_k) \qquad (2.117)$$

of ± 1, how difficult is it to determine whether or not there exists an abundant Selberg class zeta-function

$$Z(s) = \sum_{m=1}^{\infty} \frac{a(n)}{n^s} \in \mathcal{S}' \qquad (2.118)$$

such that $a(n) = a_n$ for $n = 1, \cdots, k$ and the conductor of $Z(s)$ lies in a given interval? If such a function exists, how difficult is it to construct one such zeta-function? The assumption of the GRH for $\mathcal{S}_{Kronecker}$ guarantees that there exists at most one such d.

2.6.4 *L-functions associated with elliptic curves*

In this subsection we state the L-function associated to an elliptic curve defined over $\mathbb{Q}$ following Ashnel and Goldfeld [Ashnel and Goldfeld (1997)]. See also Example 3.20. Let E be an elliptic curve defined by

$$E : y^2 = x^3 + ax + b \qquad (2.119)$$

where $a, b \in \mathbb{Q}$ with discriminant

$$\Delta_E = 4a^3 + 27b^2. \qquad (2.120)$$

Let

$$E' : y^2 = x^3 + a'x + b' \qquad (2.121)$$

be another elliptic curve defined over $\mathbb{Q}$. Two elliptic curves E and E' are isomorphic over $\mathbb{Q}$ if there exists a non-zero $u \in \mathbb{Q}$ for which $a' = u^4 a$ and $b' = u^6 b$. It follows that every elliptic curve over $\mathbb{Q}$ is isomorphic to an elliptic curve specified by a pair (a, b) with $a, b \in \mathbb{Z}$. We can make this choice canonical by requiring that the discriminant Δ_E is minimized. Associated to E there is an L-function $L_E(s)$ defined by the Dirichlet series

$$L_E(s) = \sum_{n=1}^{\infty} \frac{c_E(n)}{n^s}, \qquad (2.122)$$

where for a rational prime p (not dividing the discriminant),

$$c_E(p) = p + 1 - \sharp E(\mathbb{F}_p), \qquad (2.123)$$

and $\sharp E(\mathbb{F}_p)$ denotes the number of integer solutions (x, y) of the congruence

$$y^2 = x^3 + ax + b \quad (\bmod p) \quad \text{with} \quad 0 \le x, y \le p - 1, \qquad (2.124)$$

plus 1. The 1 refers to the additional point $\mathcal{O}$ at infinity which also lies on the curve. This formula also holds for primes dividing the discriminant, provided a more general Weierstrass minimal model is used (see [13]). For prime powers, we have the recurrence relation

$$c_E(p^{r+1}) = c_E(p)c_E(p^r) - \delta_p \cdot p \cdot c_E(p^{r-1}) \quad (r \ge 1), \qquad (2.125)$$

where $c_E(1) = 1$, and

$$\delta_p = \begin{cases} 1, & \text{if} \quad p \text{ does not divide } \Delta_E \\ 0, & \text{if} \quad p \text{ divides } \Delta_E. \end{cases} \qquad (2.126)$$

In general, if n factors into the product of prime powers $n = p_1^{e_1} \cdot p_2^{e_2} \ldots p_k^{e_k}$, then $c_E(n)$ is defined by the formula

$$c_E(n) = \prod_{j=1}^{k} c_E(p_i^{e_i}). \qquad (2.127)$$

Every elliptic curve E over $\mathbb{Q}$ gives rise to a function f that might be a weight-2 Hecke eigenform: If the L-function of the elliptic curve is given by (2.122), then the function is constructed using the same coefficients,

$$f(\tau) = \sum_{n=1}^{\infty} c_E(n) q^n, \qquad (2.128)$$

where $\operatorname{Im} \tau > 0$ and $q = e^{2\pi i \tau}$. We naturally wonder whether f is in fact an eigenform.

One of the most famous theorems of 20th century mathematics asserts that the answer is yes,

All rational elliptic curves arise from modular forms.

Since the L-functions associated with modular forms satisfy the functional equation, so does the L-function associated with an elliptic curve.

The Riemann hypothesis for L_E was first proved by Hasse [13, p. 243] and is equivalent to the bound

$$|c_E(n)| = \sqrt{n}d(n), \tag{2.129}$$

where $d(n)$ is the divisor function ((3.8)).

2.6.5   The $\mathcal{S}_{elliptic}$ class

We define the $\mathcal{S}_{elliptic}$ class to be the one of all such $L = L_E$-functions in (2.122). Then Ashnel-Goldfeld Conjecture in this case reads:

For sufficiently large $B > 0$ let $\mathcal{S}_{elliptic}^B$ denote the finite subclass of the abundant class such that the discriminant Δ_E in (2.120) lies in the interval $[B, 2B]$. For each such B, we let

$$b = [(\log B)^\mu], \tag{2.130}$$

the integral part, with $\mu > 2$. Given a list of values (2.133) with each a_j satisfying the Hasse bound $|a_j| \leq \sqrt{j}d(j)$ in (2.129), the Fourier projection function defined by (2.66)

$$F : \mathcal{S}_{elliptic}^B \to \mathbb{C}^k \tag{2.131}$$

with

$$F(L_E(s)) = (a_1, \cdots, a_b) \tag{2.132}$$

for $Z(s) \in \mathcal{S}_{elliptic}^B$ is a one-way function.

The intractable problem corresponding to Definition 2.8 reads as follows. Given a sequence

$$(a_1, \cdots, a_k) \tag{2.133}$$

of complex numbers, how difficult is it to determine whether or not there exists an abundant Selberg class zeta-function

$$Z(s) = \sum_{m=1}^{\infty} \frac{c_E(n)}{n^s} \in \mathcal{S}_{elliptic} \tag{2.134}$$

such that $c_E(n) = a_n$ for $n = 1, \cdots, k$ and the conductor of $Z(s)$ lies in a given interval? If such a function exists, how difficult is it to construct one such zeta-function?

The assumption of the RH for $\mathcal{S}_{elliptic}$ guarantees that there exists at most one such E up to isogeny.

Gauss

Chapter 3

Arithmetic functions and Stieltjes integrals

Abstract: This chapter starts by developing elementary analytic number theory, with emphasis on the use of Stieltjes integrals, thereby stating three generating functions for arithmetic functions: power series, Dirichlet series and Euler products. We give some elucidating interpretation of some results in number theory as a manifestation of the Stieltjes integration. One of the highlights is that of Lau's theorem on the summatory function associated to the convolution of two arithmetic functions. Toward the end we state a rather general principle–Abel-Tauber process, which consists of the Abelian process of forming the Riesz sums and the subsequent Tauberian process of differencing the Riesz sums, an analogue of the integration–differentiation process. In Section 3.8, we use Abel-Tauber process to establish interesting asymptotic expansion for the Riesz sums of arithmetic functions with best possible error estimate. The novelty of our method is that we incorporate the Selberg type divisor problem in this process by viewing the contour integral as part of the residual function. The novelty also lies in the uniformity of the error term in the additional parameter which varies according to the cases. Generalization of the famous Selberg Divisor problem to arithmetic progression has been made by Rieger [Rieger (1965)], Marcier [Meyer (1967)], Nakaya [Nakaya (1992)] and around the same time Nowak [Nowak (1991)] studied related subject of reciprocals of an arithmetic function and obtained asymptotic formula with the Vinogradov-Korobov error estimate with the main term as a finite sum of logarithmic terms. We shall also elucidate the situation surrounding these researches and illustrate our results by rich examples.

3.1 Introduction

There are many instances in which the partial summation formula and the integration by parts are treated as different tools. One of our aims is to show that they can be viewed as the same thing from the point of view of the Stieltjes integration. We shall also state the general Abel-Tauber process which gives an asymptotic formula for the summatory function of an arithmetic function generated by the Euler product $Z(s)$ in case it admits factoring out of the complex power $Z^*(z)$ of a certain zeta-function (the so-called Selberg-Delenge method):

$$Z(s) = Z^*(z)\Phi(s),$$

where the auxiliary factor has a wider range of absolute convergence. In existing literature, the asymptotic formulas for the summatory function of a Dirichlet convolution of two arithmetic functions and the one for the arithmetic function generated by the power of a zeta-function were treated separately, our theorem makes it possible to treat them as our Abel-Taber process at a stretch. Moreover it gives an asymptotic formula for Riesz sums of higher order.

As an application of the Stieltjes integration as well as a comparison, we shall elucidate a far-reaching theorem of Y.-K. Lau [Lau (2002)] on a precise asymptotic formula for the convolution of two arithmetic functions. One is generated by the essential factor $G_B(s)$ for which the asymptotic formula is assumed with the main term in the form of a contour integral. The other is generated by the auxiliary factor $G_B(s)$. The role of the factors is reversed in [Banerjee, Chakraborty, Kanemitsu and Magi (2016)] and [Lau (2002)]. Cf. Table 3.1 for comparison.

3.2 Arithmetical functions and their algebraic structure

This section corresponds to §3.1 **I** and we state the results from there mostly without proof.

Definition 3.1. We call any function

$$f : \mathbb{N} \to \mathbb{C}$$

an **arithmetical function** (or arithmetic function, number-theoretic function). The domain is sometimes extended to $\mathbb{Z}$ and often to $\mathbb{R}$ by 0-extension, i.e. it is zero for non-integral argument, and the range may be replaced by a general integral domain.

We write

$$R = \{f | f : \mathbb{N} \to \mathbb{C}\},$$

the set of all arithmetic functions.

For an extensive theory of arithmetic functions, we refer to [McCarthy (1986)], [Sivaramakrishnan (1988)] and [Hartman and Wintner (1938)].

Exercise 12. In R define the sum and the product of f, g by

$$(f + g)(n) = f(n) + g(n)$$

$$(fg)(n) = (f \cdot g)(n) = f(n)g(n).$$

Then $(R, +, \cdot)$ is a ring.

We introduce another product, the **Dirichlet convolution**, which is crucial in multiplicative number theory.

Definition 3.2. For $f, g \in R$, define

$$(f * g)(n) = \sum_{d|n} f\left(\frac{n}{d}\right) = \sum_{d\delta=n} f(d)g(\delta), \tag{3.1}$$

when d runs through all positive divisors of n.

Theorem 3.1. $(R, +, *)$ *is an integral domain with the (multiplicative) unity e where*

$$e(n) = \begin{cases} 1, & n = 1 \\ 0, & n \neq 1 \end{cases}; \tag{3.2}$$

$f \in R^\times$ *if and only if $f(1) \neq 0$.*

Exercise 13. Let R' be the set of all arithmetical functions $f : \mathbb{N} \cup \{0\} \to \mathbb{C}$. Define the Abel convolution $f \times g$ by

$$(f \times g)(n) = \sum_{k=0}^{n} f(k)g(n-k) = \sum_{a+b=n} f(a)g(b).$$

Then $R = (R, +, \times)$ is an integral domain with the (additive) unity e',

$$e'(n) = \begin{cases} 1, & n = 0 \\ 0, & n > 0. \end{cases}$$

As one can easily perceive, the genesis of the **Abel convolution** is the Cauchy product of power series. Suppose

$$F_1(z) = \sum_{n=0}^{\infty} f(n)z^n, \; G_1(z) = \sum_{n=0}^{\infty} g(n)z^n$$

be power series associated to f and g, respectively, which we assume are absolutely convergent in a disc $|z| < r, 0 < r$. Then the product $F_1(z)G_1(z)$ gives rise to

$$F_1(z)G_1(z) = \sum_{n=0}^{\infty} (f \times g)(n)z^n \tag{3.3}$$

in the same disc. Examples of sequences generated by power series will be given below (cf. e.g. Definition 3.5).

We now clarify the genesis of the Dirichlet convolution.

Definition 3.3. For an arithmetic function $f \in R$, we associate a Dirichlet series $F(s)$, called the **generating Dirichlet series**, which is to be, convergent in some half-plane.

$$F(s) = \sum_{n=0}^{\infty} \frac{f(n)}{n^s}, \quad \sigma > \sigma_f,$$

where σ_f indicates the **abscissa of absolute convergence**: i.e. σ_f is the infimum of σ such that

$$\sum_{n=0}^{\infty} \frac{|f(n)|}{n^\sigma} < \infty.$$

Example 3.1. If

$$f(n) \ll n^{\alpha+\varepsilon}, \quad \forall \varepsilon > 0,$$

then σ_f can be taken $\sigma_f \le \alpha$.

Given two arithmetical functions $f, g \in R$, we denote the associated generating Dirichlet series by $F(s)$ and $G(s)$, which are absolutely convergent for $\sigma > \max\{\sigma_f, \sigma_g\}$.

The product $F(s)G(s)$ also converges absolutely in the same half-plane, giving rise to $f * g$:

$$F(s)G(s) = \sum_{n=0}^{\infty} \frac{(f * g)(n)}{n^s}, \quad \sigma > \max\{\sigma_f, \sigma_g\}. \tag{3.4}$$

Formula (3.4) gives the genesis of (3.1) and these two are essential ingredients in the study of multiplicative properties of integers. (3.1) is finite and is sometimes useful to give information, while (3.4) is, though infinite, often more powerful than (3.1).

Many familiar arithmetic functions arise in this way from basic ones, which we now introduce:

$$I(n) = 1, \quad n \in \mathbb{N} \longleftrightarrow F(s) = \zeta(s), \quad \sigma > 0; \tag{3.5}$$

$$N(n) = n, n \in \mathbb{N} \longleftrightarrow F(s) = \zeta(s-1), \quad \sigma > 2, \tag{3.6}$$

where $\zeta(s)$ indicates the Riemann zeta-function to be introduced by (3.31) below.

The **Möbius function** $\mu(n)$ is introduced as the coefficients of $\zeta(s)^{-1}$ (cf. (3.12)):

$$\mu * I = e \longleftrightarrow \zeta(s)\zeta(s)^{-1} = 1$$

or

$$\sum_{d|n} \mu(d) = \begin{cases} 1 & n = 1 \\ 0 & n > 1, \end{cases} \tag{3.7}$$

which is a genesis of the most essential Möbius inversion formula, Corollary 3.4 below. For a generalization of (3.7), cf. Theorem 5.18, Chapter 5. Further examples are

$$I * I = d \longleftrightarrow \zeta(s)\zeta(s) = \zeta(s)^2 = \sum_{n=1}^{\infty} \frac{d(n)}{n^s}, \quad \sigma > 1, \tag{3.8}$$

where $d(n) = \sum_{d|n} 1$ is the **divisor function** counting the number of divisors of n (cf. (3.148) below).

$$I * N^\alpha = \sigma_\alpha \longleftrightarrow \zeta(s)\zeta(s - a) = \sum_{n=1}^{\infty} \frac{\sigma_\alpha(n)}{n^s},$$

where $\sigma_\alpha(n) = \sum_{d|n} d^\alpha$ is the **sum-of-divisor function**. We often write

$$\sigma_1 = \sigma \quad (\sigma_0 = d)$$

(3.7) is the most fundamental and forms the basis of the Möbius inversion formula. To provide a clearer picture of arithmetic functions, we make the following definition.

Definition 3.4. If $0 \neq f \in R$ satisfies

$$f(mn) = f(m)f(n) \quad (m, n) = 1, \tag{3.9}$$

then $f \in R$ is called a **multiplicative function**. If f satisfies

$$f(mn) = f(m) + f(n) \quad (m, n) = 1, \tag{3.9}'$$

then f is called an **additive function**. If (3.9) or (3.9)$'$ holds for all m, n, then f is called **completely multiplicative** (or **additive**).

Exercise 14. Prove the following

- (i) f is additive, $c > 0 \iff g(n) = c^{f(n)}$ is multiplicative;
- (ii) f is multiplicative, $f(n) > 0 \iff g(n) = \log f(n)$ is additive.

The following theorem assures that the values of multiplicative and additive function are determined by those at prime powers.

Theorem 3.2. *Let* $n = \prod_p p^{\alpha_p(n)}$ *be the canonical decomposition. Then for a multiplicative function* f *and an additive function* g, *we have*

$$f(n) = \prod_p f(p^{\alpha_p(n)}), \quad f(1) = 1 \tag{3.10}$$

$$g(n) = \sum_p f(p^{\alpha_p(n)}), \quad g(0) = 0$$

Example 3.2. The Euler function $\varphi(n) = (\mathbb{Z}/n\mathbb{Z})^\times$ ((2.32)) is multiplicative. This is because, for $(m, n) = 1$, $(\mathbb{Z}/mn\mathbb{Z})^\times = (\mathbb{Z}/m\mathbb{Z})^\times (\mathbb{Z}/m\mathbb{Z})^\times$ ((1.36), p. 33, **I**).

Since $\varphi \in G$, it suffices to know the values at prime powers p^α. Since $\varphi(p^\alpha)$ is the number of integers $\leq p^\alpha$ prime to p^α, it is clear that

$$\varphi(p^\alpha) = p^\alpha - p^{\alpha-1}.$$

Consider the convolution $\mu * N \in G$, which takes on the value $\mu(1)p^\alpha + \mu(p)p^{\alpha-1} = p^\alpha - p^{\alpha-1}$. Hence $\varphi = \mu * N$:

$$\varphi(n) = (\mu * N)(n) = \sum_{d|n} \mu\left(\frac{n}{d}\right) d. \tag{3.11}$$

For other results on $\varphi(n)$, cf. Exercise 17 below.

We now prove another algebraic structure theorem.

Theorem 3.3. *The set* G *of all multiplicative functions forms a group.*

We need a lemma which anticipates some knowledge of divisibility of integers.

Lemma 3.1. *If* $(m, n) = 1$, *then each* $d|mn$ *can be expressed uniquely in the form* $d = d_1 d_2, d_1|m_1, d_2|n, (d_1, d_2) = 1$.

Corollary 3.1. *If* $h = f * g \in G$, *then either both* $f, g \in G$, *or neither of them belong to* G.

Corollary 3.2. $\sigma_\alpha \in G$ *and in particular,* $d \in G$ *and* $\sigma \in G$.

Corollary 3.3. $\mu = I^{-1}$, *i.e.* (3.7) *holds.*

To give another proof of (3.7), it is necessary to know the values of μ, and for that purpose, the natural way is to appeal to the Euler product

$$\zeta(s)^{-1} = \prod_p (1 - p^{-s}) = \sum_{n=1}^{\infty} \frac{\mu(n)}{n^s}, \quad \sigma > 1 \tag{3.12}$$

which is a consequence of the most important Euler product for the Riemann zeta function

$$\zeta(s) = \prod_p (1 - p^{-s})^{-1} = \sum_{n=1}^{\infty} \frac{1}{n^s}, \quad \sigma > 1. \tag{3.31}$$

The Euler product of (3.31) is proved in Exercise 55, p. 93, **I**.

Assuming the validity of (3.12) for the moment, we expand the product in (3.12) to find that

$$\mu(n) = \begin{cases} 1, & n = 1 \\ 0, & p^2 | n \\ (-1)^k, & n = p_1 p_2 \cdots p_k, p_i \text{ distinct} \end{cases} \tag{3.13}$$

which is a usual definition of μ given in most textbooks.

From (3.13) it is clear that $\mu \in G$.

Proof of Corollary 3.3. By Theorem 3.3, it suffices to check the values at prime powers p^k:

$$(\mu * I)(p^k) = \mu(1) + \mu(p) + \cdots + \mu(p^k) = 1 - 1 = 0.$$

Corollary 3.4. (Möbius inversion formula) *For $f, g \in R$, the following are equivalent:*

- (i) $I * f = g$, $\quad or \sum_{d|n} f(d) = g(n), \quad n \in \mathbb{N};$
- (ii) $f = \mu * g$, $\quad or \quad f(n) = \sum_{d|n} \mu(d) g\left(\frac{n}{d}\right), \quad n \in \mathbb{N};$
- (ii)$'$ (Continuous version of (ii)) *If* $g(x) = \sum_{n \leq x} f\left(\frac{x}{n}\right)$, *then* $f(x) = \sum_{d \leq x} \mu(d) g\left(\frac{x}{d}\right), n \in \mathbb{N};$
- (iii) $\sum_{n \leq x} g(n) = \sum_{d \leq x} f(d) \left[\frac{x}{d}\right], \quad 1 \leq r \in \mathbb{R},$

where $[x]$ indicates the greatest integer $\leq x$.

Proof. That (i) and (ii) are equivalent is contained on Corollary 3.3.

(i)$\implies$ (iii) is given in Exercise below.

We give a proof which is a precursory indication of the "double to repeated" integration in §3.6.1 below. For arbitrary arithmetic functions a, c let

$$b(n) = \sum_{n \leq x} c\left(\frac{x}{n}\right). \tag{3.14}$$

Then consider the continuous convolution $a \tilde{*} b$:

$$a \tilde{*} b(x) = \sum_{d \leq x} a(n) b\left(\frac{x}{n}\right). \tag{3.15}$$

Substituting from (3.14), we deduce that

$$a \tilde{*} b(x) = \sum_{d \leq x} a(n) \sum_{\delta \leq x/d} c\left(\frac{x/d}{\delta}\right) = \sum_{d\delta \leq x} a(d) c\left(\frac{x}{d\delta}\right) = \sum_{n \leq x} c\left(\frac{x}{n}\right) \sum_{d|n} a(d). \tag{3.16}$$

Putting $c = I$ (recall I in (3.5)) and $a = f$ in (3.16) gives (iii) in view of $b(x) = [x]$ while $a = \mu$ and $b = g$ gives (ii)' in view of (3.7).

(iii)$\Longrightarrow$ (i). Reversing the argument in Exercise 15, we arrive at the first equality in (3.18):

$$\sum_{n \leq m} g(n) = \sum_{n \leq m} \sum_{d|n} f(d) \tag{3.17}$$

where we put $x = m \in \mathbb{N}$. Substituting (3.17) with m replaced by $m - 1$, we get

$$g(m) = \sum_{d|m} f(d),$$

i.e. (i) follows. This completes the proof. $\qquad\qquad\square$

Exercise 15. Give direct proofs of (ii)$'$ and the implication (i)$\Longrightarrow$ (iii).

Solution. Direct proof of (ii)' is verbatim to the proof given above. Substituting the definition of g in $\sum_{d \leq x} \mu(d) g\left(\frac{x}{d}\right)$, $n \in \mathbb{N}$ we find that it amounts to the 2-dimensional sum

$$\sum_{n \leq x} c\left(\frac{x}{n}\right) \sum_{d|n} \mu(d)$$

and we appeal to (3.7). (i)$\Longrightarrow$ (iii): Substituting the expression for g, we find that

$$\sum_{n \leq x} g(n) = \sum_{n \leq x} \sum_{d|n} f(d) = \sum_{d\delta \leq x} f(d) = \sum_{d \leq x} f(d) \sum_{\delta \leq \frac{x}{d}} 1 \tag{3.18}$$

whose innermost sum is $\left[\frac{x}{d}\right]$, completing the proof.

For applications of Möbius inversion, one may consult e.g. [Chen (2010)].

Remark 3.1. Not only the Möbius inversion formula but also Formula (3.7) appears in many other contexts. One important principle **relative primality principle** is that when we have a sum $S = S(x)$ over $n \leq x$, $(n,q) = 1$ for a given integer $q > 0$,

$$S = \sum_{\substack{n \leq x \\ (n,q)=1}} f(n), \tag{3.19}$$

then we may replace the relatively prime condition by $\sum_{d|(n,q)} \mu(d)$ and writing dn for n, we transform S into

$$S = \sum_{d|q} \mu(d) \sum_{n \leq x/d} f(dn). \tag{3.20}$$

E.g. if $f \equiv I$, then since $\sum_{n \leq x/d} 1 = \frac{q}{d}$, we deduce (3.11) again.

Similarly, if $f \equiv N$, then $\sum_{\substack{n \leq x \\ (n,q)=1}} n = \frac{1}{2} \sum_{d|q} d\mu(d) \left[\frac{x}{d}\right] \left(\left[\frac{x}{d}\right] + 1\right)$.

Another example appears in the proof of (3.41), Exercise 3.5 and Exercise 41 in Chapter 5.

Let $\Delta = \Delta_1^1$ denote the difference operator in (3.212), i.e. $\Delta f(n) = f(n+1) - f(n)$. Note that $\Delta \log n = \log\left(1 + \frac{1}{n}\right) \to 0$. We may now state Erdös theorem, which asserts that this limiting condition characterizes the logarithm function among additive arithmetic functions.

Theorem 3.4. (Erdös) *If an additive arithmetic functions f satisfies the condition*

$$\lim_{n \to \infty} \Delta f(n) = 0, \tag{3.21}$$

then we must have

$$f(n) = c \log n \tag{3.22}$$

for some constant c.

3.3   Generatingfunctionology

As already stated in Definition 3.3, the Dirichlet series generate arithmetic functions as their coefficients. As is referred to in the passage about the Abel convolution in §3.2, power series, being mostly Taylor series for certain known functions, work as generating functions of many important arithmetic functions which appear as Taylor coefficients. In this section we collect some important arithmetic functions generated by power series. Power series have been used for much longer time than Dirichlet series and there is a rich source of results obtained, especially, in the field of combinatorics. As is well-known, the region of absolute convergence of power series is a circle of finite or infinite radius (of convergence), while that of Dirichlet series is a half-plane.

Exercise 16. Prove that the function

$$f(z) = \frac{1}{e^z - 1}$$

has a simple pole at $z = 0$ with residue 1, while the function $zf(z) = \frac{z}{e^z-1}$ has a removable singularity at $z = 0$.

Solution. The first assertion follows from the Laurent series

$$\frac{1}{e^z - 1} = \frac{1}{z + \frac{z^2}{2!} + \frac{z^3}{3!} + \cdots} \qquad (3.23)$$

$$= \frac{1}{z} + f_0(z),$$

where $f_0(z) = -\frac{1}{2} + \frac{1}{12}z + \cdots$ is analytic at $z = 0$. Since $e^z = 1$ occurs for $z = 2\pi i n$, $n \in \mathbb{Z}$, the Laurent expansion (3.23) holds in the annulus $0 < |z| < 2\pi$.

The second assertion follows from the following argument. Since near $z = 0$, $zf(z) = 1 - \frac{1}{2}z + \frac{1}{12}z^2 + \cdots$, it follows that the new function

$$g(x) = \begin{cases} 1, & z = 0 \\ \frac{z}{e^z-1}, & z \neq 0 \end{cases}$$

is analytic inside the circle $|z| < 2\pi$.

Definition 3.5. The product $\frac{z}{e^z-1}e^{xz}$ is analytic in $|z| < 2\pi$ and has a Taylor expansion of the form:

$$\frac{ze^{xz}}{e^z - 1} = \sum_{n=0}^{\infty} \frac{B_n(x)}{n!} z^n, \quad |z| < 2\pi. \qquad (3.24)$$

The n-th coefficient $B_n(x)$ is called the n-th **Bernoulli polynomial**.

$B_n = B_n(0)$ is called the n-th **Bernoulli number**. They are therefore generated by the power series

$$\frac{z}{e^z - 1} = \sum_{n=0}^{\infty} \frac{B_n}{n!} z^n, \quad |z| < 2\pi, \tag{3.25}$$

whence

$$B_0 = 1, \; B_1 = -\frac{1}{2}, \; B_2 = \frac{1}{6}, \; B_4 = -\frac{1}{30}, \; B_{2k+1} = 0 \; (k = 1, 2, \cdots). \tag{3.26}$$

Apparently $\frac{z}{e^z-1} e^{xz}$ is the Abel convolution in (3.3) and since (3.25) gives the power series coefficients, it follows that

$$\frac{B_n(x)}{n!} = \sum_{k+l=n} \frac{B_k}{k!} \frac{x^l}{(l)!} \tag{3.27}$$

which leads to (3.29).

For its importance and ubiquitousness, we make special mention of the **first periodic Bernoulli polynomial** once and for all and use it freely without further mentioning:

$$\overline{B}_1(x) = x - [x] - \frac{1}{2} = -\frac{1}{\pi} \sum_{n=1}^{\infty} \frac{\sin 2\pi nx}{n}, \tag{3.28}$$

where the saw-tooth Fourier series on the far-right side is bondedly convergent and the equality holds for $x \notin \mathbb{Z}$. For $x \in \mathbb{Z}$, $\overline{B}_1(x) = -\frac{1}{2}$ while the Fourier series converges to 0 and care must be taken in this case. The saw-tooth Fourier series is often denoted by $\psi(x)$, cf. (4.25). There is still one more notation (4.48) which is however, specified to the Dedekind sums only.

$$\overline{B}_1(x) = \psi(x) = ((x)), \qquad x \notin \mathbb{Z}.$$

More generally we also define the rth periodic Bernoulli polynomial once and for all by

$$\overline{B}_n(x) = B_n(x - [x]) = \sum_{k=0}^{n} \binom{n}{k} B_{n-k} \{x\}^k \tag{3.29}$$

which has an absolutely convergent Fourier series (4.58).

In the same way, we may define the **Euler polynomials** $E_n(x)$ by

$$\frac{2e^{xz}}{e^z + 1} = \sum_{n=0}^{\infty} \frac{E_n(x)}{n!} z^n, \quad |z| < \pi \tag{3.30}$$

the series being absolutely and uniformly convergent in $|z| < \pi$ (the singularity from the origin being $z = \pm i\pi$).

As stated on [Serre (1973), p. 90, Footnote (2)], there are several definitions of Bernoulli numbers. The most commonly used ones are those in the *b*-notation in [Serre (1973)] while Leopoldt's definition differs only at one value B_1 which is defined to be $\frac{1}{2}$ rather than $-\frac{1}{2}$. Here we followed Washington's notation [Washington (1982)] and introduced the B_n by (3.25). For a systematic account of Bernoulli polynomials, we refer to [Kanemitsu (2007), Chapter 1] and we freely use the results from it. There are many generalizations of Bernoulli numbers and polynomials, some of which will be given in Chapter 5, **I**. Most of them have been introduced so as to express the special values of the relevant zeta- and *L*-functions at negative integral argument, whence at certain positive integral arguments, while the Bernoulli numbers themselves were used to express the sum of powers of natural numbers up to n, say, which were used by Euler to solve the Basler problem $\zeta(2) = \frac{\pi^2}{6}$.

Finally we mention a fundamental convergence theorem for general Dirichlet series.

Definition 3.6. Let a_n be an arbitrary sequence of complex numbers, $\{\lambda_n\}$ be an increasing sequence of real numbers with $\lambda_1 \geq 1$ and let $\lambda_n^{-s} = e^{-\log \lambda_n}$ with log meaning the principal value. Then the series of the form $F(s) = \sum_{n=1}^{\infty} a_n \lambda_n^{-s}$ are called (general) **Dirichlet series**.

Theorem 3.5. (i) *If the series $F(s)$ is convergent at a point s_0, then it is convergent in the half-plane $\sigma > \sigma_0$ and is uniformly convergent in the angular domain $|\arg(s - s_0)| \leq \theta < \frac{\pi}{2}$ and represents an analytic function there.*
(ii) *If the series $F(s)$ is absolutely convergent at a point s_0, then it is so in the half-plane $\sigma > \sigma_0$ (in which $F(s)$ is analytic).*

Proof uses the partial summation formula. Cf. e.g. [Hardy and Riesz (1915)].

3.4   Euler products

In his epoch-making paper [Riemann (1859)], the only paper on number theory in his whole career, Riemann [Riemann (1854), p. 144] introduces the **Riemann zeta-function** $\zeta(s)$ in the form

$$\zeta(s) = \prod_p (1 - p^{-s})^{-1} = \sum_{n=1}^{\infty} \frac{1}{n^s} \tag{3.31}$$

for $\sigma > 1$, where in the product, p runs through all the rational primes, and NOT the additive form first. Here we can see an extremely beautiful interplay among the giants of the whole time, the Euler product, which is one of the main contributions of Euler and its introduction to the study on distribution of primes with complex analysis by Riemann toward the establishment of the Gauss conjecture, *l'estro armonico del circollo del Euler-Gauss-Riemann*.

From the tradition starting from this very paper of Riemann, we write

$$s = \sigma + it \qquad (3.32)$$

with $\sigma, t \in \mathbb{R}$ throughout the book save for in Chapter 6, where we write $s = \sigma + j\omega$, following the tradition of electrical engineering. *The convergence of the infinite product entails the non-vanishing of it in the domain in question.* Hence the zeta-function does not vanish in the right half-plane $\sigma > 1$ and this led Riemann to come to an approach to the proof of PNT by connecting the estimate of the error term to the zero-free region. Roughly speaking, the error term will be of the form x^β, where β is the maximum of the real parts of the non-trivial zeros of the Riemann zeta-function, which are the zeros lying in the critical strip $0 < \sigma < 1$.

However, to our best knowledge at present, we cannot treat the product well and must be satisfied with the additive form. It is to be remarked that many of the most advanced zeta-functions including those of Artin, Selberg etc. introduced in Chapter 2 are defined in the form of Euler products, which are then expanded into Dirichlet series. As can be seen from Exercise 54, p. 92, **I**, which we recall below, (3.31) is analytic equivalent of the unique factorization property of the rational integers. Likewise, those zeta-functions defined by Euler products are of number-theoretic nature possessing unique factorization property.

Recall Exercise 54, p. 92, **I** which gives a proof of (3.31) by considering the finite product

$$\prod_{p \leq x}(1 - p^{-s})^{-1}, \quad \sigma > 1. \qquad (3.33)$$

We remark that the argument in that Exercise gives Euler's proof of the infinitude of primes. For if there are only finitely many primes, then (3.33) with $s = 1$ is over all primes, and (3.71), p. 92, **I** (the series form of (3.33)) with $s = 1$ is over all positive integers:

$$\prod_{p \leq x}(1 - p^{-1})^{-1} = \sum_{n=1}^{\infty} n^{-1} = \infty,$$

a contradiction. Needless to say, the first and the shortest proof of the infinitude of primes is due to Euclid, which may be stated in one line:

If $p_1, \cdots, p_n$ are all the primes, then $p_1 \cdots p_n + 1$ is also a prime.

Proof of the following theorem is verbatim to the solution to Exercise 54, p. 92, **I**.

Theorem 3.6. *If $f(n)$ is multiplicative and the series $S = \sum_{n=1}^{\infty} f(n)$ is absolutely convergent, then so is the infinite product $P = \prod \left(\sum_{k=0}^{\infty} f(p^k) \right)$ and $S = P$ holds, where p runs through all the primes.*

Proof. (3.71), p. 92, **I** should read

$$\prod_{p \leq x} \sum_{k=0}^{\infty} f(p^{-k}) = \sum_{n \in A_x}' f(n),$$

and the subsequent equation should read

$$\left| S - \prod_{p \leq x} \sum_{k=0}^{\infty} f(p^{-k}) \right| \leq \left| \sum_{n \in A_x}' f(n) \right| < \sum_{n > x} |f(n)| \to 0$$

as $x \to \infty$, whence $S = P$. Applying this result to $|f(n)|$, the absolute convergence of P follows, completing the proof. $\square$

Regarding the density of primes, much more is known. Letting, as usual

$$\pi(x) = \sum_{p \leq x} 1 = \text{the number of primes } \leq x, \qquad (3.34)$$

the **prime counting function**, then the celebrated **prime number theorem** PNT, conjectured by Gauss, states that

$$\pi(x) \sim \frac{x}{\log x}, \qquad (3.35)$$

which means that the ratio of both sides goes to 1 as $x \to \infty$ (cf. e. g. [Davenport (2000)]). To be continued on to §3.5. For the Dirichlet PNT for an arithmetic progression, cf. Theorem 2.13.

The following "**inclusion-exclusion principle**" is sometimes the underlying principle of important identities: Let $A_1, \cdots, A_r$ be the subsets of a set A and $B = A \backslash \cup_{i=1}^{r} A_i$. Then

$$\sharp B = \sharp A + \sum_{s=1}^{r} (-1)^s \sum_{\{i_1, \cdots, i_s\} \subset \{1, \cdots, r\}} \sharp(A_{i_1} \cap \cdots \cap A_{i_s}). \qquad (3.36)$$

We show below that the **Eratosthenes sieve** is a manifestation of the inclusion-exclusion principle. Before that we state

Exercise 17. For distinct primes p, q, prove by the inclusion-exclusion principle

$$\varphi(pq) = pq - p - q + 1. \tag{3.37}$$

Note that (3.37) is a basis for the RSA cryptology.

Solution. Let $X = \{1, \cdots, n\}$, $n = pq$ and let P resp. Q be the subset of X consisting of integers divisible by p resp. q. Then we want to find the number of elements in $X \backslash (P \cup Q)$. Since

$$\sharp(P \cup Q) = \sharp P + \sharp Q - \sharp(P \cap Q)$$

and $\sharp P = \left[\frac{n}{p}\right] = q$, $\sharp Q = \left[\frac{n}{q}\right] = p$, and $\sharp(P \cap Q) = 1$, it follows that

$$\sharp X \backslash (P \cup Q) = n - p - q + 1.$$

Theorem 3.7. (Eratosthenes sieve) *For $x > 0$ we have*

$$\pi(x) - \pi(\sqrt{x}) = [x] - 1 + \sum_{n|P} \mu(n) \left[\frac{x}{n}\right], \tag{3.38}$$

where P is the product of all primes $\leq \sqrt{x}$.

This is a restatement of (3.36) in the form:

$$\pi(x) - \pi(\sqrt{x}) = [x] - 1 + \sum_{s=1}^{\pi(x)} (-1)^s \sum_{\substack{p_{i_1} \cdots p_{i_s} \leq x \\ \text{distinct}}} \left[\frac{x}{p_{i_1} \cdots p_{i_s}}\right]. \tag{3.39}$$

E.g. with $x = 100$ gives $\pi(100) = 25$.

Recall that Exercise 55, p. 93, **I** uses a multiplicative version of the Eratosthenes sieve (see (3.39) below) to prove (3.36).

We add other manifestations of the above principles by the example of Dirichlet L-functions. For basic knowledge on Dirichlet characters, cf. §2.6.1 and §1.8, **I**.

Example 3.3. ([Iwasawa (1972), p. 4]) Let χ_1 and χ_2 be primitive characters with conductors f_1 and f_2, respectively. Then there is a unique primitive character with conductor f, $f \mid f_1 f_2$ such that

$$\chi(a) = \chi_1(a)\chi_2(a), \tag{3.40}$$

for $(a, f_1 f_2) = 1$. χ is called the product of χ_1 and χ_2 and denoted by: $\chi = \chi_1 \chi_2$. Note that (3.40) does not necessarily hold for $(a, f_1 f_2) > 1$.

The set of all primitive characters forms a group under the multiplication (3.40), by Exercise 57, p. 100 **I**.

The inclusion-"explosion" principle above may be applied in various contexts and the following is one example.

Example 3.4. χ be a non-principal imprimitive Dirichlet character mod q induced by the primitive character χ^* with conductor f. Prove that

$$S_q := \sum_{m=1}^{q} \chi(m)m = \frac{q}{f} \prod_{\substack{p|q \\ p\nmid f}} (1 - \chi^*(p)) \sum_{m=1}^{f} \chi^*(m)m. \qquad (3.41)$$

The following is Exercise 58, p. 101. **I**.

Example 3.5. Cf. §2.6.2. Suppose χ is a non-principal imprimitive Dirichlet character mod q induced by the primitive character χ^* with conductor f. Let $L(s, \chi)$ be the associated Dirichlet L-function ((2.83), Chapter 5). Then prove that

$$L(s, \chi) = \prod_{p|q} \left(1 - \frac{\chi^*(p)}{p^s}\right) L(s, \chi^*), \qquad (3.42)$$

where the product is over all prime divisors of q (which do not divide f).

3.5   PNT and the zero-free region

In this section we shall give a brief account of the PNT with the best known estimate for the error term, improving the first-stage approximation (3.35).

This was made possible by extracting the genuine main term in the form of the integral

$$\mathrm{li}x = \int_2^\infty \frac{\mathrm{d}t}{\log t} \qquad (3.43)$$

called the **logarithm integral**, which is sometimes defined in the form $\int_0^\infty \frac{\mathrm{d}t}{\log t} = \int_0^{1-} + \int_{1+}^\infty$. By integration by parts,

$$\mathrm{li}x = \frac{x}{\log x} + \frac{x}{\log^2 x} + \cdots \qquad (3.44)$$

which gives only the logarithmic reducing of the error term but this is gives the main term $\frac{x}{\log x}$ in (3.35).

The first proof of PNT was given in 1896 in the lines indicated by Riemann [Riemann (1859)] by Hadamard and de la Vallee Poussin independently. The latter's argument entails the error estimate

$$\pi(x) = \mathrm{li}x + O(x\delta_{VP}(x)), \tag{3.45}$$

where

$$\delta_{VP}(x) = \delta_{c_{12},1,0}(x) = e^{-c_{12}\sqrt{\log x}} \tag{3.46}$$

which we call the de la Vallee Poussin reducing factor.

After intermediate estimates due to Titchmarsh, Chudakov et al, the best known error estimate is given by the Vinogradov-Korobov reducing factor (cf. (3.196)).

$$\delta_{VK}(x) = \delta_{A,\frac{2}{3},\frac{1}{3}}(x) = \exp\left(-A(\log x)^{3/5}(\log\log x)^{-1/5}\right) \tag{3.47}$$

We note that both are special cases of (3.196).

The Vinigradov-Korobov estimate was first stated in [Walum (1991)] and in modern style in [Iwaniec (2004), §8.5]. The latter book contains very rich information on the zero-free regions for general L-function, esp. Chapter 5.

All these are unified as slowly varying (oscillating) functions cf. the passage before Theorem 3.12. Let $\lambda(n)$ denote the Liouville function [Ayoub (1963), p. 123, p. 132]:

One of its definitions is

$$\lambda(n) = (-1)^{\Omega(n)}, \tag{3.48}$$

where $\Omega(n)$ is the total number of prime factors of n. Cf. Example 3.15 and (4.26).

The PNT in terms of the Liouville function is in the same depth as that of the Möbius function and (3.45):

$$\sum_{n\leq x}\mu(n) = O(\delta_{VK}(x)) \iff \sum_{n\leq x}\lambda(n) = O(\delta_{VK}(x)). \tag{3.49}$$

To deduce the right arrow, one may appeal to the convolution argument.

(3.49) follows from Theorem 3.19 with the aid of Theorem 3.22 with Δ a constant.

In case there is another ingredient involved, typically the modulus in the case of arithmetic progressions, much more delicate analysis is needed. By our Abel-Tauber process we can obtain the Siegel-Walifisz type results, i.e. the modulus is $\ll (\log x)^A$ for an arbitrary large A. To deal with the case of bigger modulus, even the (generalized) Riemann hypothesis is not enough and zero density and other techniques are needed. Here the Riemann hypothesis means that the infinitely many non-trivial zeros of the zeta-, L-function in the **critical strip** $0 < \sigma < 1$ lies on the critical line $\sigma = \frac{1}{2}$. Here non-trivial zero means a zero in the critical strip (there are trivial zeros at $s = -2, -4, \cdots$ arising from the poles of the gamma function). Thus we formulate the celebrated **Riemann Hypothesis (RH)** once and for all: For $0 < \sigma < 1$

$$\zeta(s) = 0 \Longrightarrow \sigma = \frac{1}{2}. \tag{3.50}$$

A generalized RH for L-functions refers to non-vanishing in the critical strip of all L-functions to a fixed modulus. There are weak (or quasi) RH meaning that there exists an $a \geq \frac{1}{2}$ such that $\zeta(s) \neq 0$ in $a \leq \sigma \leq 1$. This implies for any $a < b < 1$, $\sum_{n \leq x} \mu(n) = O(x^b)$. For more information on the RH, cf. §4.1.5.

3.6   Applications of Stieltjes integrals

As is remarked, most of the results presented in §§3.2, 3.6, **I** may be unified as those for summatory functions of Dirichlet convolutions (Definition 3.2), which is presented as §3.6.1 below. In this respect, we shall show that using the notion of Stieltjes integrals we may express the results as well as the argument in a lucid way. Cf. [Apostol (1957)] and [Widder (1989)] for a rather complete theory of Stieltjes integrals.

Definition 3.7. For bounded functions f, g defined on the closed interval $I = [a, b]$ one introduces the (Riemann-)Stieltjes integral in almost verbatim to that of the Riemann integral:

$$\int_a^b f(x) \, \mathrm{d}g(x). \tag{3.51}$$

The only difference from the Riemann integral is that one uses $g(x_{j+1}) - g(x_j)$ for the difference $x_{j+1} - x_j$.

We consider a **division** Δ of I into m disjoint subintervals:

$$\Delta : a = x_0 < x_1 < \ldots < x_m = b. \tag{3.52}$$

We understand Δ means the division of I into the union of subintervals $I = \bigcup_{k=1}^{m} I_k, I_k = [x_{k-1}, x_k]$ as well as the set of all division points $\{x_0, \ldots, x_m\}$. Adding more division points gives rise to a refinement of I. We denote the maximum of the width (1-dimensional measure) of m subintervals $\mu(I_k) = x_k - x_{k-1}$ by $|\Delta|$ and call it the **size** of Δ:

$$|\Delta| = \max\{\mu(I_k) \mid 1 \leq k \leq m\}.$$

From each I_k choose $\forall \xi_k \in I_k$ and let $\Xi = (\xi_1, \ldots, \xi_m)$. Then form the finite sum

$$S(f) = S(\Delta, f) = S(\Delta, f, \Xi) := \sum_{k=1}^{m-1} f(\xi_k)(g(x_{k+1}) - g(x_k)) \qquad (3.53)$$

and call it the **Riemann-Stieltjes sum**.

If $S(\Delta, f, \Xi)$ approaches a value S say, independently of the choice of ξ_k as we make the size $|\Delta|$ of the division smaller (refining and making the size smaller), i.e.

$$\forall \varepsilon > 0, \ \exists \delta = \delta(\varepsilon) > 0 \ \forall \Delta \ \forall \Xi \ (\ |\Delta| < \delta \ \Rightarrow \ |S(\Delta, f, \Xi) - S| < \varepsilon),$$

we say that f is **integrable** (in the Riemann-Stieltjes sense) on I, and call the value S the **(Riemann-)Stieltjes integral** (3.51) of f with respect to g (with a and b lower and upper limit, respectively).

If a function f on $[a, b]$ satisfies the condition that the sum of differences is bounded:

$$\sum_{k=1}^{m-1} \|f(x_{j+1}) - f(x_j)\| = O(1) \qquad (3.54)$$

for any division (3.51), then f is said to be **of bounded variation**.

Theorem 3.8.

- (i) *The Stieltjes integral exists if f is continuous and g is of bounded variation. The role can be changed in view of Item* (ii).

- (ii) *The formula for integration by parts holds true:*

$$\int_a^b f(x)\,dg(x) = [f(x)g(x)]_a^b - \int_a^b g(x)\,df(x), \qquad (3.55)$$

provided that f is continuous and g is of bounded variation or g is continuous and f is of bounded variation.

- (iii) *If g is a step function with jumps a_n at x_n, the Stieltjes integral reduces to the sum:*

$$\int_a^b f(x)\,dg(x) = \sum_{a < x_n \leq b} f(x_n)a_n. \qquad (3.56)$$

- (iv) *If f is continuous and g is differentiable, then the Stieltjes integral reduces to the Riemann integral:*

$$\int_a^b f(x)\,\mathrm{d}g(x) = \int_a^b f(x)g'(x)\,\mathrm{d}x. \tag{3.57}$$

Remark 3.2. There are many instances where the partial summation and integration by parts are regarded as different processes. However, from the general point of view of Stieltjes integration they are exactly the same as shown below.

If in (3.56), f is differentiable, then applying (3.55) to it, we deduce that

$$\sum_{a < x_n \leq b} f(x_n)a_n = \int_a^b f(x)\,\mathrm{d}g(x) = [f(x)g(x)]_a^b - \int_a^b g(x)f'(x)\,\mathrm{d}x, \tag{3.58}$$

which is the partial summation. When f and g are differentiable of class C^1, say, we combine (3.55) and (3.57) to deduce

$$\int_a^b f(x)g'(x)\,\mathrm{d}x = [f(x)g(x)]_a^b - \int_a^b g(x)f'(x)\,\mathrm{d}x, \tag{3.59}$$

which is the formula for integration by parts.

Corollary 3.5. *We may deduce the closed formula for the sum of kth powers*

$$(k+1)\sum_{j=1}^n j^k = n(n+1)^k - \sum_{r=0}^{k-2} \binom{k}{r} \sum_{j=1}^n j^{r+1}, \tag{3.60}$$

from Theorem 3.8, where an empty sum is to be interpreted to mean 0. From (3.60) inductively, we may deduce the closed formulas for the sum of kth powers.

Proof. We evaluate the integral $I_{n+1} = I_{n+1}(k) := \int_0^{n+1} x^k \,\mathrm{d}[x]$ in two ways. First by Theorem 3.8, (iii)

$$I_{n+1} = \sum_{j=1}^{n+1} j^k. \tag{3.61}$$

By Theorem 3.8, (ii) and (iv),

$$I_{n+1} = x^k [x] \Big|_0^{n+1} - \int_0^{n+1} [x]\,\mathrm{d}x^k = (n+1)^{k+1} - \sum_{j=0}^n j \int_j^{j+1} kx^{k-1}\,\mathrm{d}x, \tag{3.62}$$

so that

$$I_{n+1} = (n+1)^{k+1} - \sum_{j=1}^{n} jx^k \big|_j^{j+1} = (n+1)^{k+1} - \sum_{j=1}^{n} j((j+1)^k - j^k)$$

$$= (n+1)^{k+1} - \sum_{j=1}^{n} j \sum_{r=0}^{k-1} \binom{k}{r} j^r = (n+1)^{k+1} - \sum_{r=0}^{k-1} \binom{k}{r} \sum_{j=1}^{n} j^{r+1}.$$

Hence it follows that

$$(k+1) \sum_{j=1}^{n} j^k + (n+1)^k = (n+1)^{k+1} - \sum_{r=0}^{k-2} \binom{k}{r} \sum_{j=1}^{n} j^{r+1}$$

or (3.60). $\qquad\square$

For $k = 1$, (3.60) reads

$$2 \sum_{j=1}^{n} j = n(n+1), \quad \sum_{j=1}^{n} j = \frac{1}{2} n(n+1). \tag{3.63}$$

For $k = 2$, (3.60) reads

$$3 \sum_{j=1}^{n} j^2 = n(n+1)^2 - \sum_{j=1}^{n} j = n(n+1)^2 - \frac{1}{2} n(n+1), \tag{3.64}$$

whence the second equality of (3.74) below.

It is customary to express the sum of kth powers by Bernoulli polynomials.

Theorem 3.9.

$$Z(n,k) := \sum_{j=1}^{n} j^k = \frac{1}{k+1} \left(B_{k+1}(n+1) - B_{k+1}(0) \right). \tag{3.65}$$

Proof. First we note

$$(n+1)^{k+1} - 1 = \sum_{l=0}^{k} \binom{k+1}{l} Z(n,l). \tag{3.66}$$

Assuming (3.65) for $l = 1, 2, \cdots, k-1$ and substituting in (3.66), we deduce that

$$(n+1)^{k+1} - 1 = \sum_{l=0}^{k-1} \binom{k+1}{l} \frac{1}{l+1} \left(B_{l+1}(n+1) - B_{l+1} \right) + (k+1)Z(n,k). \tag{3.67}$$

Note that

$$\binom{k+1}{l}\frac{1}{l+1} = \frac{1}{k+2}\binom{k+2}{l+1},$$

whence that the right-hand side of (3.67) becomes

$$\frac{1}{k+2}\sum_{l=0}^{k-1}\binom{k+2}{l+1}(B_{l+1}(n+1) - B_{l+1}) + (k+1)Z(n,k). \qquad (3.68)$$

Recalling the addition formula [Kanemitsu (2007), p. 3]

$$B_n(x+y) = \sum_{r=0}^{n}\binom{n}{r}B_{n-r}(x)y^r, \qquad (3.69)$$

we see that (3.68) becomes

$$\frac{1}{k+2}(B_{k+2}(n+2) - B_{k+2}(1)) \qquad (3.70)$$

$$- 1 - (B_{k+1}(n+1) - B_{k+1}) - \frac{1}{k+2}(B_{k+2}(n+1) - B_{k+2}) + (k+1)Z(n,k).$$

Hence, substituting (3.70) in (3.66), we conclude that

$$(n+1)^{k+1} - 1 = \frac{1}{k+2}(B_{k+2}(n+2) - B_{k+2}(n+1)) \qquad (3.71)$$

$$- 1 - (B_{k+1}(n+1) - B_{k+1}) + (k+1)Z(n,k).$$

Applying the basic difference equation

$$\frac{1}{k+2}(B_{k+2}(n+2) - B_{k+2}(n+1)) = n^{k+1}, \qquad (3.72)$$

we conclude (3.65), completing the proof. □

Exercise 18. Deduce (3.66) from (3.60).

Solution. We may rewrite (3.60) as

$$(k+1)Z(n,k) = n(n+1)^k - \sum_{l=1}^{k-1}\binom{k}{l-1}Z(n,l)$$

or

$$(n+1)^{k+1} - (n+1)^k = \sum_{l=1}^{k}\binom{k+1}{l}Z(n,l) - \sum_{l=1}^{k-1}\left(\binom{k+1}{l} - \binom{k}{l-1}\right)Z(n,l) \qquad (3.73)$$

$$= \sum_{l=1}^{k}\binom{k+1}{l}Z(n,l) - \sum_{l=1}^{k-1}\binom{k}{l}Z(n,l).$$

Hence by induction, we prove (3.66).

Exercise 19. (i) By the method of above proof, i.e. using Theorem 3.8, (ii) and (iv), deduce the following

$$\sum_{j=1}^{n} j = \frac{1}{2}n(n+1), \quad \sum_{j=1}^{n} j^2 = \frac{1}{6}n(n+1)(2n+1). \tag{3.74}$$

(ii)

$$I := \int_{0}^{n} (x^2 + 1)\,\mathrm{d}[x] = \frac{1}{6}n(2n^2 + 3n + 7). \tag{3.75}$$

Example 3.6. We derive the special case of (3.109) for the Riemann zeta-function by the Stieltjes integral:

$$\zeta(s) = \frac{1}{s-1} + \frac{1}{2} - s\int_{1}^{\infty} \overline{B_1}(x)x^{-s-1}\,\mathrm{d}x \tag{3.76}$$

valid for $\sigma > 0$. This has another application in Theorem 5.17, Chapter 6. In general suppose $f(x)$ is a continuous function on $[a, b]$. Then consider the integral

$$I = -\int_{a}^{b} \overline{B_1}(x)f(x)\,\mathrm{d}x.$$

By Theorem 3.8, (ii),

$$I = -\left[\overline{B_1}(x)f(x)\right]_{a}^{b} + \int_{a}^{b} f(x)\,\mathrm{d}\overline{B_1}(x) \tag{3.77}$$

$$= -\left[\overline{B_1}(x)f(x)\right]_{a}^{b} + \int_{a}^{b} f(x)\,\mathrm{d}x - \int_{a}^{b} f(x)\,\mathrm{d}[x].$$

Since by Theorem 3.8, (iii),

$$\int_{a}^{b} f(x)\,\mathrm{d}[x] = \sum_{n-[a]+1}^{[b]} f(n), \tag{3.78}$$

it follows that

$$\sum_{n-[a]+1}^{[b]} f(n) = \int_{a}^{b} f(x)\,\mathrm{d}x - \overline{B_1}(b)f(b) + \overline{B_1}(a)f(a) \tag{3.79}$$

$$+ \int_{a}^{b} \overline{B_1}(x)\,\mathrm{d}f(x).$$

Now choose $a = 1$, $b = N \to \infty$ and $f(x) = x^{-s}$ with $\sigma > 1$. Noting that

$$\int_1^\infty x^{-s}\, \mathrm{d}x = \frac{1}{s-1},$$

we see that (3.79) leads to

$$\zeta(s) = \frac{1}{s-1} + \frac{1}{2} + \int_1^\infty \overline{B_1}(x)\, \mathrm{d}(x^{-s}),$$

which amounts to (3.76).

Since the integral in (3.76) is absolutely convergent for $\sigma > 0$, it defines an analytic function there, giving rise to an analytic continuation of $\zeta(s)$ to the wider half-plane $\sigma > 0$. On integrating, one gets an analytic continuation to $\sigma > -1$. Then the functional equation gives the meromorphic continuation to the whole complex plane, with the simple pole at $s = 1$ with residue 1.

Example 3.7. Example 3.6 gives the Laurent expansion

$$\zeta(s) = \frac{1}{s-1} + \sum_{n=0}^\infty \frac{\gamma_n}{n!}(s-1)^n, \tag{3.80}$$

where

$$\gamma_n = (-1)^n \lim_{N\to\infty} \left(\sum_{k=1}^N \frac{\log^n k}{k} - \frac{\log^{n+1} N}{n+1} \right) \tag{3.81}$$

is the n-th **generalized Euler constant** with γ_0 indicating the Euler constant γ ([Lehmer (1975)]). We adopt Ferguson's proof [Fergusson (1963)]. In view of (3.76), we have the Taylor expansion

$$H(s) := \zeta(s) - \frac{1}{s-1} + \sum_{n=0}^\infty A_n(s-1)^n \tag{3.82}$$

around $s = 1$. Since $n! A_n = H^{(n)}(1)$, we may differentiate (3.76) under the integral sign to get

$$n! A_n = \int_1^\infty \overline{B_1}(x)\, \mathrm{d}\left(\frac{\mathrm{d}^n}{\mathrm{d}s^n} \left(e^{-s\log x} \right) \right) \Bigg|_{s=1} = (-1)^{n+1} \int_1^\infty \overline{B_1}(x)\, \mathrm{d}\left(\frac{\log^n x}{x} \right).$$

Putting $a = 1$, $b = N \to \infty$ and $f(x) = x^{-s}$ in (3.79), we find that

$$n! A_n = (-1)^n \lim_{N\to\infty} \left(\sum_{k=1}^N \frac{\log^n k}{k} - \int_1^N \frac{\log^n x}{x}\, \mathrm{d}x \right), \tag{3.83}$$

which amounts to (3.81).

Example 3.8. We elucidate [Guthman (1999)]. Let $A(x)$ denote the summatory function as in (3.88) and let $A_\varkappa(x)$ denote the $\varkappa$-th Riesz sum; in particular,

$$A_1(x) = \sum_{n \leq x} (x - n)a_n = \int_0^x A(t)\, dt. \tag{3.84}$$

nth Fourier coefficient of the cusp form f of weight k, written $f \in S_k$:

$$\varphi(x) = f(ix) = \sum_{n=0}^{\infty} a_n e^{-2\pi nx} \tag{3.85}$$

for $\operatorname{Re} x > 0$.

We consider the splitting

$$\varphi(x) = \varphi_1(\eta, x) + \varphi_{-1}(\eta, -x), \tag{3.86}$$

where

$$\varphi_1(\eta, x) = \sum_{n > \eta} a_n e^{-2\pi nx}, \varphi_{-1}(\eta, -x) = \sum_{n < \eta} a_n e^{2\pi nx}. \tag{3.87}$$

Example 3.9. (The Wintner-Inghma-Segal summability) We appeal to a continuous version of the Möbius inversion formula, Corollary 3.4, (ii)'. Let

$$A(x) = \sum_{n \leq x} a(n). \tag{3.88}$$

However, cf. (3.99) below which seems to be assumed in Segal [Segal (1975)], too or the lower limit of integration below is to read $1-$.

$$\sum_{d \leq x} \frac{\mu(d)}{d} I\left(\frac{x}{d}\right) = \frac{1}{x} \sum_{n \leq x} a_n a_n = \int_1^x t\, dA(t) = xA(x) - \int_1^x A(t)\, dt. \tag{1}$$

$$\sum_{k \leq r} a_k F\left(\frac{k}{r}\right) = \int_1^r F\left(\frac{t}{r}\right) dA(t) = -\frac{1}{r} \int_1^r F'\left(\frac{t}{r}\right) A(t)\, dt. \tag{2}$$

Deduction of [Segal (1975), (6)] is the simplest if we calculate the integral

$$I = \int_1^r F'\left(\frac{t}{r}\right) \frac{1}{t} \int_1^t A(t)\, dt\, du \tag{3.89}$$

in two different ways. One is as given by Segal and applying integration by parts as follows.

$$I = -\frac{1}{r}\left[rF\left(\frac{t}{r}\right)\frac{1}{t}\int_1^t A(u)\,du\right]_1^r + \int_1^r F\left(\frac{t}{r}\right)\left(\frac{1}{t}\int_1^t A(u)\,du\right)' dt.$$

$$(3.90)$$

Noting that the derivative inside the integral is

$$\frac{1}{t}\left(A(t) - \frac{1}{t}\int_1^t A(u)\,du\right) = \sum_{d\leq r}\frac{\mu(d)}{d}I\left(\frac{r}{d}\right),$$

by (1), we obtain

$$I = \int_1^r \sum_{d\leq t}\frac{\mu(d)}{d}I\left(\frac{r}{t}\right)\frac{1}{t}F\left(\frac{t}{r}\right)\,dt = \sum_{d\leq r}\frac{\mu(d)}{d}\int_1^r \frac{I\left(\frac{r}{t}\right)}{t}F\left(\frac{t}{r}\right)\,dt \quad (3.91)$$

on noting that $I\left(\frac{r}{t}\right) = 0$ for $t > d$.

The other way is to substitute for $\frac{1}{t}\int_1^t A(u)\,du$ from (1):

$$I = \frac{1}{r}\int_1^r F'\left(\frac{t}{r}\right)A(t)\,dt - \frac{1}{r}\int_1^r \sum_{d\leq t}\frac{\mu(d)}{d}I\left(\frac{r}{t}\right)F'\left(\frac{t}{r}\right)\,dt. \quad (3.92)$$

By (2), the first integral is $-\sum_{k\leq r} a_k F\left(\frac{k}{r}\right)$, so that it follows from (3.91) and (3.92) that

$$\sum_{d\leq r}\frac{\mu(d)}{d}\int_1^r \frac{I\left(\frac{r}{t}\right)}{t}F\left(\frac{t}{r}\right)\,dt$$

$$= -\sum_{k\leq r} a_k F\left(\frac{k}{r}\right) - \sum_{d\leq r}\frac{\mu(d)}{d}\int_1^r \frac{I\left(\frac{r}{t}\right)}{t}\frac{t}{r}F'\left(\frac{t}{r}\right)\,dt, \quad (3.93)$$

or

$$\sum_{k\leq r} a_k F\left(\frac{k}{r}\right) = \sum_{d\leq r}\frac{\mu(d)}{d}\int_1^r \frac{I\left(\frac{r}{t}\right)}{t}\left(F\left(\frac{t}{r}\right) - \frac{t}{r}F'\left(\frac{t}{r}\right)\right)\,dt. \quad (6)$$

3.6.1   *Stieltjes resultant and asymptotic formulas*

This subsection revisits §3.2, **I** and §3.6, **I** in the light of Stieltjes integrals.

Recall the divisor function $d(n) = I * I(n) = \sum_{d|n} d$, which takes on the value 2 for $n = p$ a prime and we also have an estimate $d(n) \ll n^\varepsilon$ for every $\varepsilon > 0$. We can also determine the maximal order of $d(n)$.

However, to look at the behavior of $d(n)$ at large, it is customary to consider the sum of their values up to x:

$$D(x) = \sum_{n\leq x} d(n),$$

which is called the **summary function** of $d(n)$. The main objective is to find the main term with a plausibly good error estimate. These passages apply to all arithmetic functions verbatim.

It very often happens that the arithmetic function c is given in the convolution form $a * b$. Then the summatory function $C(x) = \sum_{n \leq x} c(n)$ is the double sum

$$C(x) = \sum_{d\delta \leq x} a(d)b(\delta).$$

and the technique of "2-dimensional (double) into repeated" amounts to the Stieltjes resultant (3.98) by using results in Theorem 3.8:

$$C(x) = \sum_{d\delta \leq x} a(d)b(\delta) = \int_a^x A(x/u)\,dB(u), \tag{3.94}$$

This works when an asymptotic formula for $A(x) = \sum_{n \leq x} a(n)$ is known and $b(n)$ is smaller in a certain sense – one condition is that $b(p) = o(|a(p)|)$.

Exercise 20. Interpret Example 3.3, **I** in the light of Stieltjes resultant: We have the asymptotic formula

$$\Phi(x) = \sum_{n \leq x} \varphi(n) = \frac{1}{2\zeta(2)} x^2 + xE(x) + O(x), \tag{3.95}$$

where

$$E(x) = -\sum_{n \leq x} \frac{\mu(n)}{n} \overline{B}_1\left(\frac{x}{n}\right) = O(\log x). \tag{3.96}$$

Solution. Some typos in Example 3.3, **I** are corrected. Since

$$\Phi(x) = \sum_{mn \leq x} m\mu(n) = \int_1^x A(x/u)\,dB(u)$$

we have

$$A(x) = \sum_{n \leq x} n = \frac{1}{2}[x]([x] + 1) = \frac{1}{2}x^2 - x\overline{B}_1(x) + O(1)$$

on applying

$$[x] = x - \overline{B}_1(x) - \frac{1}{2}. \tag{3.97}$$

Hence it follows that

$$\Phi(x) = \int_1^x \left(\frac{1}{2}\left(\frac{x}{u}\right)^2 - f\overline{B}_1\left(\frac{x}{u}\right) \right) \frac{x}{u} \, dB(u) + O(1)$$

$$= \frac{x^2}{2} \sum_{n \le x} \frac{\mu(n)}{n^2} - xE(x) + O(x).$$

Writing

$$\sum_{n \le x} \frac{\mu(n)}{n^2} = \sum_{n=1}^{\infty} \frac{\mu(n)}{n^2} - \sum_{n > x} \frac{\mu(n)}{n^2} = \frac{1}{\zeta(2)} + O(x),$$

we conclude (3.95).

Exercise 21. Interpret Example 3.4, **I** in the light of Stieltjes resultant: The summatory function of the divisor function $d(n)$ admits the asymptotic formula

$$D(x) = \sum_{n \le x} d(n) = x \log x + O(x).$$

For an improvement, cf. Exercise 22 below.

Solution. Indeed,

$$D(x) = \int_1^x A(x/u) \, dB(u)$$

and $A(x) - [x]$, so that

$$D(x) = \int_1^x \left[\frac{x}{u}\right] dB(u) = \sum_{n \le x} \left[\frac{x}{n}\right] = x \sum_{n \le x} \frac{1}{n} + O(x)$$

by (3.97).

Using the asymptotic formula (3.106) with the error $O(x)$ we conclude the assertion.

In literature there are general theorems on convolutions, notably in the books [Ivić (1985)], [Postnikov (1988)].

J. P. Tull ([Tull (1958)]-[Tull (1961)]) developed a general method for obtaining asymptotic formulas for the summatory function of the convolution of two arithmetic functions $a(n)$ and $b(n)$ whose summatory functions $A(x)$ and $B(x)$ satisfy asymptotic formulas. Indeed, his method is more general than just for summatory function but can also treat the Stieltjes resultant: Given two functions A and B defined for $x \ge 1$ of bounded variation on each bounded interval, one defines the **Stieltjes resultant** C of A and B by

$$C(x) = (A \times B)(x) = \int_a^x A(x/u) \, dB(u), \tag{3.98}$$

whenever the integral exists and for all x, $C(x)$ lies between the limits $\lim_{h\to\mp0} C(x \pm h)$. If $A(1) = B(1) = 0$, then (3.98) may be also written as $C(x) = (B \times A)(x) = \int_a^x B(x/u)\,dA(u)$. Hence it is better to define the summatory function as

$$A(x) = \sum_{n<x} a(n). \tag{3.99}$$

Cf. Widder [Widder (1989), pp. 83-91]. I,e., in what follows we assume

$$A(1) = 0 \tag{3.100}$$

for all summatory functions A.

We may now elucidate the hyperbola method developed in §3.6, **I**. The main ingredient is the following dissection based on the inclusion-exclusion principle. Let $\rho = \rho(x)$, $0 < \rho \le x$ to be chosen suitably and consider the dissection of the sum

$$C(x) = \sum_{mn\le x} a(m)b(n) \tag{3.101}$$

$$= \sum_{m\le\rho} a(m)B\left(\frac{x}{m}\right) + \sum_{n\le\rho^{-1}x} b(n)A\left(\frac{x}{m}\right) - A(\rho)B\left(\frac{x}{m}\right).$$

We divide the integral in (3.98) into two with $a = 1$: $\left[1, \frac{x}{\rho}\right] \cup \left[\frac{x}{\rho}, x\right]$. The first integral amounts to $\sum_{n\le\rho^{-1}x} b(n)A\left(\frac{x}{m}\right)$. We apply the formula for integration by parts (3.55) in the following form.

$$\int_{\rho^{-1}x}^x A(x/u)\,dB(u) = [A(x/u)B(u)]_{\rho^{-1}x}^x - \int_{\rho^{-1}x}^x B(u)\,dA(x/u). \tag{3.102}$$

The first term on the right-hand side is $-A(\rho)B\left(\frac{x}{m}\right)$ while the last integral becomes on writing $\frac{x}{u} = v$,

$$\int_1^\rho B(x/u)\,dA(u) = \sum_{m\le\rho} a(m)B\left(\frac{x}{m}\right). \tag{3.103}$$

Thus we have proved

Theorem 3.10. *The hyperbola method is the dissection of the Stieltjes resultant with integration by parts.*

In order to improve the error term for the Dirichlet divisor problem, we introduce the **hyperbola method** due to Dirichlet himself. This in fact is *another manifestation of the inclusion-exclusion principle* (3.36) as with Exercise 17.

The method will turn out to be very useful in deriving asymptotic formulas for the sum of other arithmetic functions, too.

Exercise 22. Elucidate §3.6, **I**. In particular, deduce the formula

$$D(x) = \sum_{n \leq x} d(n) = \sum_{mn \leq x} 1 = x \log x + (2\gamma - 1)x + \Delta(x), \qquad (3.104)$$

where

$$\Delta(x) = \sum_{m \leq \rho} \overline{B}_1\left(\frac{x}{m}\right) + \sum_{n \leq \rho^{-1}x} \overline{B}_1\left(\frac{x}{n}\right) + O\left(\frac{x}{\rho^2} + \frac{\rho^2}{x}\right) \qquad (3.105)$$

$$= -2 \sum_{n \leq \sqrt{x}} \overline{B}_1\left(\frac{x}{n}\right) + O(1)$$

for $\rho = \sqrt{x}$.

Solution. Since (3.101) amounts to

$$D(x) = \sum_{m \leq \rho} \left[\frac{x}{m}\right] + \sum_{n \leq \rho^{-1}x} \left[\frac{x}{n}\right] - [\rho]\left[\rho^{-1}x\right],$$

we apply (3.97) and then

$$\sum_{n \leq x} \frac{1}{n} = \log x + \gamma - B_1(x)\frac{1}{x} + O\left(\frac{1}{x^2}\right). \qquad (3.106)$$

To make $\frac{x}{\rho} = \rho$ hold, we are to choose $\rho = \sqrt{x}$.

$$D(x) = x \log x + (2\gamma - 1)x + O(\sqrt{x}),$$

i.e. Dirichlet's result.

Using the Fourier series for $\overline{B}_1(x)$ and the estimate for trigonometrical sums, we can prove that

$$\Delta(x) = O(x^{\frac{1}{3}+\varepsilon})$$

or of smaller order. The problem of determining the true order of the error term is called the Dirichlet divisor problem. It is conjectured that the exponent is to be $\frac{1}{4} + \varepsilon$ in view of the Ω-result but remains unproven.

In Exercises 20, 21 and Exercise 22, we appealed to asymptotic formulas for the summatory function of powers of natural numbers $\sum_{n \leq x} n, \sum_{n \leq x} 1, \sum_{n \leq x} \frac{1}{n}$.

These are not isolated and can be unified in the following far-reaching theorem, (3.108) (cf. [Kanemitsu (2007), Chapter 3]).

Theorem 3.11. (Integral Representations, [Kanemitsu (2007), Theorem 3.1, p. 55]) *For any $l \in \mathbb{N}$ with $l > \operatorname{Re} u + 1$ and for any $x \geq 0$, we have the integral representation*

$$L_u(x,a) := \sum_{n \leq x} (n+a)^u = \sum_{r=1}^{l} \frac{\Gamma(u+1)}{\Gamma(u+2-r)} \frac{(-1)^r}{r!} \overline{B_r}(x)(x+a)^{u-r+1}$$

$$+ \frac{(-1)^l}{l!} \frac{\Gamma(u+1)}{\Gamma(u+1-l)} \int_x^\infty \overline{B_l}(t)(t+a)^{u-l} dt$$

$$+ \begin{cases} \dfrac{1}{u+1}(x+a)^{u+1} + \zeta(-u,a), & u \neq -1 \\ \log(x+a) - \psi(a), & u = -1 \end{cases},$$

$$(3.107)$$

where $\Gamma(s)$ and $\psi(s)$ are the gamma function and Gauss' digamma function, respectively and $\zeta(s,a)$ is the Hurwitz zeta-function defined by (4.71).

Also the asymptotic formula

$$L_u(x,a) = \sum_{r=1}^{l} \frac{(-1)^r}{r} \binom{u}{r-1} \overline{B_r}(x)(x+a)^{u-r+1} + O\left(x^{\operatorname{Re}(u)-l}\right)$$

$$+ \begin{cases} \dfrac{1}{u+1}(x+a)^{u+1} + \zeta(-u,a), & u \neq -1 \\ \log(x+a) - \psi(a), & u = -1 \end{cases}$$

$$(3.108)$$

holds true as $x \to \infty$.

Furthermore, the integral representation

$$\zeta(-u,a) = a^u - \frac{1}{u+1}a^{u+1} - \sum_{r=1}^{l} \frac{(-1)^r}{r} \binom{u}{r-1} B_r a^{u-r+1}$$

$$+ (-1)^{l+1} \binom{u}{l} \int_0^\infty \overline{B_l}(t)(t+a)^{u-l} dt,$$

$$(3.109)$$

which follows from (3.107) by putting $x = 0$, holds true for all complex $u \neq -1$, where l can be any natural number subject only to the condition that $l > \operatorname{Re} u + 1$.

We may expect to obtain an asymptotic formula for the Stieltjes resultant C as a generalization of Lau's results [Lau (2002)] which is expounded in the next section.

3.7   Lau's theorem

We elucidate Lau's theorem in the light of Stieltjes integration. Although it looks like there are some miraculous cancellations occurring in the process, we shall show that the cancellations are necessitated by the argument based on Stieltjes integration.

Suppose for simplicity (the upper limit 1 may be replaced by any $b \geq 1$)

$$0 \leq a < 1. \tag{3.110}$$

Suppose two summatory functions are given as

$$A(x) = \sum_{n \leq x} a(n) = M_A(x) + E_A(x), B(x) = \sum_{n \leq x} b(n) = M_B(x) + E_B(x)$$

$$\tag{3.111}$$

and the error terms satisfy some estimates that allow the intermediate procedures.

Let $G_A(s)$ (resp. $G_A(s)$) indicate the generating function of $\{a(n)\}$ (resp. $\{b(n)\}$). These generating functions satisfy some conditions as described by Lau. Given fixed numbers $0 \leq a < \eta < 1$ and $0 \neq \alpha < a$,

(1) $G_A(s)$ (resp. $G_B(s)$) is absolutely convergent for $\sigma > 1$ (resp. $\sigma > \alpha$).

(2) $G_A(s)$ has analytic continuation on $\sigma > \alpha - \epsilon$ (resp. $\sigma > \alpha$) and satisfies

$$G_A(s) \ll |t|^{1-\varepsilon}, \quad \alpha \leq \sigma \leq 1 - . \tag{3.112}$$

(3) $G_B(s)$ has analytic continuation in a wider region and it is analytic on $U^- = U^-(a, \delta, Y)$ and satisfies

$$G_B(s) \ll \delta^{-c}, \quad |s - a| = \delta, s \in U^-, \tag{3.113}$$

where $\delta < Y$ satisfies $0 < \delta < \eta - a$, so that

$$a + \delta < \eta, \quad \alpha < a - Y. \tag{3.114}$$

Also the main term is given as a contour integral

$$M_B(x) = \Phi(x) = \frac{1}{2\pi i} \int_{\mathcal{H}_Y(a, \delta)} G_B(s) x^s \frac{ds}{s}. \tag{3.115}$$

Under the above conditions, the summatory function for the convolution $a * b$:

$$C(x) = (A \times B)(x) = \sum_{n \leq x} (a * b)(n) = \sum_{mn \leq x} a(m) b(n). \tag{3.116}$$

Let $\mathcal{N}(x)$ indicate a **slowly varying function** or a slowly oscillating function appearing in the precise form of the prime number theorem etc., §3.5.

Also cf. de Haan [Haan (1970)], Seneta [Seneta (1976)], Ivič. [Ivić (1985)]. c with or without suffix indicates some unspecified positive constants.

Theorem 3.12. (Lau's theorem) *For sufficiently small $\epsilon' > 0$ let*

$$y = y(x) = x \exp(-c(\epsilon')\mathcal{N}(x)). \tag{3.117}$$

Then we have

$$C(x) = \sum_{n=1}^{\infty} b(n)M_A\left(\frac{x}{n}\right) + \frac{1}{2\pi i}\int_{\mathcal{H}} G_B(s)G_A(s)x^s\frac{ds}{s} \tag{3.118}$$

$$+ \sum_{n\leq y} b(n)\left(E_A\left(\frac{x}{n}\right) - G_{A,\alpha}(0)\right) + E_C(x),$$

where

$$E_C(x) = O(x^a\delta(x)), \quad \delta(x) = \delta_c(x) = \exp(-c\mathcal{N}(x)). \tag{3.119}$$

Recall the formula (3.55) for integration by parts.

Lemma 3.2. *Suppose*

$$M_A(x) = O(x(\log x)^{c'}), x \to \infty \quad M_A(x) = O(x^\eta), x \to +0 \tag{3.120}$$

and that

$$E_B(x) = O(x^a\delta_{c_0}(x)), \quad \delta_{c_0}(x) = \exp(-c_0\mathcal{N}(x)). \tag{3.121}$$

Then all the error terms $E_j(x)$ are absorbed in the error term (3.119).

$$E_1(x) = \int_y^{\infty} M_A\left(\frac{x}{t}\right)\,dE_B(t), \quad E_2(x) = \sum_{m\leq x/y} a(m)O\left(E_B\left(\frac{x}{m}\right)\right), \tag{3.122}$$

$$E_4(x) = \frac{1}{2\pi i}\int_{(\alpha)} \frac{G_A(w)}{w(w-s)}\left(\frac{x}{y}\right)^{w-s}\,dw.$$

Proof

$$E_1(x) = \int_y^{\infty} M_A\left(\frac{x}{t}\right)\,dO(x^a\delta_{c_0}(x)) \ll \frac{x}{y}y^a\exp(-c_0\mathcal{N}(y)) \ll \delta_{c_1}(x)$$

by (3.117).

$$E_2(x) = x^a\delta_{c_0}(x)\sum_{m\leq x/y} \frac{|a(m)|}{m^a}. \tag{3.123}$$

Noting that for $e > 0$

$$S(x) = \sum_{m \le t} \frac{|a(m)|}{m^{1*r}} = O(1),$$

the sum in (3.123) is $<< O(1)\left(\frac{x}{y}\right)^{r+1-a}$. Hence

$$E_2(x) \ll x^a \delta_{c_2}(x).$$

Finally, by (3.113) with $\delta = \frac{1}{\log x}$

$$E_4(x) = x^\alpha y^{a-\alpha} (\log y)^{c'} \int_{(\alpha)} \frac{|G_A(w)|}{|w|^2} |dw|. \tag{3.124}$$

By (3.112), the integral is finite and so by (3.117), $E_4(x) = x^a \delta_{c_6(\epsilon)}(x)$.

We note that the procedure is remindful of that of Ivič [Ivić (1985), 395-429], Postnikov [Postnikov (1988), pp. 289-300].

Proof of Theorem 3.12.

First we shall derive (3.133) and (3.134) with $y > 0$ and with admissible error. y is to be chosen later as (3.117).

We write

$$C(x) = \int\int_{uv \le x} dA(u) dB(v), \tag{3.125}$$

and divide the integral into two parts $n \le y$ and $n > y$ to deduce that

$$
\begin{aligned}
C(x) &= \int_{v \le y} \int_{u \le \frac{x}{v}} + \int_{y < v \le x} \int_{u \le \frac{x}{v}} \\
&= \int_{v \le y} dB(v) \int_{u \le \frac{x}{v}} dA(u) + \int_{u \le \frac{x}{y}} dA(u) \int_{y < v \le \frac{x}{u}} dB(v)
\end{aligned} \tag{3.126}
$$

whence we have

$$C(x) = S_1 + S_2, \tag{3.127}$$

where

$$S_1 = S_1(x, y) = \sum_{n \le y} b(n) A\left(\frac{x}{n}\right), \, S_2 = S_2(x, y) = \sum_{m \le x/y} a(m)\left(B\left(\frac{x}{m}\right) - B(y)\right). \tag{3.128}$$

Writing

$$S_1 = \sum_{n=1}^{\infty} b(n) M_A\left(\frac{x}{n}\right) - \sum_{n > y} b(n) M_A\left(\frac{x}{n}\right) + \sum_{n \le y} b(n) E_A\left(\frac{x}{n}\right), \tag{3.129}$$

we see that the second sum is

$$\int_y^\infty M_A\left(\frac{x}{t}\right) dB(t) = \int_y^\infty M_A\left(\frac{x}{t}\right) dM_B(t) + E_1(x),$$

so that (3.129) becomes

$$S_1 = \sum_{n=1}^\infty b(n) M_A\left(\frac{x}{n}\right) + \sum_{n\le y} b(n) E_A\left(\frac{x}{n}\right) - \int_y^\infty M_A\left(\frac{x}{t}\right) dM_B(t) + E_1(x),$$

(3.130)

where here and in what follows $E_j(x)$ indicates an error term defined and estimated in Lemma 3.2.

On the other hand, by (3.55),

$$S_2 = S_2(x,y) = \sum_{m\le x/y} a(m)\left(B\left(\frac{x}{m}\right) - B(y)\right) + E_2(x) \qquad (3.131)$$

$$= \int_{1-}^{x/y}\left(B\left(\frac{x}{t}\right) - B(y)\right) dA(t) + E_2(x)$$

$$= \left[\left(B\left(\frac{x}{t}\right) - B(y)\right) A(t)\right]_{1-}^{x/y} - \int_{1-}^{x/y} A(t)\, d\left(B\left(\frac{x}{t}\right) - B(y)\right) + E_2(x)$$

$$= -\int_{1-}^{x/y} M_A(t)\, dM_B\left(\frac{x}{t}\right) - \int_{1-}^{x/y} E_A(t)\, dM_B\left(\frac{x}{t}\right) + E_2(x).$$

Hence adding (3.130) and (3.131), we see that the integrals boil down to

$$-\int_{1-}^{x/y} E_A(t)\, dM_B\left(\frac{x}{t}\right) + \int_0^{1-} M_A(t)\, dM_B\left(\frac{x}{t}\right).$$

Hence

$$C(x) = \sum_{n=1}^\infty b(n) M_A\left(\frac{x}{n}\right) + \sum_{n\le y} b(n) E_A\left(\frac{x}{n}\right) - \int_{1-}^{x/y} E_A(t)\, dM_B\left(\frac{x}{t}\right)$$

(3.132)

$$+ \int_0^{1-} M_A(t)\, dM_B\left(\frac{x}{t}\right) + E_3(x),$$

where $E_3(x) = \max\{E_1(x), E_2(x)\}$ which amounts to the formula [Lau (2002), (2.8), p. 38].

Now note that for $0 < t < 1$, we have $A(t) = 0$ and so $M_A(t) = -E_A(t)$ for $0 < t < 1$. Hence the last two integrals combine to give the integral over $(0, x/y)$, so that

$$C(x) = \sum_{n=1}^\infty b(n) M_A\left(\frac{x}{n}\right) + \sum_{n\le y} b(n) E_A\left(\frac{x}{n}\right) + J + E_3(x), \qquad (3.133)$$

where

$$J = -\int_0^{x/y} E_A(t)\, \mathrm{d}M_B\left(\frac{x}{t}\right). \tag{3.134}$$

At this point we appeal to the special feature of the main term $M_B(x)$ as the contour integral (3.115) to obtain the integral expression

$$\Phi'(x) = \frac{1}{2\pi i} \int_{\mathcal{H}} G_B(s) x^{s-1}\, \mathrm{d}s. \tag{3.135}$$

Hence the integrator in (3.134) becomes

$$\mathrm{d}M_B\left(\frac{x}{t}\right) = -x\Phi'\left(\frac{x}{t}\right)\frac{\mathrm{d}t}{t^2} = -x\frac{1}{2\pi i}\int_{\mathcal{H}} G_B(s)\left(\frac{x}{t}\right)^{s-1}\mathrm{d}s\frac{\mathrm{d}t}{t^2}. \tag{3.136}$$

Substituting this in (3.134), we derive that

$$J = \frac{1}{2\pi i}\int_{\mathcal{H}} G_B(s) x^s \left\{ \int_0^{x/y} E_A(t) t^{-s-1}\, \mathrm{d}t \right\} \mathrm{d}s. \tag{3.137}$$

If we follow Lau's argument on p. 39, we apply the Riesz sum of order 1 referred to as Perron's formula in the form

$$A(x) = \frac{1}{2\pi i}\int_{2-i\infty}^{2+i\infty} G_A(w) x^w\, \mathrm{d}w. \tag{3.138}$$

The application of Perron's formula in infinite integral form is to be understood as in Davenport [Davenport (2000), p. 105], i.e. as the truncated one applied with a proper error term which however can be absorbed in the error term. Indeed, in Lau the infinite integral is replaced by finite integral

Under these conditions, for $\alpha \leq \lambda \leq \eta$

$$E_A(t) = \frac{1}{2\pi i}\int_{\lambda-iT}^{\lambda+iT} G_A(w) t^w \frac{\mathrm{d}w}{w} + G_{A,\lambda}(0) + I(\lambda, T, t), \tag{3.139}$$

where $G_{A,\lambda}(0) = G_A(0)$ if $\lambda < 0$ and is 0 otherwise.

Substituting (3.139) in (3.137) and interchanging the integration, we are to evaluate the integral

$$\int_0^{x/y} t^{w-s-1}\, \mathrm{d}t \tag{3.140}$$

which can be improper at $t = 0$ and is proper only for $\lambda = \eta$ in view of (3.114). However we are to choose $\lambda = \alpha$ for extracting part of the main term, and so we need to divide the integral inside the curly bracket in

(3.137) into two parts as in Lau. There is no need to split at $t = 1$ and at any midpoint $t = t_0 > 0$, we have

$$\int_{t_0}^{x/y} E_A(t) t^{-s-1} \, dt = \frac{1}{2\pi i} \int_{\lambda - iT}^{\lambda + iT} G_A(w) \frac{dw}{w} \int_{t_0}^{x/y} t^{w-s-1} \, dt \quad (3.141)$$

$$+ G_{A,\lambda}(0) \int_{t_0}^{x/y} t^{-s-1} \, dt + \int_{t_0}^{x/y} I(\lambda, T, t) t^{-s-1} \, dt$$

$$= \frac{1}{2\pi i} \int_{\lambda - i\infty}^{\lambda + i\infty} G_A(w) \left(\left(\frac{x}{y}\right)^{w-s} - t_0^{w-s} \right) \frac{dw}{w(w-s)}$$

$$+ s^{-1} G_{A,\lambda}(0) \left(t_0^{-s} - \left(\frac{y}{x}\right)^s \right)$$

after letting $T \to \infty$, which corresponds to [Lau (2002), (2.11)].

Now w lies on the contour $\mathcal{H}_Y(a, \delta)$, so that $a - Y \le \mathrm{Re}\, s \le a + \delta$. Hence if $\mathrm{Re}\, w = \lambda = \eta$, then $G_{A,\eta}(0) = 0$ and $\mathrm{Re}\, w - \mathrm{Re}\, s = \eta - a - \delta > 0$ by (3.114). Hence in (3.141) we may regard $t_0 = 0$, so that

$$\int_0^{t_0} E_A(t) t^{-s-1} \, dt = \frac{1}{2\pi i} \int_{\eta - i\infty}^{\eta + i\infty} G_A(w) \frac{dw}{w(w-s)} \quad (3.142)$$

on regarding x/y as t_0. Hence adding (3.141) with $\lambda = \alpha$ and (3.142) leads to

$$\int_0^{x/y} E_A(t) t^{-s-1} \, dt \quad (3.143)$$

$$= \frac{1}{2\pi i} \int_{(\alpha)} \frac{G_A(w)}{w(w-s)} \left(\frac{x}{y}\right)^{w-s} \, dw + s^{-1} G_{A,\alpha}(0) \left(t_0^{-s} - \left(\frac{y}{x}\right)^s \right)$$

$$+ \frac{1}{2\pi i} \left\{ \int_{(\eta)} - \int_{(\alpha)} \right\} \frac{G_A(w)}{w(w-s)} \, dw.$$

Since the last summand equals

$$s^{-1} G_A(s) - s^{-1} G_{A,\alpha}(0), \quad (3.144)$$

(3.143) with $t_0 = 1$ becomes

$$\int_0^{x/y} E_A(t) t^{-s-1} \, dt \quad (3.145)$$

$$= \frac{1}{2\pi i} \int_{(\alpha)} \frac{G_A(w)}{w(w-s)} \left(\frac{x}{y}\right)^{w-s} \, dw + s^{-1} \left(G_A(s) - G_{A,\alpha}(0) \left(\frac{y}{x}\right)^s \right).$$

Substituting (3.145) in (3.137), we find that

$$J = \frac{1}{2\pi i} \int_{\mathcal{H}} G_B(s) \left(G_A(s) x^s - G_{A,\alpha}(0) y^s \right) \frac{ds}{s} + E_4(x), \quad (3.146)$$

where $E_4(x)$ is absorbed in the error term (3.119).

Substituting (3.146) in (3.133) and noting

$$\frac{1}{2\pi i} \int_{\mathcal{H}} G_B(s) y^s \frac{ds}{s} = \sum_{n \leq y} b(n) \qquad (3.147)$$

completes the proof of Lau's theorem up to the error estimates, which is done in Lemma 3.2. It follows also that y should be chosen almost x, i.e. as in (3.117).

3.8   Abel-Tauber process

In mathematics it is very often the case that we are to consider the average of data rather than the datum itself. In number theory we consider the average form of an arithmetic function $\{a_n\}$ as a (Riesz) sum with as good an estimate of the error term as possible. The most common procedure for attaining this aim is to use a generating function $Z(s) = \sum_{n=1}^{\infty} \frac{a_n}{\lambda_n^s}$ absolutely convergent in some half-plane, where $s = \sigma + it$ is the complex variable. There is a large class of arithmetic functions whose generating function has an Euler product. The generating function for the so-called Selberg type divisor problem is a complex power of a zeta-function and so belongs to this class. If that is the case, one may appeal to the factoring-out method in which the main procedure is factoring out the essential zeta-function $Z(s)^*$ from the Euler product form of $Z(s)$, leaving an inessential factor $\Phi(s)$ convergent in a wider region and negligible in the process of estimation of the error term. The essential factor is usually the product of zeta-functions and is very often in the form of a certain complex power of a suitable zeta-function.

Here the problem pertains to the arithmetic function $d_z(n)$ defined by

$$\zeta(s)^z = \prod_p \left(1 - \frac{1}{p^s}\right)^{-z} = \sum_{n=1}^{\infty} \frac{d_z(n)}{n^s} \qquad (3.148)$$

for $\sigma > 1$. This includes among others the divisor function cf. (3.8) and the more general Plitz divisor function (6.89).

From a generating function $Z(s)$ with an Euler product we factor the essential factor $Z^*(s)$ out as a product of known zeta-functions. For example, if the coefficient $z(p)$ of the first order prime is $z(p) = z + O(p^{-\delta})$ (with $z \neq 0$ a constant, $\delta > 0$, $O(p^{-\delta})$ being an appropriate allowance element), then the factor is (3.148).

If $z(p) = 0$ occurs for a finite set S of primes, then we factor out the finite part $\Phi_0(s) = \prod_{p \in S}$ so that the coefficient of the first order prime in $Z(s)/\Phi_0(s)$ is always non-zero. Cf. (3.160) and Example 3.10.

We may apply the Abel-Tauber process to the essential factor $Z^*(s) = \sum_{n=1}^{\infty} \frac{a_n^*}{\lambda_n^s}$ and deduce the asymptotic formula for the summatory function of a_n^*. If the remaining factor $\Phi(s) = \frac{Z(s)}{Z^*(s)} = \sum_{n=1}^{\infty} \frac{b_n^*}{\lambda_n^s}$, then a_n is given as the Dirichlet convolution $a = a^* * b^*$ and we may establish some interesting results for the original arithmetic function. This study will be conducted elsewhere from a more general point of view.

3.8.1 *The Rieger-Nowak-Nakaya results*

A great number of papers have been published on the Selberg type divisor problem including [Delange (1959)], [Delange (1971)], [Friedlander (1972)], [Hasse and Suetuna (1931)], [Nowak (1991)], [Nowak (1991)], [Nowak (1992)], [Scourfield (1991)], [Suetuna (1931)], [Suetuna (1950)], [Tennenbaum (1995)], etc. and references therein. The error term in these papers is, however, not as good as that of the prime number theorem (PNT).

The best method is due to Balasubramanian and Ramachandra [Balasubramanian and Ramachandra (1998)] and Nowak [Nowak (1991)]-[Nowak (1992)]. The former was generalized by Nakaya [Nakaya (1991)]. These three authors treated the proper main term *in the form of a contour integral* and were able to exhibit the error estimate as good as that for the PNT and then showed that it can be expanded into an asymptotic formula. Among them, Nowak mainly considers the average order of the *reciprocals* of arithmetic functions (in the lines of [Koninck and Ivić (1980)]) including the number $a(n)$ of non-isomorphic Abelian groups ([Nowak (1991)]) and $d_{\varkappa,k}(n)$ in [Nowak (1991)], cf. Theorem 3.13. And in [Nowak (1992)], the average order of the function $\tau_{\mathbf{B},\mathcal{A}}$ is considered, which counts the number of natural numbers n belonging to both $\mathbf{B}$ and $\mathcal{A}$. Here $\mathbf{B}$ is the set of natural numbers which are equal to the norm of an ideal in $\mathfrak{o}_k$ and $\mathcal{A} = \mathcal{A}(a,q)$ is the arithmetic progression $n \equiv a \pmod{q}$. As an intriguing and typical result of Nowak, we state

Theorem 3.13. (Nowak, [Nowak (1991)]) *Let k be a number field of degree l over $\mathbb{Q}$ and let $d_{\varkappa,k}(n)$ be the number of $\varkappa$ ideals $\mathfrak{A}_i \in \mathfrak{o}_k$, the norm of whose product is n. Then we have*

$$\sideset{}{'}\sum_{n \le x} d_{m,k}(n)^{-1} = x \sum_{j=0}^{M(x)} A_j (\log x)^{\frac{1}{ml^2} - j - 1} + O(x \delta_{VK}(x)), \qquad (3.149)$$

where δ_{VK} is the Vinogradov-Korobov reducing factor in (3.47), which gives the best known error estimate and

$$M(x) = \left[a(\log x)^{3/5}(\log\log x)^{-6/5}\right] = \left[-\log(\delta_{VK}(x))(\log\log x)^{-1}\right].$$
(3.150)

The summatory functions in [Nowak (1991)] and [Nowak (1991)] may be thought of as special cases of the sum $B_{1,c}(x) = B_1(x) = \sum_{n \le x} g(n)^c$ in Example 3.10, with the best possible error term. Suetuna [Suetuna (1931)] imposed the condition that c is a fixed positive constant on p. 155 but this just comes from the fact that Ramanujan considered such cases of divisor functions. These are 'small functions' in the terminology of Scourfield [Scourfield (1991)]. In [Balasubramanian, Kanemitsu and Ramachandra (2003)], c is considered as an arbitrary constant.

We note that both $A_1(x) = \sum_{ng(n) \le x} 1$ and $B_1(x)$ (cf. (3.163) below) are already considered in [Suetuna (1931)]. In [Balasubramanian and Ramachandra (1998), Remark 4], there is a close relationship between $A_1(x)$ and $B_{1,-1}(x) = \sum_{n \le x} g(n)^{-1}$, i.e. $A_1(x) \sim B_{1,-1}(x)$, which is again noticed in [Scourfield (1991), (1.14)]. This explains the reason why these asymptotic formulas hold true simultaneously, reminiscent of several equivalent statements of the PNT.

The next step would be to consider the same problem for an arithmetic progression or the (narrow) ideal class in an algebraic number field.

In this direction Rieger [Rieger (1965)] generalized a result of Selberg [Selberg (1956)] to an arithmetic progression $n \equiv l \,(\text{mod } k)$. Although the main result [Rieger (1965), Satz 1] states an asymptotic formula for $D_z(x; k, l) := \sum_{\substack{n < x \\ n \equiv l \,(\text{mod } k)}} d_z(n)$ with logarithmic reduction, he has reached the highest summit given as [Rieger (1965)] in the following Siegel-Walfisz type theorem.

Theorem 3.14. (Rieger, Satz 1)

(i) For $A > 0$, $|z| < A$ with $(k, l) = 1$, we have

$$D_z(x; k, l) = \frac{1}{\varphi(k)} L_z(x; k) + O\left(\frac{x^{\beta(k)}}{\varphi(k)}(\log 2k)^{-c_1 A}\right)$$
$$+ O_A(x\delta_{VP}(x)),$$
(3.151)

where $\varphi(k)$ is the Euler function and $\delta_{VP}(x)$ is the de la Vallee Poussin reducing factor (3.46).

Here $L_z(x; k)$ is the integral of the function $\frac{1}{2\pi i} Z^*(s) \frac{x^s}{s}$ along the contour $C_r = C_r(k)$ and the contour goes along the lower edge of the real axis from $A_0 = 1 - \frac{c_6}{\log 8k}$ to $1 - r$, encircling the circle with center at the logarithmic singularity at $s = 1$ and small radius $r > 0$ in clockwise direction, returning to the point A_0 along the upper edge of the real axis.

(ii) Under the condition, similar to (3.225),

$$k \ll \log^B x, \quad B > 0 \tag{3.152}$$

we have

$$D_z(x; k, l) = \frac{1}{\varphi(k)} L_z(x; k) + O_{A,B}(x\delta_{VP}(x)), \tag{3.153}$$

where the O-constant is ineffective.

For the auxiliary sum which is naturally associated to all the arithmetic progression type issues

$$D(x; \chi) := \sum_{n < x} \chi(n) d_z(n),$$

where χ is a Dirichlet character of modulus k. For this purpose Rieger started investing the formula for the Riesz sum of order 1, cf. (3.198),

$$\int_1^x D_z(v; x) \, dv = \frac{1}{2\pi i} \int_{2-i\infty}^{2+i\infty} (L(s, \chi))^z \frac{x^{s+1}}{s(s+1)} \, ds \tag{Rie, 3}$$

and obtained intermediate result with the de la Vallee Poussin error term.

Lemma 3.3. [Rieger (1965), Hilfssatz 5] *For $A > 0$, $|z| < A$, $x > c_2(A) > 0$, $k < e^{\sqrt{\log x}}$ we have*

$$D_z(x; \chi) = E(\chi) L_z(x; k) + E_1(\chi) O(x^{\beta(k)} (\log 2k)^{c_{11} A}) + O_A(x\delta_{VP}(x)), \tag{3.154}$$

where

$$E(\chi) = \begin{cases} 1 & if \chi = \chi_0 \\ 0 & if \chi \neq \chi_0 \end{cases} \quad and \quad E_1(\chi) = \begin{cases} 1 & if \chi = \chi_1 \\ 0 & if \chi \neq \chi_1, \end{cases} \tag{3.155}$$

where χ_0 is the principal character and χ_1 is the exceptional character as defined in §3.8.9.

Thus we may say that Rieger's argument is a precursor to all subsequent papers including [Banerjee, Chakraborty, Kanemitsu and Magi (2016)] which constitutes the main body of §3.8, i.e., one first establishes an asymptotic formula for the Riesz sum of order $\varkappa$ and then applies (an equivalent to) the differencing method to deduce a Tauberian result with the reducing factor as in the prime number theorem.

Rieger's result has been generalized by Mercier [Meyer (1967)] to the case of a multiplicative function with logarithmic reduction. Although Mercier considers the two cases where $(k, l) = 1$ and $(k, l) = k$, the latter case leads to the progression $k(\frac{l}{k} + n)$, $n \in \mathbb{N}$ and will not be considered here. The relatively prime case amounts to Rieger's.

Prior to Mercier, Nakaya [Nakaya (1992)] made a generalization of Rieger's result and obtained an equivalent assertion to the (generalized) RH. He also studied a generalization of the prime counting function (3.156) and obtained a similar assertion. His method is the same as Rieger's starting from the Riesz sum formula [Rieger (1965), eq. 3] of degree 1 and applying the differencing. The first half of his main theorem is the same as Rieger's Satz 1 above with the de la Vallee Poussin error estimate.

Our method may be described as a generalized elaboration of the Rieger-Nakaya-Nowak method i.e., first we obtain an Abelian result as asymptotic formulas for the higher order Riesz sums of the arithmetic function generated by $Z^*(s) = \sum_{n=1}^{\infty} \frac{a_n^*}{\lambda_n^s}$ which are interesting in their own right. Then applying Landau's differencing argument we deduce an asymptotic formula for the summatory function of a_n^*. We may describe the latter as the Tauberian process. We note that Landau's differencing method goes back to [Landau (1915)]. The novelty of our method is the incorporation of the Selberg type divisor problem in the process by viewing the contour integral of the form $\frac{1}{2\pi i} \int_{C_r} Z^*(s) \frac{x^s}{s} \, ds$ as part of the residual function.

Another novelty of our results lies in the uniformity of the error term in the additional parameter Δ, which varies according to the cases. A typical case is $\Delta = N\mathfrak{f} \cdot |d|$ as in Theorem 3.21.

3.8.2   *Other complex powers of L-functions*

As is discussed in §3.8.1, the complex powers of zeta- and L-functions have been studied rather extensively as the essential factor of the given generating functions with Euler product to extract crucial information therefrom. As a typical example, in [Selberg (1954)] Selberg's motivation for considering the complex power was to obtain an asymptotic formula for the sum

$$\pi_k(x) = \sum_{\substack{n \le x \\ n = p_1 \cdots p_k}} 1. \tag{3.156}$$

The generating function associated with this sum is

$$Z(s) = \prod_p \left(1 + \frac{z}{p^s}\right) \tag{3.157}$$

and the essential factor $Z^*(s)$ is evidently (3.148).

Another example is the generating function considered by Delange [Delange (1971)]. Let $f(n)$ be an additive function which takes only positive integer values and $\chi(n)$ is a bounded multiplicative function satisfying

$$\sum_{k=1}^{\infty} \frac{|\chi(p^k)||z|^{f(p^k)}}{p^{k\sigma}} < \infty \tag{3.158}$$

with $\chi(p) = 0$ only for those $p \in S$, where S is a finite set. Then

$$Z(s) := \sum_{k=1}^{\infty} \frac{\chi(n)z^{f(n)}}{n^s} = \Phi_0(s) \prod_{p \notin S} \left(1 + \sum_{k=1}^{\infty} \frac{\chi(p^k)z^{f(p^k)}}{p^{ks}}\right). \tag{3.159}$$

Here $\Phi_0(s) = \prod_{p \in S}$ is a finite part. This assures the condition $\chi(p) \neq 0$ in the second product.

The essential factor $Z^*(s) = Z^*(s, z)$ of $Z(s)$ in this case is

$$Z^*(s) = \zeta(s)^z$$

since

$$\Phi(s) = \Phi(s, z) = \frac{Z(s)}{Z^*(s)} = \Phi_0(s) \prod_p \left(1 + \sum_{k=2}^{\infty} \frac{a(p)}{p^{ks}}\right) \tag{3.160}$$

is absolutely convergent for $\sigma > \sigma_0(|z|)$. The essential factors which are to be raised to a complex power may belong to the Selberg class $\mathcal{S}$ (cf. §3.8.11) which contains the Riemann zeta-, Dirichlet L- and cusp form zeta-functions among others. Also complex powers of various zeta functions are interesting in their own right. One may consider the product

$$P_1(s)^{z_1} P_2(s)^{z_2} P_3(s)^m$$

as in [Balasubramanian and Ramachandra (1998)], [Ramachandra (1976)]. Here P_1 is the product of Selberg class zeta-functions, P_2 is the product of their derivatives and P_3 is their logarithm.

As is expounded at the end of the preceding section, our aim is two-fold. The first is to derive results on the asymptotic formula for Riesz sums (of order $\varkappa \geq 0$) of the coefficients of a generating function $F(s) = Z^*(s)$, with the error term as good as the one in the PNT incorporating complex powers of zeta-functions, i.e. allowing a singularity up to a logarithmic one in addition to poles. In the case of the Selberg type divisor problem, the main term is in the form of a contour integral which can be expanded into an asymptotic formula. This will be treated as part of the residual function.

The second aim is to make the dependence explicit of the error estimate on the additional parameter Δ.

The following example ([Balasubramanian, Kanemitsu and Ramachandra (2003)]) illustrates the factoring out of the essential zeta-function in a rather general setting.

Example 3.10. (Ideal-function-like function) Let K/k be an algebraic extension of number fields of degree n. Let K^* be the smallest Galois extension of k containing K, let $\mathfrak{G} = \mathrm{Gal}(K^*/k)$ be the corresponding Galois group with $\#\mathfrak{G} = g$, and let $\mathfrak{o}_K$ be the ring of integers in K. We regard $\mathfrak{G}$ as a (transitive) subgroup of the n-th symmetric group. Let

$$\mathfrak{G} = \bigcup_{j=1}^{h} \mathfrak{C}_j$$

be the decomposition of $\mathfrak{G}$ into the disjoint union of conjugate classes $\mathfrak{C}_j$, with $\mathfrak{C}_1 \ni 1$ and $\#\mathfrak{C}_j = h_j$.

By a theorem of Dedekind, those prime ideals $\mathfrak{p}_0$ which ramify in $\mathfrak{o}_K$ are those dividing the discriminant $D_{K/k}$ of K/k. All unramified prime ideals may be classified into h classes corresponding to conjugate classes $\mathfrak{C}_j$ in the sense that if $\sigma \in \mathfrak{G}_j$ decomposes into the product of e_j disjoint cycles consisting of $f_{j_1} \cdots f_{je_j}$ symbols, then

$$\mathfrak{p}_j = \mathfrak{P}_{j1} \cdots \mathfrak{P}_{je_j}, \quad N_{K/k}\mathfrak{P}_{ji} = \mathfrak{p}^{f_{ji}}.$$

We say that $\mathfrak{p}_j$ belongs to $\mathfrak{C}_j$ for $1 \leq j \leq h$.
Given $h+1$ numbers $\lambda_j > 0$, $1 \leq j \leq h$, assume that

$$g(\mathfrak{p}_j) = \frac{1}{\lambda_j} + O\left(\exp\left(-c_1(\log N\mathfrak{p}_j)^a\right)\right) \tag{3.161}$$

with $c_1 > 0$, $a > \frac{3}{2}$. Then define the **ideal-function-like function** $g(\mathfrak{a})$ multiplicatively. Further assume that

$$g(\mathfrak{a}) \gg (N\mathfrak{a})^{-\alpha}, \quad \alpha > 0. \tag{3.162}$$

In this setting the summatory function

$$A_1(x) = \sum_{N\mathfrak{a}g(\mathfrak{a})\leq x} 1, \quad B_1(x) = \sum_{N\mathfrak{a}\leq x} g(\mathfrak{a})^c \tag{3.163}$$

was considered in [Balasubramanian, Kanemitsu and Ramachandra (2003)]. The generating function for $A_1(x)$ is

$$Z(s) = \sum_{\mathfrak{a}} (N \mathfrak{a} g(\mathfrak{a}))^{-s}, \quad \sigma > 1. \tag{3.164}$$

By multiplicativity of $g(\mathfrak{a})$, we may express $Z(s)$ as

$$Z(s) = \prod_{\mathfrak{p}} \left(1 + \frac{1}{(N\mathfrak{p}g(\mathfrak{p}))^s} + \frac{1}{(N\mathfrak{p}g(\mathfrak{p}))^{2s}} + \cdots \right). \tag{3.165}$$

According to the classification of prime ideals into $h+1$ classes, we obtain

$$Z(s) = \Phi_0(s) \prod_{j=1}^{h} Z_j(s), \tag{3.166}$$

where $\Phi_0(s)$ is the product (3.165) restricted to ramified primes $\mathfrak{p}_0$ and $Z_j(s)$ is the product over all prime ideals $\mathfrak{p}_j$ belonging to $\mathfrak{C}_j, 1 \leq j \leq h$. The product for $\Phi_0(s)$ is absolutely convergent for $\sigma > 0$ and can be neglected.

Let $\chi_j (1 \leq j \leq h)$ be the simple characters of $\mathfrak{G}$ (with χ_1 principal), and let $L(s, \chi_j) = L(s, \chi_j; K^*/k)$ be the corresponding Artin L-function. Cf. §2.3 for Artin L-functions. Let

$$\tau_i(s) = \frac{1}{g} \sum_{j=1}^{h} \lambda_j^{-s} h_j \chi_i(\mathfrak{G}_j^{-1}) = \frac{1}{g} \sum_{G \in \mathfrak{G}} \lambda_G^{-s} \chi_i(G^{-1}), \tag{3.167}$$

where $\lambda_G = \lambda_j$ if $G \in \mathfrak{C}_j$.

The main result in [Balasubramanian, Kanemitsu and Ramachandra (2003)] is

Theorem 3.15. ([Balasubramanian, Kanemitsu and Ramachandra (2003), Theorem 1]) *The generating Dirichlet series* (3.164) *admits the representation*

$$Z(s) = Z^*(s)\Phi(s), \quad Z^*(s) = \exp\left(\sum_{j=1}^{\hbar} \tau_j(s) \log L(s, \chi_j) \right) \tag{3.168}$$

where $\Phi(s)$ *is regular and non-vanishing for* $\sigma > \frac{1}{2}$ *and is bounded in* $\sigma \geq \frac{1}{2} + \delta$ ($\delta > 0$).

From this expression, one may obtain a general result with the best known error estimate by the convolution argument.

Note that **Suetuna's divisor function**

$$T(\mathfrak{a}) = T_{\varkappa}(\mathfrak{a}) = \sum_{\mathfrak{A}_1 \cdots \mathfrak{A}_{\varkappa} = \mathfrak{a}} 1, \tag{3.169}$$

the number of expressions of $\mathfrak{a} \subset \mathfrak{o}_k$ as a product of $\varkappa$ ideals $\mathfrak{A}_j$ in $\mathfrak{o}_K$, is a typical example of $g(\mathfrak{a})$.

Example 3.11. This example depends on the last part of [Suetuna (1925)]. We consider a special case of Example 3.10 in that K/k is an Abelian extension $k/\mathbb{Q}$ of degree ℓ with the Dedekind zeta-function

$$\zeta_k(s) = \sum_{n=1}^{\infty} \frac{F(n)}{n^s}, \quad \sigma > 1, \tag{3.170}$$

where $F(n)$ is the ideal-function counting the number of ways in which n can be expressed as a product of norms of ideals. Cf. Remark 1.3. We consider the $\varkappa$-dimensional divisor function as a special case of (3.169).

$$T(n) = T_\varkappa(n) = \sum_{N\mathfrak{a}_1 \cdots N\mathfrak{a}_\varkappa = n} 1, \tag{3.171}$$

which is generated by

$$\zeta_k(s)^\varkappa = \sum_{n=1}^{\infty} \frac{T_\varkappa(n)}{n^s}, \quad \sigma > 1. \tag{3.172}$$

Since $T(n)$ and $T_n^r(n)$ are multiplicative, we have similarly to (3.165)

$$\zeta_k(s)^\varkappa = \prod_p \left(1 + \frac{T(p)}{p^s} + \frac{T(p^2)}{p^{2s}} + \cdots \right) \tag{3.173}$$

and

$$Z(s) = \sum_{n=1}^{\infty} \frac{T_\varkappa^r(n)}{n^s} = \prod_p \left(1 + \frac{T^r(p)}{p^s} + \frac{T^r(p^2)}{p^{2s}} + \cdots \right) \tag{3.174}$$

for any $r \in \mathbb{R}$. From (3.173) and the Euler product for the Dedekind zeta-function, we conclude that

$$T_\varkappa(p) = \varkappa F(p) \tag{3.175}$$

for every prime p. Viewing k as a class field, it follows that $F(p) = 0$ or $F(p) = \ell$ according as p belongs to the principal class or not. Hence save for finite number of primes, (3.175) gives

$$T_\varkappa^r(p) = \varkappa^r \ell^{r-1} F(p). \tag{3.176}$$

Hence the essential factor is $Z^*(s) =)\zeta_k(s) \varkappa^r \ell^{r-1}$ and the auxiliary factor is $\prod_p \left(1 + O\left(\frac{|T^r(p^2)|}{p^{2\sigma}} \right) \right)$, which is absolutely convergent for $\sigma > \frac{1}{2}$. From this Suetuna deduced an asymptotic formula for the sum $\sum_{n \leq x} T_\varkappa^r(n)$ for $\varkappa^r \ell^{r-1}$ being an integer and $\varkappa^r \ell^r \geq 4$ using the 4th power moment of the Dirichlet L-functions. In the case k is a quadratic field ($\ell = 2$) and $\varkappa = 1$, $r = 2$, the explicit decomposition

$$Z(s) = \frac{\zeta_k(s^2)}{\prod_{p|d}(1 + p^{-s})\zeta(2s)} \tag{3.177}$$

holds true, where d is the absolute value of the discriminant of the field. From this he deduces

$$\sum_{n \leq x} F^2(n) = ax \log x + bx + O\left(x^{\frac{1}{2}+\varepsilon}\right) \tag{3.178}$$

with explicit expression

$$a = \frac{24h^2}{w^2 d \prod_{p|d}\left(1 + \frac{1}{p}\right)} \tag{3.179}$$

where h and w indicate the class number and the number of roots of unity discussed in Chapter 2.

3.8.3 *Riesz sums*

For the theory of Riesz sums, cf. Hardy and Riesz [Hardy and Riesz (1915)], Chandrasekharan and Minakshisundaram [Chandrasekharan and Minakshisundaram (1952)], [Kanemitsu (2007), Chapter 6] and references therein.

Example 3.12. Given an increasing sequence $\{\lambda_k\}$ of real numbers and a sequence $\{\alpha_k\}$ of complex numbers, we let

$$A_\lambda(x) = A_\lambda^0(x) = {\sum_{\lambda_k \leq x}}' \alpha_k \tag{3.180}$$

be the summatory function of a sequence $\{\alpha_k\}$, where the prime on the summation sign means that when $\lambda_k = x$, the corresponding term is to be halved. Then we have

$$\varkappa \int_0^x (x - t)^{\varkappa - 1}\, \mathrm{d}A_\lambda(t) \tag{3.181}$$

$$= \varkappa \int_0^x (x - t)^{\varkappa - 1} A_\lambda(t)\mathrm{d}t$$

$$= {\sum_{\lambda_k \leq x}}' (x - \lambda_k)^{\varkappa} \alpha_k$$

$$= A^{\varkappa}(x) = A_\lambda^{\varkappa}(x),$$

say, where $A_\lambda^{\varkappa}(x)$ is called the Riesz sum of order $\varkappa$ as in [Chandrasekharan and Minakshisundaram (1952), p. 2] and [Hardy and Riesz (1915), p. 21]. The Riesz means, or sometimes typical means, were introduced by M. Riesz and have been studied in connection with summability of Fourier series and of Dirichlet series [Chandrasekharan and Minakshisundaram (1952)] and [Hardy and Riesz (1915)].

(3.181) or rather normalized $\frac{1}{\Gamma(\varkappa+1)}A^\varkappa(x)$ which appears in [Kanemitsu (2007), (G-8-2)] is called the **Riesz sum** of order $\varkappa$. If $\frac{1}{\Gamma(\varkappa+1)}A^\varkappa(x)$ approaches a limit A as $x \to \infty$, the sequence $\{\alpha_k\}$ is called Riesz summable or $(R, \varkappa, \lambda)$ summable to A, which is called the **Riesz mean** of the sequence. Sometimes the negative order Riesz sum is considered, in which case the sum is taken over all n which are not equal to x.

3.8.4   *Abelian process*

Definition 3.8. Let $\varkappa$ be a real number (mostly we assume that it is nonnegative) and let $\{\lambda_n\}_{n=1}^\infty$, $\{\ell_n\}_{n=1}^\infty$ be arbitrary sequences of real numbers strictly increasing to infinity such that $\lambda_1 \geq 1$, $\ell_1 \geq 0$. Let $\{a_n\}_{n=1}^\infty$ be any sequence of complex numbers. Then we write

$$\begin{cases} A_\lambda^\varkappa(x) = \frac{1}{\Gamma(\varkappa+1)} \sum_{\lambda_n \leq x} a_n(x - \lambda_n)^\varkappa \\ A_\ell^\varkappa(x) = \frac{1}{\Gamma(\varkappa+1)} \sum_{\ell_n \leq x} a_n(x - \ell_n)^\varkappa \end{cases} \tag{3.182}$$

and refer to $A_\lambda^\varkappa(x)$ (resp. $A_\ell^\varkappa(x)$) as the Riesz sum of order $\varkappa$ of the second (resp. first) kind associated to the series $\sum_{n=1}^\infty a_n \lambda_n^{-s}$ (resp. $\sum_{n=1}^\infty a_n e^{-s \log \ell_n}$), where absolute convergence of the series is assumed in some half-plane $\mathrm{Re}\, s = \sigma > \sigma_a$. For the special choice of λ (resp. ℓ), i.e. $\lambda_n = n$ or $= N\mathfrak{a}$, $N\mathfrak{a}$ denoting the norm of the integral ideal $\mathfrak{a}$ (resp. $\ell_n = \log n$ or $= \log N\mathfrak{a}$), we denote the corresponding Riesz sum $A_\lambda^\varkappa(x)$ (resp. $A_\ell^\varkappa(\log x)$) by $A_a^\varkappa(x)$ (resp. $A_l^\varkappa(x)$) and refer to it as the arithmetic (resp. logarithmic) Riesz sum of order $\varkappa$ associated to the series $\sum_{n=1}^\infty a_n n^{-s}$ or $\sum_{\mathfrak{a}} a_{\mathfrak{a}} N\mathfrak{a}^{-s}$.

Settings for the generating function:
Consider the generating function $F(s)$ of the sequence $\{a_n\}_{n=1}^\infty$ as a Dirichlet series

$$F(s) = \sum_{n=1}^\infty a_n \lambda_n^{-s} \tag{3.183}$$

with abscissa of absolute convergence σ_a, without loss of generality we may assume σ_a to be positive. Also let

$$B(b^*) = \sum_{n=1}^\infty |a_n| \lambda_n^{-b^*} \tag{3.184}$$

be one of its majorants, *a fortiori* $b^* > \sigma_a$. We assume the following conditions on F:

(i) $F(s)$ can be continued analytically to a *meromorphic function* in some (possibly thin) vertical strip-shape domain R' with the right-hand line $\sigma = b^*$ and bounded on the left by a piecewise smooth Jordan curve

$$\Gamma : \sigma = f(t), \quad 0 < f(t) < b^*, \tag{3.185}$$

where $f(t)$ is given by (3.190). By abuse of language, by meromorphy we allow up to logarithmic singularities and refer to them as poles. All the poles up to logarithmic singularities of $F(s)$ lying in R' are contained in the finite part R of R' and are not on Γ, where the subdomain R is described in (ii). We denote the sum of the residues of $\frac{\Gamma(s)}{\Gamma(s+\varkappa+1)}F(s)x^{s+\varkappa}$ in R by $Q_\varkappa(x)$:

$$Q_\varkappa(x) = \frac{1}{2\pi i}\int_{\partial R}\frac{\Gamma(s)F(s)}{\Gamma(s+\varkappa+1)}x^{s+\varkappa}\,\mathrm{d}s. \tag{3.186}$$

In case there is a logarithmic singularity we assume it to be at $s = \beta$ for the sake of simplicity and *by abuse of notation* we understand the residual function $Q_\varkappa(x)$ to mean the sum of residues in R plus the integral along the contour along the contour C_r described in Theorem 3.14 (or similar to it). Here the contour C_r may be given explicitly as

$$C_r : s = \sigma, \beta - \psi(0) \le \sigma < \beta - r; s = \beta + re^{i\theta}, 0 \le \theta \le 2\pi; \tag{3.187}$$
$$s = \sigma, \beta - \psi(0) \le \sigma < \beta - r.$$

(ii) R is a subdomain of R' whose boundary consists of $\overline{DA} : f(-T) \le \sigma \le b^*, t = -T, \overline{BC} : f(T) \le \sigma \le b^*, t = T, \overline{AB} : \sigma = b^*, -T \le t \le T$, and that part of DC of Γ with $|t| \le T$ with T large enough for all the poles of $F(s)$ in R' to be contained in R (T will be taken as in (3.192)).

(iii) Suppose that $F(s)$ satisfies the following growth conditions: there exists a constant $\varkappa > \mu - 1$ such that

$$F(s) = O\left(T^{\mu+\varepsilon}\right) \qquad \text{on } \overline{DA} \text{ and } \overline{BC}; \tag{3.188}$$
$$F(s) = O\left(|t|^r V(t)\right) \qquad \text{on } \Gamma \text{ if } |t| \ge t_0;$$
$$F(s) = O\left(W(f(t), t_0)\right) \qquad \text{on } \Gamma \text{ if } |t| \le t_0$$

where V, W are positive, integrable and $V(y) = o(y^\varepsilon)$ as $y \to \infty$, and $t_0 > 0$ is some constant. Let

$$\mathcal{L} = \mathcal{L}(t) = \log \Delta(|t| + 2) \tag{3.189}$$

and

$$f(t) = \beta - \psi(t) \geq \eta > 0; \quad \psi(t) = A\mathcal{L}^{-a}(\log \mathcal{L})^{-b}, \quad |t| \geq t_0 \quad (3.190)$$

with constants $a \geq 0, b \geq 0, A > 0, \Delta > 0$ and $\beta > 0$.

There is another type of zero-free region as in Hinz's Theorem 3.22, i.e.

$$f(t) = \beta - \mathcal{M}(\Delta, t)^{-1} > 0, \quad |t| \geq t_0, \quad (3.191)$$

$$\mathcal{M}(\Delta, t) = \max\{\Delta, A\mathcal{L}^a(\log \mathcal{L})^b\}$$

and $\Delta = |d| \log N\mathfrak{f}$ (in Hinz's result $a = \frac{2}{3}$, $b = \frac{1}{3}$).

Remark 3.3. The left boundaries (3.190) and (3.191) arise from the zero-free region of the relevant zeta-function. They are not essentially different on the ground that for $A, B > 0$

$$\max\{A, B\} \leq A + B \leq 2\max\{A, B\}.$$

In applications, we assume $\beta = \sigma_a$ and choose $b^* = \beta + \frac{1}{\log x}$. Also all the zeta-functions $Z(s)$ that appear satisfy the condition

$$Z(s) = O(e^{\varepsilon|t|})$$

for every $\varepsilon > 0$ in every strip $\sigma_1 \leq \sigma \leq \sigma_2$, the growth conditions (3.188) in Theorem 3.16 amount to $F(s) = O(|t|^\tau V(t))$ by the Phragmén-Lindelöf principle. We tacitly assume the known estimates for the reciprocals of zeta- and L-functions, e.g. for the Hecke L-function $L_K(s, \chi)$, we have ([Fogels (1965)])

$$L_K(s, \chi)^{-1} = O(\log^n \Delta(|t| + 2))$$

in the zero-free region corresponding to (3.189):

$$\sigma > \beta - \frac{c}{\mathcal{L}}.$$

This is absorbed in the V-function while in most of the applications $W = O(1)$; care must be taken when one tries to maintain uniformity in the parameter Δ.

Now we state our main theorem in [Banerjee, Chakraborty, Kanemitsu and Magi (2016)].

Theorem 3.16. *Suppose F is a Dirichlet series which satisfies the conditions in the setting and that*

$$T = x^\alpha \quad (3.192)$$

with a constant $\alpha > 0$. Also suppose that

$$\log \Delta \ll (\log x)^{\frac{1}{1+a}-\eta} \quad (\eta > 0). \tag{3.193}$$

Then for any $\varkappa > \tau$ we have

$$A_\lambda^\varkappa(x) = Q_\varkappa(x) + R_{\lambda,I}^\varkappa(x), \tag{3.194}$$

where $R_{\lambda,I}^\varkappa$ is the error term satisfying

$$R_{\lambda,I}^\varkappa(x) = O\left(x^{\varkappa+b^*}\left(x^{\alpha(\mu+\varepsilon-\varkappa-1)} + x^{-\alpha\varkappa}B(b^*)\right)\right) \tag{3.195}$$
$$+ O\left(x^{\varkappa+u}W(\beta,\Delta)\right) + O\left(x^{\varkappa+\beta}\delta(x)\right),$$

where $u = \max_{|t| \le |t_0|} f(t)$ and $\delta = \delta(x)$ is the reducing factor such that

$$\delta(x) = \delta_{A,a,b}(x) = \exp\left(-A(\log x)^{1/(a+1)}(\log \log x)^{-b/(a+1)}\right). \tag{3.196}$$

In case the left boundary is given by (3.191), *we still have the same asymptotic formula with the uniform estimate for the error term provided that*

$$\log \Delta \ll (\log x)^{a/(a+1)}(\log \log x)^{b/(a+1)}. \tag{3.197}$$

If $\varkappa \le \tau$, *then the last term* $O\left(x^{\varkappa+\beta}\delta_A(x)\right)$ *in* (3.195) *is to be replaced by* $O\left(x^{\varkappa+\beta}x^{\alpha(\varkappa-\tau)}\log x\right)$.

Theorem 3.17. *Under the same conditions as in Theorem 3.16, suppose* $f(t)$ *is given by*

$$f(t) = \beta = constant.$$

Then

$$A_\lambda^\varkappa(x) = Q_\varkappa(x) + R_{\lambda,II}^\varkappa(x), \tag{3.198}$$

where

$$R_{\lambda,II}^\varkappa(x) = O\left(x^{\varkappa+b^*}\left(x^{\alpha(\mu+\varepsilon-\varkappa-1)} + x^{-\alpha\varkappa}B(b^*)\right)\right) + O\left(x^{\varkappa+u}W(\beta,\Delta)\right).$$

Similar results hold for the logarithmic Riesz sums with the following replacement to be made:
- $Q_\varkappa(x)$ is replaced by $P_\varkappa(x)$ which is the sum of the residues of $\frac{1}{s^\varkappa}F(s)e^{xs}$ in R and $B(b^*)$ in (3.184) is replaced by

$$B^*(b^*) = \sum_{n=1}^\infty |a_n|\exp\left(-b^*\ell_n\right); \quad T = e^{\alpha x}.$$

Remark 3.4. (3.193) gives only the Siegel-Walfisz type result in Theorem 3.21.

Example 3.13. Let K be an algebraic number field of degree χ with discriminant d. Let $\mathfrak{o}_K$ be the ring of algebraic integers in K and let $\mathfrak{f}$ be an arbitrary fixed non-zero ideal of $\mathfrak{o}_K$. Let $A_\mathfrak{f}$ be the group of all fractional ideals with numerators and denominators relatively prime to $\mathfrak{f}$ and $H^*(\mathfrak{f})$ denote the ray class group of K, i.e. the quotient of $A_\mathfrak{f}$ modulo the group $S_\mathfrak{f}$ of principal ideals (α) with totally positive α such that $\alpha \equiv 1 \pmod{\mathfrak{f}}$. We define the Möbius function $\mu(\mathfrak{a})$ on ideals in the same manner as in the rational case. Let

$$M_\ell^\varkappa(x) = \frac{1}{\varkappa!} \sum_{N\mathfrak{a} \leq x} \frac{\mu(\mathfrak{a})}{N\mathfrak{a}} \left(\log \frac{x}{N\mathfrak{a}} \right)^\varkappa. \tag{3.199}$$

In this particular case

$$F(s) = Z(z) = Z^*(s) = \zeta_K(s+1)^{-1}$$

where $\zeta_K(s) = \sum_{\mathfrak{a} \neq 0} \frac{1}{N\mathfrak{a}^s}$ is the Dedekind zeta-function of K. Cf. Remark 1.3. Clearly

$$P_\varkappa(\log x) = \sum_{n=1}^\varkappa \frac{a_n}{(\varkappa - n)!} (\log x)^{\varkappa - n} \tag{3.200}$$

where a_n's are the Laurent coefficients given by

$$\zeta_K(s)^{-1} = \sum_{n=1}^\infty a_n (s-1)^n, \quad s \to 1.$$

Then by Theorem 3.16 for logarithmic Riesz sums we have

$$M_\ell^\varkappa(x) = \sum_{n=1}^\varkappa \frac{a_n}{(\varkappa - n)!} (\log x)^{\varkappa - n} + O(\delta_{VK}(x)), \tag{3.201}$$

where $\delta_{VK}(x)$ is the Vinogradov-Korobov reducing factor (cf. (3.196)).

Example 3.14. Let $q_K(n)$ denote the characteristic function of K-free integers (K-freie Zahlen) and let K be a fixed set of positive integers with the least element $k' \geq 2$. Also let Q_K be the set of all K-void integers, i.e. those integers in whose canonical decompositions there is no exponent from K. For example, if $K = \{n \in \mathbb{N} | n \geq 2\}$, then the K-void integers are nothing but square-free integers.

We write $S = S_K = \{n | k' \leq n, n \notin K\}$ and put $k = k_K = \infty$ or $\min S$ according as S is empty or not. Also writing $T = \emptyset$ or $T = T_K = \{n | k \leq n \in K\}$ according as $k = \infty$ or not, we put $q = q_K = \infty$ or $\{\min S\}$ according as $T = \emptyset$ or not.

Clearly $k < q$ when k is finite. The generating function of $q_K(n)$ is given by (for $\sigma > 1$)

$$Z(s) = \sum_{n=1}^{\infty} \frac{q_K(n)}{n^s} = \zeta(s) \prod_p \left(1 - (1 - p^{-s}) \sum_{a \in K} p^{-as} \right). \qquad (3.202)$$

It is known that

$$Z(s) = Z^*(s)\Phi(s), \quad Z^*(s) = \frac{\zeta(s)\zeta(ks)}{\zeta(k's)} \qquad (3.203)$$

where

$$\Phi(s) = \Phi_k(s) = \prod_p \left(1 - p^{-2ks} \right.$$
$$\left. + \left[(1 - p^{2ks})(1 - p^{-k's})^{-1} (p^{-(k+k')s} - (1 - p^{-s}) \sum_{k < a \in K} p^{-as}) \right] \right)$$

since the product is absolutely convergent for $\sigma > \max\{\frac{1}{k+k'}, \frac{1}{q}\}$.

When $k = \infty$ one needs to replace $\zeta(ks)$ and $\Phi(s)$ by 1 and $1 + (1 - p^{-k's})^{-1}$ respectively. Thus for any real t and any $\sigma > b_1 > \max\{\frac{1}{k+k'}, \frac{1}{q}\}$

$$|\Phi(\sigma + it)| \leq \exp\left(2 \sum_p p^{-b_1 \min\{k+k', q\}} \right)$$

$$\sim \exp\left(-2 \log[b_1 \min\{k + k', q\}] \right)$$

$$= \left(\frac{1}{b_1 \min\{k + k', q\}} \right)^2.$$

By Theorem 3.16 for arithmetic Riesz sums we have

Theorem 3.18. *Let* $\varkappa > 1 - \frac{k+1}{2k'}$. *Then*

$$\frac{1}{\Gamma(\varkappa + 1)} \sum_{n \leq x}' (x - n)^\varkappa q_K(n) = \alpha_K x^{\varkappa + 1} + O_{K,\varkappa}\left(x^{\varkappa + \frac{1}{k'}} \delta_{VK}(x) \right)$$

where $\delta_{VK}(x)$ *is the reducing factor in (3.47) with an absolute constant* A *such that* $0 < A < \frac{k}{k'(k+k')}(\log 3)^{2/3}(\log \log 3)^{1/3}$.

3.8.5   *Proof of Theorem 3.16*

We apply the Cauchy residue theorem to $F(s)$ in the domain R. In case there is a logarithmic singularity, the contour ∂R is to be modified by connecting the contour C_r in (3.187). Then we obtain (3.194), where the error term is the integrals along ∂R. Horizontal integrals are apparently negligible and so it is enough to estimate the vertical integral

$$\frac{1}{2\pi i} \int_\Gamma \frac{\Gamma(s)}{\Gamma(s+\varkappa+1)} F(s) x^{s+\varkappa} \, ds \tag{3.204}$$

and for that purpose, the problem amounts to the estimate of

$$\int_{t_0}^{T} x^{-\psi(t)} t^{\tau-\varkappa-1} \, dt \tag{3.205}$$

since $\frac{d}{dt}(f(t)+it) \ll 1$. We divide the interval $[t_0, T]$ into two $[t_0, Y] \cup [Y, T]$, where $Y = \exp\{A(\log \Delta x)^{\frac{1}{1+a}}(\log\log \Delta x)^{-\frac{b}{1+a}}\}$ is chosen.

With this choice of Y, the above calculation shows that the integral along $[t_0, Y]$ is $O\left(x^{\varkappa+\beta}\delta(x)\right)$ in (3.196).

On the other hand, the integral (3.205) over $[Y, T]$ is

$$\ll \int_{Y}^{\infty} t^{\tau-\varkappa-1} \, dt \ll Y^{\tau-\varkappa}, \tag{3.206}$$

which is negligible.

We now turn to the case where $f(t)$ is given by (3.191). As is remarked in Remark 3.3, this case is similar to the previous case and the problem amounts to the estimate of

$$I = I_T := \int_{t_0}^{T} x^{-\mathcal{M}(\Delta,t)^{-1}} t^{\tau-\varkappa-1} \, dt \tag{3.207}$$

corresponding to (3.205). We distinguish two cases.

(I) $\log \Delta > A\log(T+2)^a \log\log(T+2)^b$. Since $\mathcal{M}(\Delta,t) = \log \Delta$, we have

$$I \ll x^{\beta-(\log \Delta)^{-1}} \int_{t_0}^{T} t^{\tau-\varkappa-1} \, dt \ll x^\beta \exp\left(-\frac{\log x}{\log \Delta}\right) \ll x^\beta \delta(x) \tag{3.208}$$

in view of (3.197).

(II) $\log \Delta \le A\log(T+2)^a \log\log(T+2)^b$. We divide the range of integration into $[t_0, T_0] \cup [T_0, T]$ with the solution $T_0 = T_0(\Delta, t)$ of the equation $A\mathcal{L}^a(\log \mathcal{L})^b = \log \Delta$. It follows that $\exp(\log \Delta^{\frac{1}{a+1}}) \ll T_0 \ll \exp(\log \Delta^{\frac{1}{a}})$.

There are two cases: if $t \in [t_0, T_0]$, then $\mathcal{M}(\Delta, t) = A\mathcal{L}^a (\log \mathcal{L})^b$ and if $t \in [T_0, T]$, then $\mathcal{M}(\Delta, t) = \log \Delta$. Hence

$$I = x^\beta \int_{t_0}^{T_0} \exp(-A\mathcal{L}^{-a}(\log \mathcal{L})^{-b}) t^{\tau - \varkappa - 1} \, dt + x^{\beta - (\log \Delta)^{-1}} \int_{T_0}^{T} t^{\tau - \varkappa - 1} \, dt.$$
(3.209)

With the same Y from (3.206) we divide the range as $[t_0, Y] \cup [Y, T_0]$ and estimate the first integral as in the case of (3.190) treated above (with $\Delta = 1$). We can easily see that the first integral $\ll x^\beta \delta(x)$. The second integral can be estimated as in case of (I) with t_0 replaced by T_0. So the second integral becomes $\ll x^\beta \exp\left(-\frac{\log x}{\log \Delta}\right) T_0^{\tau - \varkappa} \ll x^\beta \delta(x)$.

Hence altogether, the integral in (3.207) is $\ll x^\beta \delta(x)$, completing the proof.

3.8.6 *Tauberian process*

Theorem 3.19. *Suppose that the conditions of Theorem 3.16 are satisfied. Let q be the maximum of the real parts of poles of $F(s)$ in R and let r be the maximum order of poles with real parts q. Also define θ' to be 1 or 0 according as $a_n \geq 0$ or not. Then, for $0 < y \leq \frac{1}{\varkappa} x$*

$$A_0^\varkappa(x) = Q_0(x) + \theta' O \left(\sum_{x < \lambda \leq x + \varkappa y} |a_n| \right) + O(y x^{q-1} \log^{r-1} x) \quad (3.210)$$

$$+ O(y^{-\varkappa} R_{\lambda, i}^\varkappa(x))$$

where $i = I$ or II according to the choice of $f(t)$. In case there is a logarithmic singularity as $s = \beta$, the main term Q_0 is to be interpreted as an integral along the limit of the contour C_r.

Theorem 3.20. *Suppose that the conditions of Theorem 3.16 are satisfied and that for some constant $C > 0$*

$$\sum_{\ell_n \leq x} |a_n| = x^{q'} \log^{r' - 1} x (C + o(1)) \quad (3.211)$$

holds true. Then

$$A_0^\varkappa(x) = \theta P_0(\log x) + \theta' O\left(\delta x^{q'} \log^{r'-1} x\right) + \theta O(\delta^{1/\varkappa} x^q \log^{r-1} x)$$

$$+ O\left(\delta^{-1} x^{-\alpha \varkappa} (b - q')^{[r']+1} + x^{b + (\mu + \varepsilon - \varkappa - 1)\alpha}\right) + O(\delta^{-1} x^u W(\beta, \Delta))$$

$$+ O(x^\beta \delta),$$

where all δ^c's amount to the reducing factor δ with possibly different constants A, a etc. in (3.196).

3.8.7   *Proof of Theorem 3.19*

The general formula for the *difference operator* of order $\alpha \in \mathbb{N}$ with difference $y \geq 0$ is given by

$$\Delta_y^\alpha f(x) = \sum_{\nu=0}^\alpha (-1)^{\alpha-\nu} \binom{\alpha}{\nu} f(x + \nu y). \tag{3.212}$$

If f has the α-th derivative $f^{(\alpha)}$, then

$$\Delta_y^\alpha f(x) = \int_x^{x+y} dt_1 \int_{t_1}^{t_1+y} dt_2 \cdots \int_{t_{\alpha-1}}^{t_{\alpha-1}+y} f^{(\alpha)}(t_\alpha)\, dt_\alpha. \tag{3.213}$$

We apply the difference operator to (3.198) and we will get

$$\Delta_y^\varkappa A_\lambda^\varkappa(x) = \sum_{\nu=0}^\varkappa (-1)^{\varkappa-\nu} \binom{\varkappa}{\nu} \frac{1}{\Gamma(\varkappa+1)} \sum_{\lambda_n \leq x+\nu y} a_n(x + \nu y - \lambda_n)^\varkappa. \tag{3.214}$$

We divide the inner sum into two: $\lambda_n \leq x$ and $x < \lambda_n \leq x + \nu y$ and note that the first sum amounts

$$\frac{1}{\Gamma(\varkappa+1)} \sum_{\lambda_n \leq x} a_n \Delta_y^\varkappa (x - \lambda_n)^\varkappa$$

and we can check

$$\frac{1}{\Gamma(\varkappa+1)} \Delta_y^\varkappa (x - \lambda_n)^\varkappa = y^\varkappa.$$

Using these and estimating the second sum trivially, we infer that

$$\Delta_y^\varkappa A_\lambda^\varkappa(x) = y^\varkappa A_0^\varkappa(x) + O\left(y^\varkappa \sum_{x < \lambda_n \leq x+\nu y} |a_n| \right). \tag{3.215}$$

Now differentiating the defining equation (3.186) under the integral sign, we obtain

$$\frac{d^\varkappa}{dx^\varkappa} Q_\varkappa(x) = \frac{1}{2\pi i} \int_{\partial R} \frac{F(s)}{s} x^s\, ds = Q_0(x). \tag{3.216}$$

We have for any t_ν in the interval $x \leq t_\nu \leq x + \nu y$

$$|Q_0(t_\nu) - Q_0(x)| = \left| \int_x^{t_\nu} Q_0'(y)\, dy \right| \leq \varkappa y \max_{x \leq y \leq x + \varkappa x} |Q_0'(y)|. \tag{3.217}$$

Letting r_ξ denote the order of the pole of $F(s)$ at $s = \xi \in R$, we have

$$\operatorname{Res}_{s=\xi} \frac{F(s)}{s} x^s = x^\xi \sum_{\nu=1}^{r_\xi} \frac{c_{-\nu}}{(\nu-1)!} \log^{\nu-1} x,$$

where $c_{-\nu}$ are Laurent coefficients of $F(s)$ at $s = \xi$. Hence

$$Q_0(x) = \sum_{\xi} x^{\xi} \sum_{\nu=1}^{r_{\xi}} \frac{c_{-\nu}}{(\nu-1)!} \log^{\nu-1} x. \tag{3.218}$$

From (3.217) and (3.218) we obtain

$$Q_0(t_{\nu}) = Q_0(x) + O(yx^{q-1} \log^{r-1} x). \tag{3.219}$$

Substituting (3.219) in (3.213), we derive that

$$\Delta_y^{\varkappa} Q_{\varkappa}(x) = y^{\varkappa} Q_0(x) + O(y^{\varkappa+1} x^{q-1} \log^{r-1} x). \tag{3.220}$$

From (3.212) it follows that

$$\Delta_y^{\varkappa} R_{\lambda,i}^{\varkappa}(x) = O(\max_{0 \le \nu \le \varkappa} |R_{\lambda,i}^{\varkappa}(x+\nu y)|) = O(|R_{\lambda,i}^{\varkappa}(x)|) \tag{3.221}$$

if only $y \ll x$.

From (3.215), (3.220) and (3.221) we conclude (3.210), thereby completing the proof of Theorem 3.19.

Example 3.15. Let $\lambda(n)$ denote the Liouville function defined by (3.48). Then we have for $\sigma > 1$

$$Z(s) = \sum_{n=1}^{\infty} \frac{\lambda(n)}{n^s} = Z^*(s)\Phi(s), \quad Z^*(s) = \frac{1}{\zeta(s)}, \quad \Phi(s) = \zeta(2s). \tag{3.222}$$

Since $\Phi(s)$ is absolutely convergent for $\sigma > \frac{1}{2}$, the problem amounts to the essential factor $\frac{1}{\zeta(s)} = \sum_{n=1}^{\infty} \frac{\mu(n)}{n^s}$ in (3.12). Cf.

Example 3.16. Notation being the same as in Example 3.13 and for $\mathfrak{C} \in H^*(\mathfrak{f})$ we put

$$M(x, \mathfrak{C}) = \sum_{\substack{N\mathfrak{a} \le x \\ \mathfrak{a} \in \mathfrak{C}}} \mu(\mathfrak{a}). \tag{3.223}$$

We have the following version of the Siegel-Walfisz prime ideal theorem.

Theorem 3.21. *Let*

$$\Delta = N\mathfrak{f} \cdot |d| \tag{3.224}$$

and suppose that

$$\Delta \ll \log^A x \tag{3.225}$$

with an arbitrary large constant A. Then

$$M(x, \mathfrak{C}) = O_{n,A}(x\delta_{VK}(x)) \tag{3.226}$$

where $\delta_{VK}(x) = \delta_a(x)$ is the Vinogradov-Korobov reducing factor (3.47) with a constant $a = a(n, A) > 0$ dependent (at most) on n and A and so being the O-constant.

The theorem follows on appealing to Theorem 3.22 and Siegel's estimate (3.241).

Example 3.17. Let $f(n)$ be the number of representations of the square-free integer n as norms of products of completely splitting primes and ramified primes. Then

$$Z(s) := \sum_{n=1}^{\infty} \frac{f(n)}{n^s} = \prod_{p \mid d}(1 + p^{-s}) \prod_{\chi_d(p)=1}(1 + 2p^{-s}) \qquad (3.227)$$

and we have the decomposition [Kanemitsu (2010)], [Zhai (2005)]

$$Z(s) = Z_1(s)Z_2(s)\zeta_K(3s), \qquad (3.228)$$

where

$$Z_1(s) = \frac{\zeta_K(s)\zeta(4s)}{\zeta_K(4s)^2}\Phi(s) = \frac{\zeta_K(s)}{\zeta_K(4s)L(4s,\chi_d)^2}\Phi(s), \quad \sigma > 1 \qquad (3.229)$$

and

$$Z_2(s) = \frac{L(2s,\chi_d)}{\zeta_K(2s)^2} = \frac{1}{\zeta(2s)^2 L(2s,\chi_d)}, \qquad \sigma > \frac{1}{2}. \qquad (3.230)$$

This decomposition is not exactly of the form considered so far because it is designed to deduce the summatory function of $f(n)$.

Using Theorem 3.21 we may prove generalizations of asymptotic formulas in [Buschman (1959)] sharp estimate on the error term, which will be done elsewhere.

The following examples do not fall into the class considered above but rather into the class of Rankin-Selberg type zeta-functions whose study will be conducted elsewhere (cf. §3.8.10).

Example 3.18. Let β be a multiplicative function defined by $\beta(1) = 1$, $\beta(n) = \alpha_1 \cdots \alpha_r$ for $n = p_1^{\alpha_1} \cdots p_r^{\alpha_r}$. Given any Dirichlet character mod k, Knopmacher [Knuth (1997)] studied the generating function

$$Z(s) = \sum_{n=1}^{\infty} \frac{\chi(n)\beta(n)}{n^s} = Z^*(s)\Phi(s),$$

$$Z^*(s) = L(s,\chi)L(2s,\chi^2)L(3s,\chi^3), \Phi(s) = \frac{1}{L(6s,\chi^6)}.$$

Example 3.19. ([Hasse and Suetuna (1931), Satz 11]) Notation being the same as in Example 3.13 with extension degree 2, one considers a generalization of the Dirichlet divisor problem. The generating Dirichlet series $Z(s)$ has the representation

$$Z(s) = Z^*(s)\Phi(s), \quad Z^*(s) = \zeta_k(s)^2\zeta_K(s), \quad \Phi(s) = \frac{1}{\zeta_K(2s)}. \qquad (3.231)$$

Example 3.20. For basics on elliptic curve, cf. §2.6.4. Let E be an elliptic curve defined over $\mathbb{Q}$ with conductor N_E and let

$$L_E(s) = \prod_{p \nmid N_E} \left(1 - \frac{a_p}{p^s} + \frac{1}{p^{2s-1}} \right)^{-1} \prod_{p \mid N_E} \left(1 - \frac{a_p}{p^s} \right)^{-1} = \sum_{n=1}^{\infty} \frac{a(n)}{n^s} \quad (3.232)$$

be the L-function associated with E for $\sigma > 2$. It can be continued meromorphically over the whole plane in the case considered including the case where the curve has complex multiplication. It has a decomposition

$$L_E(s) = Z^*(s)\Phi(s), \quad Z^*(s) = \frac{1}{L_h(s)}, \quad \Phi(s) = \zeta(s)\zeta(s-1) \quad (3.233)$$

where $L_h(s)$ is the Hecke L-function with Groössencharacter. Cf. [Butzer and Stark (1986)].

Let χ be a primitive character of conductor f. The twist of L-function is defined by

$$L_E(s, \chi) = \sum_{n=1}^{\infty} \frac{a(n)\chi(n)}{n^s} \quad (3.234)$$

which can be continued analytically over the whole place and satisfies the functional equation

$$\Lambda_E(s, \chi) := \left(\frac{2\pi}{f\sqrt{N_E}} \right)^{-s} \Gamma(s) L_E(s, \chi) = c\Lambda_E(2 - s, \chi), \quad (3.235)$$

where c is a certain constant.

Let $N(x)$ denote the number of cubic characters whose conductor $\leq x$ and consider the Z-function

$$Z(s) = \frac{1}{N(x)} \sum_{f \leq x} |L_E(s, \chi)|^{2k}. \quad (3.236)$$

Let $L(E \otimes E, s) = \sum_{n=1}^{\infty} \frac{a^2(n)}{n^{s+1}}$ be the Rankin-Selberg L-function associated with $L_E(s)$. Then we have a decomposition

$$Z(s) = Z^*(s)\Phi^*(s), \quad Z^*(s) = L(E \otimes E, 2(s - 2))^{k^2} \quad (3.237)$$

and $\Phi^*(s)$ is analytic for $\sigma > \frac{1}{3}$. Cf. [David, Fearnley, and Kisilevsky (2004)].

Remark 3.5. Although in above examples the essential factor is a real number power of a certain zeta-function, but as we have seen, in many cases as in §3.8.2, it is a complex power and there arises a logarithmic singularity. The remarkable feature of our Abel-Tauber theorem(s) is that this case is also included as mentioned at the end of "Settings for the generating function", (i) by viewing the residual function contains a contour integral along (3.187). We provide a table of the essential factor $Z^*(s)$ and the auxiliary factor $\Phi(s)$ in our theorems and Lau's theorem:

paper	ess. factor	aux. factor
Abel-Tauber	$Z^*(s)$ $(\sigma > 1)$	$\Phi(s)$ $(\sigma > a)$
Lau	$G_B(s)$ $(\sigma > a)$	$G_A(s)$ $(\sigma > 1)$

Table 3.1. Essential and auxiliary factors (domain of absolute conv.)

We note that the generating Dirichlet series $F(s)$ in our Abel-Tauber theorems is the essential factor $Z^*(s)$. Although we do not impose any further conditions on the auxiliary factor, we may impose a similar condition as in Lau's theorem and we can establish a theorem on convolution generated by $F(s)\Phi(s)$. Giving details of this procedure is rather instructive, which we will give elsewhere.

3.8.8 *Quellenangaben*

The Riesz sums may be thought of as integration or 'Abelian process' ([Briggs (1962)]) while 'differencing' the Riesz sum to deduce a formula for the Riesz sum of order 0 corresponds to differentiation or 'Tauberian process'. The logarithmic Riesz sums also appeared in [Berndt and Kim (2015)] and for which the generating function satisfies the functional equation.

General modular relations are treated in [Chandrasekharan and Raghavan Narasimhan (1962)], [Berndt and Knopp (2008)] and most comprehensively in [Kanemitsu (2007), Chapt. 6].

Below we indicate some available information and new plausible research problems.

3.8.9 *Zero-free region for L-functions*

We collect some known results on the Dirichlet L-function $L(s,\chi)$ and Hecke L-function $L_K(s,\chi)$ although the former may be thought of as a special case of the latter.

(i) For $\sigma > 0$,

$$L(s,\chi) - E(\chi)\frac{\varphi(k)}{k}\frac{1}{s-1}$$

is regular.

For real t, $0 < c_1 < \frac{1}{8}$ and for all but one characters χ (mod k) one has $L(s,\chi) \neq 0$ in

$$\sigma > 1 - \frac{c_1}{\log k(|t| + 2)}$$

with a possible exception of the real simple zero $\beta(k)$ corresponding to a possible exceptional character χ_1. For this Siegel's theorem ensures that for each $\varepsilon > 0$ there exists a $c_1(\varepsilon)$ such that

$$\beta(k) < 1 - c_1(\varepsilon)k^{-\varepsilon}. \tag{3.238}$$

(ii) Let K be a number field with discriminant d and let $L_K(s, \chi)$ be the Hecke zeta-function, where χ indicates a character of the narrow ideal class modulo a fixed ideal $\mathfrak{f}$ in K as is defined in Landau [Landau (1918)]. Hinz's result [Hinz (1980)] is the best known zero-free region explicit with respect to the modulus, given by the following

Theorem 3.22. [Hinz (1980), Satz 1.1'] *For a suitable $c = c(K) > 0$, in the domain*

$$\sigma \geq 1 - \frac{c}{\mathcal{M}(\Delta, t)} \tag{3.239}$$

there are no zeros of $L_K(\sigma + it, \chi)$ except for a possible real simple exceptional zero $\beta_1 = \beta_1(\mathfrak{f}) < 1$, where

$$\mathcal{M}(\Delta, t) = \max\{\Delta, (\log(|t| + 3))^{2/3}(\log\log(|t| + 3))^{1/3}\} \tag{3.240}$$

and $\Delta = |d| \log N\mathfrak{f}$ is defined in (3.224).

The estimate for β_1 (by Siegel's method) is known in ([Fogels (1963)]). There is an ineffective constant $c(\varepsilon, K) > 0$ such that

$$\beta_1 < 1 - c(\varepsilon, K)(N\mathfrak{f})^{-\varepsilon} \tag{3.241}$$

for arbitrarily small $\varepsilon > 0$.

Using this zero-free region, we may obtain an estimate as strong as Hinz's [Hinz (1980), Satz 4.1, p. 242]. We should have the main term in the form of a contour integral as in [Balasubramanian and Ramachandra (1998)] and [Balasubramanian, Kanemitsu and Ramachandra (2003)] as well as in Nakaya [Nakaya (1992)]. Then the asymptotic formula is obtained with logarithmic reduction, by taking a few terms in the asymptotic expansion of the contour integral. Balasubramanian and Ramachandra [Balasubramanian and Ramachandra (1998)] showed a finitely main term with the error estimate as good as in the PNT but they took as many as $(\log x)^{4/5}$ terms.

Another interesting aspect is the dependence on the magnitude of the discriminant. Hinz's result is explicit in the modulus $\mathfrak{f}$ only but by elaborating his method in the context of Fogels [Fogels (1963)], [Fogels (1965)], one can take Δ into account as defined in (3.224).

Remark 3.6. It seems that Hinz follows the method expounded in [Proskurin (1979)] to deduce the PNT from the zero-free region (3.239).

The method gives the intermediate estimate of Čudakov but the argument works with the zero-free region (3.239) to arrive at the Vinogradov-Korobov estimate.

3.8.10 *Case of infinitely many poles*

We note that in most of the above examples, the generating function has *infinitely* many poles in the critical strip on the left of $\sigma = \beta$. They arise from the Z^*-function and for which the distribution law of zeros is known. Such a case has been treated in [Hafner and Stopple (2000)], [Roy (2016)] and [Chakraborty, Kanemitsu and Maji (2016)]. The authors in the last two papers move the line of integration to the left of the critical strip and apply the functional equation to derive a modular relation with residual function as the infinite sum over non-trivial zeros in the critical strip. The convergence of the infinite sum over poles is assured by the known gap between successive ordinates of the zeros of the relevant Z^*-functions. In this method one applies the contour integration to a rectangle with height $T > 0$ each time and show that the integrals along the horizontal lines as well as those along the vertical lines are convergent as $T \to \infty$.

Another plausible way is to establish an explicit formula in terms of non-trivial poles of the Z^*-function. Here we need an integral expression for the Z^*-function which is valid in the critical strip. There are a few known integral representations which might be used.

One is the Mellin transform of the Hurwitz zeta-functions, which seems to have been first considered by Mikolás [Mikolás (1956), Satz 2]. His result has been elucidated as [Chakraborty, Kanemitsu and Tsukada (2009), Theorem 7.6, p. 172].

Another is Koshlyakov's expression [Koshlyakov (1954), (6.2)]

$$\zeta_\Omega(s) = 2\sin\left(\frac{\pi s}{2}\right)\int_0^\infty \sigma(x)x^{-s}\,\mathrm{d}x. \tag{3.242}$$

3.8.11 *Modified Selberg class functions*

The factoring out method depends essentially on the Euler product expression for the generating function which may not satisfy the functional equation. In §3.8.10, we mentioned a possibility of passing over the critical strip and apply the functional equation. An extensive theory of functional equations has been developed as modular relations in [Kanemitsu (2007)], where the Euler product was not assumed on the ground that the zeta-function with Euler product is rather limited to number-theoretic situation. For the present problem, the zeta-function is to have both functional equation and Euler product. Such a class of functions is known as the modified Selberg class of zeta-functions, modified because one assumes the Euler product $\mathcal{B}(I)$ below (in Selberg's original definition [Selberg (1989)], a substitute of Euler product was assumed). For the Selberg class zeta-functions, cf. Definition 2.5. We consider the sequence $\{c(n)\}$ with $c(1) = 1$ such that

$$c(n) = O(n^\delta) \tag{3.243}$$

for some fixed $\delta > 0$. The generating Dirichlet series

$$L(s) = \sum_{n=1}^{\infty} \frac{c(n)}{n^s} \tag{3.244}$$

is absolutely convergent for $\sigma > \delta + 1$. We say that $L(s)$ belongs to the *(modified) Selberg class $\mathcal{S}_1$* of functions [Bombieri and Hejal (1995)] in the sense that

- $\mathcal{B}(I)$: It has an Euler product

$$L(s) = \prod_p \left(1 - \frac{\alpha_1(p)}{p^s}\right)^{-1} \cdots \left(1 - \frac{\alpha_d(p)}{p^s}\right)^{-1} \tag{3.245}$$

with

$$|\alpha_{kp}| \le p^\vartheta, \quad k = 1, \cdots, d \tag{3.246}$$

for some fixed $\vartheta \in [0, \frac{1}{2})$.

- $\mathcal{B}(II)$:

$$\sum_{p \le X} \sum_{k=1}^{d} |\alpha_{kp}|^2 = O(X^{1+\varepsilon}) \tag{3.247}$$

for every $\varepsilon > 0$. Thus the series

$$L(s) = \sum_{n=1}^{\infty} \frac{a_n}{n^s} \tag{3.248}$$

as well as the Euler product are absolutely convergent in the same right half-plane $\sigma > 1$. Also $L(s)$ does not vanish there.

- $\mathcal{B}(III)$: $L(s)$ can be analytically continued over the whole plane to a meromorphic function of finite order with a finite number of poles, all on the line $\sigma = 1$ and it satisfies the functional equation

$$Q^s G(s) L(s) = \Phi(s) = \omega \bar{\Phi}(1 - s), \tag{3.249}$$

where $G(s)$ is the product of Gamma functions and ω is a complex number with $|\omega| = 1$. In order to make $L(s)$ suit our purpose, we are to assume the distribution of zeros in the critical strip. This enables one to raise $L(s)$ to the complex power z ([Selberg (1989)]):

$$L(s)^z = \exp\left(z \log L(s)\right) = \sum_{n=1}^{\infty} \frac{g_z(n)}{n^s} \qquad (3.250)$$

in the same half-plane (logarithm takes the principal value there). If $d_z(n)$ signifies the Selberg divisor function generated by $\zeta^z(s)$, then

$$g_z(p^k) = \sum_{k_1 + \cdots + k_d = k} d_z(p^{k_1}) \alpha_1(p^{k_1}) \cdots d_z(p^{k_d}) \alpha_d(p^{k_d}). \qquad (3.251)$$

Thus to recover the 'Selberg type' divisor problem itself one needs to consider

$$L(s)^{\boldsymbol{w}} = \prod_p \left(1 - \frac{\alpha_1(p)}{p^s}\right)^{-w_1} \cdots \left(1 - \frac{\alpha_d(p)}{p^s}\right)^{-w_d} \qquad (3.252)$$

where $\boldsymbol{w} = (w_1, \cdots, w_d)$. Further developments will be given elsewhere, cf. e.g. [Chakraborty, Kanemitsu and Laurinčikas].

Chapter 4

Number theory in the unit disc

Abstract: Our attention has been drawn to the intriguing identity for arithmetical Fourier series. On one hand, there appears the Liouville function which is a prime number-theoretic entity and on the other hand, Riemann's example of a nowhere differentiable function. The integrated identity can be derived from the functional equation only. But to differentiate it, one needs the estimate for the error term for the Liouville function which is as deep as the PNT. [Murty (2004)] gives a very instructive proof of the quadratic reciprocity from the theta-transformation formula, which in turn as is well known (to Riemann) is equivalent to the functional equation. The situation is suggestive of Ingham's treatment [Ingham (1964)] of the prime number theorem, i.e. first apply the Abelian process (integration) and then Tauberian process (differencing) which needs more information. A big advantage of our treatment is that while the theta-function $\Theta(\tau)$ is a dweller in the upper-half plane, its integrated form $F(z)$ is a dweller in the extended upper half-plane including the real line, thus making it possible to consider the behavior under the increment of the real variable, where the integration is along the horizontal line. The elliptic theta-function $\theta(s) = \Theta(-i\tau)$ is a dweller in the right half-plane $\{\sigma > 0\}$, where the integration is along the vertical line. In terms of Lambert series, there is an idea of Wintner to deal with limiting behavior of the Lambert series on the circle of convergence, i.e. radial integration. Our case corresponds to integration along an arc.

It seems that some of the results in [Itatsu (1981)] are not true and since in Murty-Pacelli's treatment [Murty (2004)] should give the correct expression for the $R(p, q)$, we stick to it in our presentation.

Here one can perceive two far-away-looking subjects merging on the real line as limiting behaviors of zeta- and theta-functions.

4.1 Lambert series

4.1.1 *Boundary functions of certain Lambert series*

First we follow [Wintner (1944)] to introduce the Lambert series. Let $\{a_n\} \subset \mathbb{C}$ be such that

$$\limsup |a_n|^{1/n} \le 1, \tag{4.1}$$

i.e. such that the power series $\sum_{n=1}^{\infty} a_n z^n$ is absolutely convergent in $|z| < 1$. Then the **Lambert series**

$$f(z) = \sum_{n=1}^{\infty} a_n \frac{z^n}{1-z^n} \tag{4.2}$$

is absolutely convergent in $|z| < 1$ and represents an analytic function and moreover the power series of this function can be obtained by formal rearrangement of (4.2), i.e.

$$f(z) = \sum_{n=1}^{\infty} b_n z^n \quad (|z| < 1), \tag{4.3}$$

where

$$b_n = \sum_{d|n} a_d. \tag{4.4}$$

Alongwith (4.2), we may consider

$$f_+(z) := \sum_{n=1}^{\infty} a_n \frac{z^n}{1+z^n}. \tag{4.5}$$

Riemann's posthumous Fragment [Riemann (1854)], based on Jacobi's Fundamenta Nova, §40 [Jacobi (1829)], consists of two parts, Fragment **I** and Fragment **II**, the latter of which contains only formulas and almost no text. Dedekind, entrusted with editing and publishing the fragments, succeeded in elucidating the genesis of all the formulas in Fragment **II** by introducing the most celebrated Dedekind eta-function defined by (4.44) below. All the results in Fragment **II** deal with the asymptotic behavior of those modular functions from Jacobi's Fundamenta Nova [Jacobi (1829)], for which the variable tends to rational points on the unit circle. After Dedekind, several authors including Smith [Smith (1972)], Hardy [Hardy (1903)], Rademacher [Rademacher (1931)] made some more incorporations of Fragment **II**. In 2004, Arias-de-Reyna [Arias-de-Reyna (2004)] analyzed all the formulas in Fragment **II** again by applying ad hoc methods to each of the formulas.

For the moment we shall dwell on Wintner's paper [Wintner (1944)], elucidating Riemann's procedure in Fragment **I**. Riemann divides the Lambert series by z and integrate the result from 0 to $z = re^{i\theta}$, $r = |z| < 1$, and then puts $r = 1$ to obtain his results, i.e. "Fourier expansions", in a formal way. Wintner's legitimation reads as follows. Slightly modifying Riemann's procedure by the trivial factor $e^{i\theta}$, which makes the function $F = F(r, \theta)$ a function in z, one obtains

$$F(re^{i\theta}) = \int_0^r \frac{f(re^{i\theta})}{r}\,dr = \sum_{n=0}^{\infty} \frac{b_n}{n}(re^{i\theta})^n,$$

or

$$F(z) = \int_0^r \frac{f(re^{i\theta})}{r}\,dr = \sum_{n=0}^{\infty} c_n z^n, \qquad (4.6)$$

with

$$c_n = \frac{1}{n}b_n = \frac{1}{n}\sum_{d|n} a_n. \qquad (4.7)$$

In the case of (4.2), we should have

$$c_n = \frac{1}{n}\sum_{d|n} (-1)^n a_n. \qquad (4.8)$$

Then Wintner asks if $F(z)$ tends to a measurable boundary function $F(e^{i\theta})$ as $r \to 1$ (within the Stolz path) and if so, then whether or not the boundary function is of class L^p, so that Riemann's formal trigonometric series actually is the Fourier series of the boundary function.

Use is made not only of the Riesz-Fischer condition [Chakraborty, Kanemitsu and Tsukada (2009), Theorem 7.1, p. 139]

$$\sum_{n=-\infty}^{\infty} |c_n|^2 < \infty \qquad (4.9)$$

but also a weaker condition

$$\sum_{n=-\infty}^{\infty} |n|^{\varepsilon}|c_n|^2 < \infty, \qquad (4.10)$$

which lies between (4.9) and the L^p-condition

$$\sum_{n=-\infty}^{\infty} |c_n|^{p/(p-1)} < \infty \qquad (p \geq 2) \qquad (4.11)$$

for an f to be of class L^p for some $p > 2$ (for L^p-space cf. Definition 5.2). Condition (4.10) implies that the trigonometric series $\sum_{n=-\infty}^{\infty} c_n e^{in\theta}$ is convergent a.e. (almost everywhere) and represents the values of the function a.e. We refer to this as the Hausdorff-Paley extension of Fischer-Riesz theorem to be used again in §6.3.4. The contents of this subsection has a close relationship to that of §6.3.4.

Theorem 4.1. (Wintner) *Suppose*

$$a_n = O(n^{\lambda - \delta}) \tag{4.12}$$

for some $\delta > 0$ and a fixed $0 < \lambda \le \frac{1}{2}$. Then the boundary function $F(e^{i\theta})$ exists and is measurable such that

$$F(re^{i\theta}) \to F(e^{i\theta}) \quad a.e., \text{ as } r \to 1 \tag{4.13}$$

along the Stoltz path. If $\lambda < \frac{1}{2}$, then $F(e^{i\theta})$ is of class $L^{1/\lambda}$ and if in (4.12), the exponent can be taken arbitrarily small, then it is of class L^∞.

Proof. Recall that the divisor function, which is the special case of the sum-of-divisors function, satisfies the estimate

$$d(n) = \sum_{d \mid n} 1 = O(n^\varepsilon) \tag{4.14}$$

for every $\varepsilon > 0$ (e.g. [Hartman and Wintner (1938)]). Hence it follows that

$$c_n = O(n^{\lambda - 1 - \delta}) \tag{4.15}$$

for some $\delta > 0$. Hence if $\lambda < \frac{1}{2}$, then the L^p-condition (4.11) is satisfied and if $\lambda = \frac{1}{2}$, then the series for $F(re^{i\theta})$ is Cauchy in L^2 and so there exists a function $F(e^{i\theta})$ of class L^2 such that

$$F(e^{i\theta}) \sim \sum_{n=-\infty}^{\infty} c_n e^{in\theta}. \tag{4.16}$$

This together with the condition (4.10) implies that

$$F(e^{i\theta}) = \sum_{n=-\infty}^{\infty} c_n e^{in\theta} \quad a.e. \tag{4.17}$$

Hence (4.13) follows by Abel's continuity theorem. $\square$

Example 4.1.

(i) In the case $a_{2n} = 0, a_{2n+1} = 4(-1)^{n+1}$, we obtain the Lambert series

$$f(z) = \frac{2K}{\pi} = 4 \sum_{n=1}^{\infty} \frac{(-1)^{n+1} z^{2n+1}}{1 - z^{2n+1}} + 1, \tag{4.18}$$

which in the notation of (4.3) reads

$$f(z) = \frac{2K}{\pi} = 4 \sum_{\ell,m,n=0}^{\infty} d_{4,1}(n) z^{2^{\ell}(4m-1)^2 n} + 1, \qquad (4.19)$$

with $d_{4,1}(n)$ denoting the number of divisors of n of the form $4k+1$. Hence

$$F(z) = \sum_{n=0}^{\infty} \frac{d_{4,1}(n)}{n} z^n, \quad |z| < 1. \qquad (4.20)$$

Hence by Theorem 4.1, the boundary function $F(e^{i\theta})$ exists and

$$F(e^{i\theta}) = \sum_{n=1}^{\infty} \frac{d_{4,1}(n)}{n} e^{in\theta} \quad \text{a.e.} \qquad (4.21)$$

(ii) In the case $a_n = 1$, we obtain the Lambert series considered by Lambert himself ([Lambert (1771)])

$$f(z) = \sum_{n=1}^{\infty} \frac{z^n}{1-z^n} = \sum_{n=1}^{\infty} d(n) z^n, \quad |z| < 1. \qquad (4.22)$$

Hence

$$F(z) = \sum_{n=1}^{\infty} \frac{d(n)}{n} z^n, \quad |z| < 1. \qquad (4.23)$$

Hence by Theorem 4.1, the boundary function $F(e^{i\theta})$ exists and

$$F(e^{i\theta}) = \sum_{n=0}^{\infty} \frac{d(n)}{n} e^{in\theta} \quad \text{a.e.} \qquad (4.24)$$

Following Wintner [Wintner (1937)], [Wintner (1944)], we denote the Fourier series (3.28) by $\psi(x)$. Then

$$\psi(x) = \begin{cases} -\frac{1}{\pi} \sum_{m=1}^{\infty} \frac{\sin 2\pi m x}{m}, & x \notin \mathbb{Z} \\ 0, & x \in \mathbb{Z} \end{cases} \qquad (4.25)$$

is the **saw-tooth Fourier series**. Indeed, this was also established as one of the earliest instances of the modular relation by A. Z. Walfisz [Walfisz (1923)].

Exercise 23. Prove that the imaginary part of (4.24) gives the expansion treated in [Wintner (1937)], which in turn gives rise to the series considered by Riemann [Riemann (1854)].

Indeed, this was proved by Chowla and Walfisz [Chowla and Walfisz (1935)]. Cf. [Kanemitsu (2015), §6.3.4].

We note that the estimate (4.14) gives rise to one for the sums over all divisors including the estimate for the sum-of-divisors function in the proof of (4.42) and for further estimation of (2.129).

One may also consider other Lambert series, e.g. the one studied by Titchmarch [Titchmarsh (1938)].

4.1.2 *Riemann's legacy*

This section is intimately connected with [Kanemitsu (2007), §6.3.4, pp. 206-209] which contains various remarks on Fourier series involving arithmetical functions, especially in connection with Riemann's legacy. Indeed, Riemann left three legacies and the only paper on number theory has been expounded in §3.4. The second one is related to Fourier series which he gave as some examples of the threshold of discontinuous and integrable function and a continuous nowhere differentiable function in 1854 and around 1861, respectively. We shall touch on this presently. The third is his posthumous fragments which has been expounded in §4.1.1 above. Since the main ingredient in Riemann's fragments are the study on the boundary functions as a radial limit of Lambert series, we notice that Riemann is the most ancient precursor of number theory and *a fortiori* control theory in the unit disc.

$$\sum_{n=1}^{\infty} \frac{\lambda(n)}{n} \psi(nx) = -\frac{1}{\pi} \sum_{n=1}^{\infty} \frac{\sin 2\pi n^2 x}{n^2}, \qquad (4.26)$$
$$\lambda(n) = (-1)^{\Omega(n)},$$

where $\Omega(n)$ is the total number of prime factors of n.

We note that the series on the right of (4.26) is Riemann's example of a continuous, "non-differentiable" function. According to Weierstrass, Riemann asserted (around 1861 [Neuenschwander (1978)], [Segal (1978)]) that the function

$$f(x) = \sum_{n=1}^{\infty} \frac{\sin n^2 x}{n^2}$$

is nowhere differentiable. Weierstrass tried to prove this assertion but failed. In 1872 he constructed his own example of a continuous nowhere differentiable function which is of the form

$$\sum_{n=0}^{\infty} \cos(b^n \pi x)$$

where $0 < a < 1$ and b is a positive integer satisfying

$$ab > 1 + 3/2\pi.$$

It is interesting to note that the above sine series is the imaginary part of the series

$$\sum_{n=1}^{\infty} \frac{e^{2\pi i n^2 x}}{n^2} \tag{4.27}$$

considered by Itatsu [Itatsu (1981)], Luther [Luther (1986)] et al, to prove the "non-differentiability" of its imaginary part.

The real part of (4.27) is

$$f(x) = \sum_{n=1}^{\infty} \frac{\cos 2\pi n^2 x}{n^2} \tag{4.28}$$

which is another example of Riemann discussed by Casorati and Prym in 1865 ([Neuenschwander (1978)]).

This contrast between the real and imaginary parts of (4.27) explains why Riemann was not interested in the real part in his posthumous Fragment **II** ([Riemann (1953)] (cf. [Wintner (1937), p. 633], [Arias-de-Reyna (2004), p. 59, 63]), see also below.

In §6 of his Habilitationsschrift [Riemann (1854)] (1854), Riemann gives

$$\sum_{n=1}^{\infty} \frac{\psi(nx + \frac{1}{2})}{n^2}$$

as an example of a discontinuous, nonetheless integrable function.

As remarked by [Arias-de-Reyna (2004), p. 59], this example has its origin in the two functions considered by him in Fragment **I**:

$$\sum_{n=1}^{\infty} \frac{(-1)^n}{n^2} \psi\left(\frac{nx}{2\pi} + \frac{1}{2}\right), \qquad \sum_{k=0}^{\infty} \frac{1}{(2k+1)^2} \psi\left(\frac{(2k+1)x}{2\pi} + \frac{1}{2}\right).$$

By employing a method in [LiMZ (2010)] we now prove

Theorem 4.2. *The trigonometric identity*

$$\sum_{n=1}^{\infty} \frac{1}{n^z} \overline{B}_l\left(nx + \frac{1}{2}\right) = -\frac{l!}{2\pi i} \sum_{\substack{n=-\infty \\ n \neq 0}}^{\infty} \frac{\sigma_{l-z}^*(n)}{n^l} e^{2\pi i n x} \tag{4.29}$$

holds true a.e. whose special case reduces to Riemann's formula.

Proof. Recalling

$$\sum_{n=1}^{\infty} \frac{(-1)^{n-1}}{n^s} = \left(1 - 2^{1-s}\right) \zeta(s), \ \text{Re} s = \sigma > 1, \tag{4.30}$$

we see immediately that

$$\left(1 - 2^{1-s-z}\right) \zeta(s+z)\zeta(s+l) = \sum_{n=1}^{\infty} \frac{\sigma_{l-z}^*(n)}{n^{s+l}}, \tag{4.31}$$

where

$$\sigma_a^*(n) = \sum_{d|n} (-1)^d d^a$$

is an analogue of the sum-of-divisors function.

We use (4.31) as the generating Dirichlet series [LiMZ (2010), (3.3)]. The remaining part of the proof is almost the same as that of [LiMZ (2010), Theorem 2]. Indeed, we consider the integral

$$-\frac{(-1)^l l!}{2\pi i} \int_{(c)} \frac{\Gamma(s)}{\Gamma(s+l)} \left(1 - 2^{1-s-z}\right) \zeta(s-z)\zeta(1-s-l)x^{-s} ds, \ x > 0 \tag{4.32}$$

where $0 < c < 1$. As long as $\text{Re} z > 1 - c$, the series for $\zeta(s+z)$ being absolutely convergent, we may change the order of summation and integration to conclude that (4.31) is equal to

$$-(-1)^l l! \sum_{n=1}^{\infty} \frac{1}{n^z} \frac{1}{2\pi i} \int_{(c)} \frac{\Gamma(s)}{\Gamma(s+l)} (1 - 2^{1-s-z})\zeta(1-s-l)(nx)^{-s} ds \tag{4.33}$$

$$= -(-1)^l l! \sum_{n=1}^{\infty} \frac{1}{n^z} \frac{1}{2\pi i} \int_{(c)} \frac{\Gamma(s)}{\Gamma(s+l)} \zeta(1-s-l)(nx)^{-s} ds$$

$$- (-1)^l l! \sum_{n=1}^{\infty} \frac{1}{(2n)^z} \frac{1}{2\pi i} \int_{(c)} \frac{\Gamma(s)}{\Gamma(s+l)} \zeta(1-s-l)(2nx)^{-s} ds.$$

Recalling [LiMZ (2010), (2.1)]

$$\overline{B}_l(x) = -\frac{(-1)^l l!}{2\pi i} \int_{(c)} \frac{\Gamma(s)}{\Gamma(s+l)} \zeta(1-s-l)x^{-s} ds, \tag{4.34}$$

we find that

$$\sum_{n=1}^{\infty} \frac{B_l(nx) - B_l(2nx)}{n^z}$$

$$= -\frac{(-1)^l l!}{2\pi i} \int_{(c)} \frac{\Gamma(s)}{\Gamma(s+l)} (1 - 2^{1-s-z})\zeta(s+z)\zeta(1-s-l)x^{-s} ds.$$

Using

$$\zeta(1 - s - l)x^{-s} = \frac{1}{(2\pi)^l} \left((-1)^{-s} + (-1)^{-l}\right) i^l (2\pi i x)^{-s} \Gamma(s + l)\zeta(s + l)$$

and (4.31), we conclude that

$$\Phi_l^*(z, x) = -\frac{(-1)^l l!}{(2\pi i)^{1+l}} \int_{(c)} \Gamma(s) \frac{1}{(2\pi i)^s} \left((-1)^{-s} + (-1)^{-l}\right) x^{-s} \sum_{n=1}^{\infty} \frac{\sigma_{l-z}^*(n)}{n^{s+l}} \mathrm{d}s$$

$$= -\frac{(-1)^l l!}{(2\pi i)^{l+s}} \left((-1)^{-s} + (-1)^{-l}\right) e^{-x} \sum_{n=1}^{\infty} \frac{\sigma_{l-z}^*(n)}{n^{s+l}}$$

by [Erdélyi (1953), p. 49]

$$\frac{1}{2\pi i} \int_{(c)} \Gamma(s) z^{-s} \mathrm{d}s = e^{-z}, \quad \mathrm{Re}\, z \geq 0. \tag{4.35}$$

□

He also asserted [Riemann (1854)] that the trigonometrical series

$$\sum_{n=1}^{\infty} \frac{1}{n} \psi\left(nx + \frac{1}{2}\right) \tag{4.36}$$

converges to the "Fourier series"

$$\sum_{n=1}^{\infty} \frac{c(n)}{n} \sin 2\pi n x, \tag{4.37}$$

for $x \in \mathbb{Q}$, where $c(n) = \sum_{d|n}(-1)^d$.

More generally, let $\Phi_j(s, x)$ be the series considered by Hardy and Littlewood [Hardy and Littlewood (1923)]

$$\Phi_j(s, x) = \sum_{n=1}^{\infty} \frac{\bar{B}_j(nx)}{n^s} \quad (1 \leq j \in \mathbb{N}) \tag{4.38}$$

and its special case

$$\Phi_s(x) = \Phi_1(s, x) = \sum_{n=1}^{\infty} \frac{\psi(nx)}{n^s}, \tag{4.39}$$

where $\overline{B}_j(x)$ is the j-th periodic Bernoulli polynomial defined by (3.29).

Wintner [Wintner (1937)] proved that $\Phi_1(x) = \Phi_1(1, x)$ is integrable a là Lebesgue, has the trigonometrical series (4.37) as its Fourier series and is such that $\Phi_1(x)$ is of class L^q for every $q > 0$. The convergence of (4.39) to $\Phi_1(x)$ is a consequence of the Hausdorff-Paley extension of Fischer-Riesz theorem in §4.1.1.

Since

$$\psi\left(x+\frac{1}{2}\right) = \psi(2x) - \psi(x)$$

([Wintner (1937), p. 633]), this implies Riemann's assertion above, a.e. Hartman and Wintner [Hartman and Wintner (1938)] note that the method of Chowla and Walfisz is not applicable to $\Phi_s(x)$ for $\sigma < 1$ and they apply the method of [Wintner (1937)] to prove the a.e. convergence of the Fourier series to $\Phi_j(s,x)$:

$$\Phi_j(s,x) \sim -\frac{j!}{(2\pi i)^j} \sum_{k=-\infty}^{\infty} \frac{\sigma_{j-s}(k)}{k^j} e^{2\pi i k x} \tag{4.40}$$

for $\frac{1}{2} < \Re s \leq 1$ ([Hartman and Wintner (1938), p. 116]).

Although Hartman and Wintner [Hartman and Wintner (1938)] state the case of the divisor functions, their theorem is true for general arithmetic functions satisfying the order conditions in L^p-context and it reads

Theorem 4.3 (Hartman-Wintner). *Given a Dirichlet series $\Phi(s)$ defined by [Kanemitsu (2015), (6.30), §6.1.6], where $\sigma_\Phi < 2$ and $\gamma_n = n$, we have the identity valid for $\lambda + \frac{1}{2} < \sigma < \lambda + 1$ and a.a. values of x*

$$-\sum_{n=1}^{\infty} \frac{a_n}{n^s} \psi(xn) = \frac{1}{\pi} \sum_{n=1}^{\infty} \frac{A_n}{n^s} \sin 2\pi nx, \tag{4.41}$$

provided that

$$\sum_{d|n} |a_d| \ll n^{\lambda+\varepsilon} \tag{4.42}$$

for any $\varepsilon > 0$ (4.41) means that the Fourier series $\frac{1}{\pi}\sum_{n=1}^{\infty} \frac{A_n}{n^s}\sin 2\pi nx$ converges a.e. to the L^q-class function $-\Phi_s(x)$ for

$$2 \leq q < \frac{1}{1+\lambda-\sigma}. \tag{4.43}$$

Proof of this theorem follows almost verbatim to that of [Hartman and Wintner (1938), p. 115]. Indeed, the Fourier series for the (Riemann integrable) partial sum $\Phi_s^N(x) = -\sum_{n=1}^{N} \frac{a_n}{n^s}\psi(xn)$ is

$$-\sum_{n=1}^{\infty} \frac{A_n^N}{n} \sin 2\pi xn,$$

where $A_n^N = \sum_{d|n, d\leq N} a_d d^{1-s}$. Hence $|A_n^N| \ll n^{\lambda+1-\sigma+\varepsilon}$ by (4.42). Hence it suffices to have $\frac{q}{q-1}(\sigma - \lambda) > 1$ or the condition (4.43).

4.1.3 *Lambert series and Dedekind sums*

In §4.1.3 we make a list (hopefully rather complete) of the existing proof of the eta-transformation formula in which the Dedekind sums appear. In §4.1.4 we shall derive the transformation formula for certain Lambert series under the action of an arbitrary modular transformation, from the functional equation of Hecke's type for the corresponding zeta-functions. They are the sum and the difference of some Hurwitz zeta-function, thereby establishing the result corresponding to Apostol's, which complements Goldstein-de la Torre's result.

For $\tau \in \mathcal{H} = \{\tau \in \mathbb{C} \mid \operatorname{Im} \tau > 0\}$, the upper half-plane, let $q = e^{2\pi i \tau}$. Then the celebrated **Dedekind eta function** is defined by

$$\eta(\tau) = q^{\frac{1}{24}} \prod_{n=1}^{\infty} (1 - q^n), \tag{4.44}$$

which Dedekind introduced in his elucidation of Riemann's posthumous Fragment **II** as stated in §4.1.1.

Dedekind [Dedekind (1930)] proved the transformation formula for $\eta(\tau)$ under the general modular substitution $\sigma = \begin{pmatrix} a & b \\ c & d \end{pmatrix} \in \mathrm{SL}_2(\mathbb{Z})$:

$$\log \eta(\sigma\tau) = \log \eta(\tfrac{a\tau+b}{c\tau+d})$$
$$= \log \eta(\tau) + \tfrac{1}{2} \log \tfrac{c\tau+d}{i} + \tfrac{\pi i}{12} \Phi(\sigma), \tag{4.45}$$

where we take the principal branch of the logarithm,

$$\Phi(\sigma) = \begin{cases} \frac{a+d}{c} - 12 s(c, d), & c \neq 0 \\ -\frac{b}{d}, & c = 0 \end{cases}, \tag{4.46}$$

and where

$$s(h, k) = \sum_{m=1}^{k} \left(\!\left(\frac{m}{k}\right)\!\right) \left(\!\left(\frac{hm}{k}\right)\!\right) \tag{4.47}$$

is the Dedekind sum, $k \in \mathbb{N}$, $(h, k) = 1$ and

$$((x)) = -\frac{1}{\pi} \sum_{m=1}^{\infty} \frac{\sin 2\pi m x}{m} \tag{4.48}$$

is the saw-tooth Fourier series defined by (4.25) which differs from the first periodic Benoulli polynomial (defined by (3.28) and appearing in (4.53) below) at integer arguments only. This is a specified notation for the Dedekind sums only. (4.45) shows that $\eta(\tau)$ is a cusp form of weight 12 for $SL_2(\mathbb{Z})$ and indeed,

$$\Delta(\tau) := \eta(\tau)^{24} = q \prod_{n=1}^{\infty} (1 - q^n)^{24} \tag{4.49}$$

is the well-known Ramanujan Δ-function (discriminant function). Denoting the normalized Eisenstein series of weight 4 and 6 for $SL_2(\mathbb{Z})$ by E_4 and E_6, we have

$$\Delta(\tau) = \frac{1}{1728}(E_4^3(\tau) - E_6^2(\tau)) \tag{4.50}$$

$$= \sum_{n=1}^{\infty} \tau(n)q^n,$$

where $\tau(n)$ is the famous Ramanujan function. The deduction of the q-expansion from the automorphic property (4.50) of Δ (and therefore of η) is due to Jacobi and pursued by many authors thereafter as we shall describe later.

The special case of (4.45) for one of the generators of $SL_2(\mathbb{Z})$ reads

$$\eta\left(-\frac{1}{\tau}\right) = \sqrt{\frac{\tau}{i}}\eta(\tau), \qquad \left(\left|\arg\frac{\tau}{i}\right| < \frac{\pi}{2}\right). \tag{4.51}$$

Many proofs have been given for (4.51) or (4.45) and we shall give a chronological list of existing proofs before [Yifan (2004)].

First we shall give an expression for $s(h,k)$ in terms of the cotangent function ([Apostol (1951)])

$$s(h,k) = \frac{1}{4k} \sum_{m=1}^{k-1} \cot\frac{m}{k}\pi \cot\frac{hm}{k}\pi. \tag{4.52}$$

(4.52) follows from (4.47) by way of Eisenstein's formula ([Eisenstein (1844)], [Mathews (1986)])

$$\overline{B}_1\left(\frac{\mu}{k}\right) = -\frac{1}{2k} \sum_{\nu=1}^{k-1} \sin\frac{2\mu\nu}{k}\pi \cot\frac{\nu}{k}\pi. \tag{4.53}$$

Substituting (4.53), we see that

$$(h,k) = \frac{1}{8k^2} \sum_{\lambda=1}^{k-1}\sum_{\nu=1}^{k-1} \cot\frac{\nu}{k}\pi \cot\frac{\lambda}{k}\pi$$

$$\times \sum_{\mu=1}^{k} \left(\cos\frac{2\mu(\lambda - \nu h)}{k}\pi - \cos\frac{2\mu(\lambda + \nu h)}{k}\pi\right).$$

But, since

$$\sum_{\mu=1}^{k} \cos\frac{2\mu(\lambda \pm \nu h)}{k}\pi = \begin{cases} k, & \lambda \pm \nu h \equiv 0 \,(\mathrm{mod}\,k); \\ \\ 0, & \text{otherwise,} \end{cases}$$

it follows that

$$s(h,k) = \frac{1}{8k}\sum_{\nu=1}^{k-1}\cot\frac{\nu}{k}\pi\cot\frac{h\nu}{k}\pi - \frac{1}{8k}\sum_{\nu=1}^{k-1}\cot\frac{\nu}{k}\pi\cot\frac{-h\nu}{k}\pi,$$

which is (4.52).

The Dedekind sum obeys the reciprocity law. To state it, it is convenient to remove the positivity condition from the definition of $s(h,k)$ and define for $(h,k) = 1$,

$$\begin{aligned} s(h,k) &= \sum_{m\not\equiv 0(\,\mathrm{mod}\,k)} \left(\!\!\left(\frac{m}{k}\right)\!\!\right)\left(\!\!\left(\frac{hm}{k}\right)\!\!\right) \\ &= \frac{1}{4k}\sum_{m\not\equiv 0(\,\mathrm{mod}\,k)}\cot\frac{m}{k}\pi\cot\frac{hm}{k}\pi. \end{aligned} \tag{4.54}$$

Then the **reciprocity law** reads

$$s(h,k) + s(k,h) = \frac{1}{12}\left(\frac{h}{k} + \frac{1}{hk} + \frac{k}{h}\right) - \frac{1}{4}\mathrm{sign}(hk), \tag{4.55}$$

where $\mathrm{sign}(x) = \frac{x}{|x|}$ means the sign of $x \neq 0$. Dedekind [Dedekind (1930)] proved the more general reciprocity law

$$\sum_{j=1}^{3}\Phi(\sigma_j) = -3\mathrm{sign}(c_1, c_2, c_3) \quad \sigma_j = \begin{pmatrix} * & * \\ c_j & * \end{pmatrix}. \tag{4.56}$$

With $\sigma_1 = \begin{pmatrix} 0 & -1 \\ 1 & 0 \end{pmatrix}$, (4.56) gives (4.55).

There are many generalizations of Dedekind sums (cf. [Holschneider (1991)], [Asano (2003)]). We state just one of them:

$$s_{q,r}(h,k) = \sum_{m\,\mathrm{mod}\,k}\overline{B}_{q+1-r}\left(\frac{m}{k}\right)\overline{B}_r\left(\frac{hm}{k}\right) \tag{4.57}$$

for q odd and $0 \leq r \leq q+1$ obeys a reciprocity formula, where $\overline{B}_\nu(x)$ signifies the ν-th periodic Bernoulli polynomial having the Fourier expansion

$$\overline{B}_\nu(x) = -\frac{\nu!}{(2\pi i)^\nu}\sum_{\substack{n=-\infty \\ n\neq 0}}^{\infty}\frac{e^{2\pi i x}}{n^\nu}. \tag{4.58}$$

For $p \geq 1$ *odd* and $|x| < 1$, let

$$\widetilde{G}_p(x) = \sum_{m=1}^{\infty} \sum_{n=1}^{\infty} m^{-p} x^{mn} = \sum_{m=1}^{\infty} \frac{1}{m^p} \frac{x^m}{1 - x^m} \qquad (4.59)$$

be the Lambert series and let for $x > 0$

$$G_p(x) = \widetilde{G}_p(e^{-2\pi x}) = \sum_{m=1}^{\infty} \frac{m^{-p}}{e^{2\pi mx} - 1} = \sum_{m=1}^{\infty} \sigma_{-p}(m) e^{-2\pi mx}, \qquad (4.60)$$

$\sigma_a(n) = \sum_{d|n} d^a$ being the sum-of-divisors function. Then we have the transformation formula for the Lambert series for $p > 1$, odd

$$G_p\left(\frac{x - ih}{k}\right) + \frac{1}{2}\zeta(p)\left(1 - (-1)^{\frac{p-1}{2}} x^{p-1}\right)$$

$$+ \frac{(2\pi x)^p}{2(p+1)!} \sum_{m=0}^{p+1} (p + 1m)(ix)^{-m} s_{p,m}(h, k) \qquad (4.61)$$

$$= (-1)^{\frac{p-1}{2}} x^{p-1} G_p\left(\frac{1/x - iH}{k}\right),$$

where H is an integer such that $hH \equiv -1 (\bmod\ k)$. This corrected version of Apostol's [Apostol (1950), Theorem 2] was first given by Mikolás [Mikolás (1957), p. 106] and shortly thereafter by Iseki [Iseki (1957)]. There are several other proofs (cf. Berndt [Berndt (1973)]).

We note that the special case of (4.61) with $k = 1$, $h = H = 0$, $p = 2n + 1$, gives rise to, when divided by $(\pi x)^n$:

$$(\pi x)^{-n} G_{2n+1}(x) + \frac{1}{2}\zeta(2n+1)\left((\pi x)^{-n} - (-1)^n \left(\frac{x}{\pi}\right)^n\right)$$

$$+ \frac{2^{2n+1}(\pi x)^{n+1}}{2(2n+2)!} \sum_{m=0}^{2n+1} \binom{2n+2}{m} (ix)^{-m} B_{2n+1-m} B_m$$

$$= (-1)^n \left(\frac{x}{\pi}\right)^n G_{2n+1}\left(\frac{1}{x}\right),$$

which becomes, on writing $\alpha = \pi x$, $\beta = \frac{\pi}{x}$,

$$\alpha^{-n}\left(\frac{1}{2}\zeta(2n+1) + G_{2n+1}\left(\frac{\alpha}{\pi}\right)\right)$$

$$= (-\beta)^n \left(\frac{1}{2}\zeta(2n+1) + G_{2n+1}\left(\frac{\beta}{\pi}\right)\right) \qquad (4.62)$$

$$- 2^{2n} \sum_{j=0}^{n+1} (-1)^j \frac{B_{2j}}{(2j)!} \frac{B_{2n+2-2j}}{(2n+2-2j)!} \alpha^{n+1-j} \beta^j,$$

the famous formula of Ramanujan for the values of the Riemmann zeta-function at positive odd arguments $s = 2n + 1$ [Berndt (1989), Chapter 14, Entry 21].

That (4.61) implies (4.62) was first noticed by Apostol (Letter to E. Grosswald, January 24, 1973) and Berndt [Berndt (1973)] independently. As noted by Berndt [Berndlt (1977)], many proofs of Ramanujan's formula were given thereafter.

Now we shall see how Ramanujan was led to the proof of (4.51) (cf. [Berndlt (1977), p. 160], [Berndlt (1991), p. 43]). Using (4.60) and applying [Kanemitsu (2003)], we get a slightly more general form of (4.62):

$$\sum_{k=1}^{\infty} \sigma_{-2n-1}(k)e^{-2\pi kx} + (-1)^{n+1}x^{2n}\sum_{k=1}^{\infty}\sigma_{-2n-1}(k)e^{-\frac{2\pi k}{x}} = P(x), \quad (4.63)$$

where

$$P(x) = \sum \text{Res}(2\pi)^{-s}\Gamma(s)\zeta(s)\zeta(s + 2n + 1)x^{-s}. \quad (4.64)$$

Formula (4.63) is valid also for $n = 0$, in which case it reads

$$G(\tau) = G\left(-\frac{1}{\tau}\right) + \frac{\pi}{12}\left(\frac{i}{\tau} - \frac{\tau}{i}\right) + \frac{1}{2}\log\frac{\tau}{i} \quad (4.65)$$

where

$$G(\tau) = -\sum_{k=1}^{\infty}\log(1 - q^k) = \sum_{k=1}^{\infty}\sigma_{-1}(k)q^k. \quad (4.66)$$

Since

$$G(\tau) = \frac{\pi i}{12}\tau - \log\eta(\tau),$$

(4.65) amounts to (4.51).

We remark that as the cotangent function plays an important role in the context of $L(1, \chi)$, χ odd, and of the Dedekind sum (4.52), so does it in the transformation formula (4.61) for the Lambert series. Indeed, Iseki's proof [Iseki (1957)] depends on the partial fraction expansion

$$2\pi i\left(\frac{e^{2\pi\alpha mz}}{e^{2\pi mz} - 1} - \frac{\delta(\alpha)}{2}\right) = -\sum_{n=-\infty}^{\infty}\frac{e^{2\pi i\alpha n}}{n + imz}, \quad (4.67)$$

which is valid for $z \in \mathbb{C} \setminus \mathbb{Z}$ and $\delta(\alpha) = 1$ or 0 according as $\alpha = 0$ or not, and reduces to (4.55) in the special case of $\alpha = 0 : LHS = -\pi\cot\pi imz$.

We shall now give a list of references related to the eta transformation formula in chronological order with brief comments. Among the references, [Iseki (1957)], [Goldstein and Torre (1974)], [Berndt (1973)],

[Sczech (1978)], [Elstoldt], [Balasubramanian, Ding, Kanemitsu and Tanigawa (2008)] contain detailed comments on earlier works.

Jacobi [Jacobi (1829), §40] was the starting point for Riemann's investigation, uncompleted during his lifetime. Dedekind [Dedekind (1930)] succeeded in elucidating Riemann's Fragment II (cf. [Arias-de-Reyna (2004)] for some contraversial arguments on [Dedekind (1930)]) and laid a sound foundation for further developments (cf. Wintner [Wintner (1944)] for lucid comments on Fragment I). The direction of argument deducing the q-expansion (4.49) from the automorphic property (4.50) of Δ (and therefore of η) is due to Jacobi and pursued by Hurwitz [Hurwitz (1881)], Hecke [Hecke (1970)], Lang [Landau (1918)], Petersson [Petersson (1982)] et al, which we abbreviate "automorphy $\to$ q-expansion".

After Hecke [Hecke (1970)], there appeared an influential paper of Rademacher [Rademacher (1931)] in which he developed the Mellin transform method (i.e. the modular relation method [Kanemitsu (2003)] to prove (4.61) with $p = 1$ and which became one of the main resources for the later developments by Apostol [Apostol (1950)], [Apostol (1964)], Iseki [Iseki (1961)] and Goldstein and de la Torre [Goldstein and Torre (1974)]. (cf. the comments in [Goldstein and Torre (1974)] and [Rademacher (1931)]).

In principle, we do not mention the papers on the transformation formula in connection with partitions, leaving it for another occasion, but we refer to only one paper by Schoenfeld [Schoenfeld (1944)] here, which deals with such a problem by the Mellin transform method.

Then there appeared another influential paper of Apostol [Apostol (1950)] mentioned above, in which he first defined a generalization $s_{p,1}(h, k)$ (cf. (4.57)) of the Dedekind sum (4.54) and proved the transformation formula (4.61) for an odd integer $p > 1$ (cf. the above remark).

Fischer [Fischer (1951)] gave a direct proof of

$$\eta(\sigma\tau) = \varepsilon \left(\frac{c\tau + d}{i} \right)^{1/2} \eta(\tau).$$

K. Iseki [Iseki (1952)] gave a proof of the case $p = 1$ using real integrals and a formula in the theory of theta-functions.

Koecher's paper [Koecher (1953)] occupies a unique position as one giving a new proof of the Kronecker limit formula for the Eisenstein series through its characterization in terms of its characteristic properties and thereby proving the q-product expansion for the Δ-function.

Siegel [Siegel (1954)] gave a simple proof of (4.51) using the complete interpretation of the cotangent functions, which method was elaborated by

Rademacher [Rademacher (1931)] to cover the case of the general substitution $\sigma \in \mathrm{SL}_2(\mathbb{Z})$.

After [Iseki (1957)] in which he used the partial fraction formula for

$$\frac{e^{2\pi\alpha m\tau}}{e^{2\pi m\tau} - 1},$$

Iseki [Iseki (1961)] applied the method of Rademacher [Rademacher (1931)] to treat the case $p = 1$ and Apostol [Apostol (1964)] treated the case $p > 1$ odd by the same method applied to

$$F_p(\alpha, \beta, s) = \frac{\Gamma(s)}{(2\pi)^s} \{ \zeta(s, \alpha) l_{p+s}(\beta) + \zeta(s, 1 - \alpha) l_{p+s}(1 - \beta) \} \tag{4.68}$$

$$= i^{p-1} F_p(1 - \beta, \alpha, s).$$

Meyer [Meyer (1967)] used the functional equation for the incomplete theta-function to prove (4.51).

In the same year 1968 there appeared two papers which elucidate the appearance of the modular relation. The paper of Weil [Weil (1968)] is exposed in Ogg [Ogg (1969)] and is better known than Schoenberg [Schoenberg (1968)] in both of which they use the generating Dirichlet series $\zeta(s)\zeta(s+1)$ to attain (4.51). Schoenberg also deduced the infinite product expansion for $\eta(\tau)$ by employing the generating Dirichlet series $L(2s, \chi_{12})$, where χ_{12} is a primitive character mod 12.

Then in 1972 there appeared three papers [Berndlt and Venkatachaliengar (2001)], [Grosswald (1972)] and [Lewittes (1972)] referring to (4.51) and Lambert series. Of these, Groswald's papers brought Ramanujan's formula to light and Lewitter method was improved by Bernalt [Ber1] to give a transformation formula for generalized Eisenstein series.

In 1973 two most widely read books of Lang [Lang (1973)] and Serre [Sczech (1978)] appeared. [Landau (1918), Chapter 18, §4] gives a proof of (4.49) using the theory of elliptic functions and product expansion of the σ-function. [Serre (1973), Chap VIII, §4] is an expansion of Hurwitz's theory [Hurwitz (1904)].

Goldstein and de la Torre [Goldstein and Torre (1974)] give a proof of the η-transformation (4.45) based on the modular relation for the generating zeta-function

$$F_{p,q;\varepsilon}^{(1)}(s) = \left(\frac{2\pi}{c}\right)^{-s} \Gamma(s)\zeta_\varepsilon(s, p, c)\zeta_\varepsilon(s + 1, q, c), \tag{4.69}$$

where $c \in \mathbb{N}$ and $\zeta_\varepsilon(s, h, c)$ signifies the partial zeta-function with signature.

$$\zeta_\varepsilon(s, h, c) = \sum_{\substack{n=-\infty \\ n \equiv h (\bmod c)}}^{\infty} {}' \frac{\nu_\varepsilon(n)}{|n|^s}, \tag{4.70}$$

for $\sigma > 1$ and for $x \neq 0$,

$$\nu_\varepsilon(x) = \left(\frac{x}{|x|}\right)^\varepsilon, \qquad \varepsilon = 0 \text{ or } 1,$$

is the signature character, the prime on the summation sign indicating that the term $n = 0$ is to be omitted.

For more details on this subject and generalizations thereof cf. §4.1.4.

Schoenberg's book [Schoenberg (1974)] contains a proof of (4.49) based on a summation method for Eisentein series of weight 2.

Berndt's [Berndt (1975)] is a complete form of [Berndt (1973)] and gives an explicit transformation formula for generalized Eisenstein series with incorporation of generalized Dedekind sums.

In 1976, other two books appeared which were widely appreciated. [Apostol (1976), Chap 3.2] contains a proof of (4.51) by the method of siegel.

[Weil (1976)] resurrects the theory of Eisenstein and Kronecker and is a must.

[Berndt (1973)] contains the most detailed exposition of Ramanujan's formula (4.62) and related results and is very useful.

Walker [Walker (1978)] uses the partial fraction expansion for the cotangent function to transform $(\frac{\tau}{i})^{1/4}\eta(\tau)$ in the form of a double infinite product which is invariant under the substitution $\tau \to -1/\tau$. Also used is the Euler-Maclaurin formula.

Sczech [Sczech (1978)], incorporating [Hurwitz (1904)], gives a proof of the general transformation formula. He uses the partial fraction expansion for the cotangent function (as in [Iseki (1957)]) and Eisenstein summation formula [Weil (1976)].

Peterson's book [Petersson (1982)] has an appendix suggesting the proof of (4.49) based on some transformative properties of the Weierstrass zeta-function. A simplified version appears in Elstoldt [Elstoldt].

Chan's paper [Chan (1988)] contains a proof ofeqrefdedekindetatrans using the Weierstrass elliptic function and comparing two values at half-periods, which idea is close to [Petersson (1982)] (cf. [Elstoldt]). Basu [Basu (1999)] is a repetition of [Iseki (1957)] using the same Fourier expansion (4.67).

The paper [Kanemitsu (2009)], which constitutes the main body of this section, may be thought of as a continuation to the paper [Kanemitsu (2003)], where as one example, it is shown that [Weil (1968)] is the singular case $n = 0$ of (4.63).

Kohnen [Kotre (1995)] uses averaging procedure which may be viewed as a multiplicative version of the classical Hecke operator.

Elstoldt [Elstoldt] is in manuscript form simplifying [Petersson (1982)] by the partial fraction expansion for the cotangent function.

[Berndlt and Venkatachaliengar (2001)] refers to (4.63) in the setting of (4.62) and gives its proof using the partial fraction expansion for the coth, being in the same veins as in [Sczech (1978)] et al. They also refer to the differentiated form.

The paper [Balasubramanian, Ding, Kanemitsu and Tanigawa (2008)] which was written without the knowledge of [Berndlt and Venkatachaliengar (2001)], elucidated [Apostol (1950)], [Apostol (1964)], [Iseki (1957)], [Iseki (1961)], [Goldstein and Torre (1974)], [Sczech (1978)], [Basu (1999)] and others by showing that the eta-transformation is the special case $n = 0$ of Ramanujan's formula (4.63).

4.1.4 *Transformation formula for certain Lambert series*

The purpose of this subsection is to elaborate on the Goldstein-de la Torre argument, deducing the transformation formula for the Lambert series $G_{p,q;\varepsilon}^{(r)}(w)$ ((4.83) below) under the action of the full modular group. They deduce it from the functional equation for the product $F_{p,q;\varepsilon}^{(r)}(s)$ ((4.84) below) of partial zeta-functions defined in (4.70). Thus, we shall play the role of Apostol [Apostol (1964)] dealing with the case $r > 1$, with the role of [Iseki (1961)] played by [Goldstein and Torre (1974)] for $r = 1$. We shall find out that the Dedekind sum $s_{q,r}(h, k)$ (cf. (4.57)) takes place in the case of r being odd, which we are going to study. We hope to study the case of r being even, in anticipation of finding a counterpart of the Dedekind sum built of Bernoulli polynomials elsewhere.

Let $\zeta(s, x)$ denote the Hurwitz zeta-function defined for $x > 0$ by

$$\zeta(s, x) = \sum_{n=0}^{\infty} \frac{1}{(n + x)^s}, \quad \sigma > 1. \tag{4.71}$$

This is continued meromorphically over the whole plane by the functional equation (4.79) with a unique simple pole at $s = 1$ (cf. e.g. [Kanemitsu (2007)]).

Let $\{x\}$ denote the fractional part of the real number x. Then for any $h \in \mathbb{Z}$, $c\{\frac{h}{c}\}$ is the least non-negative residue $\bmod c$ of h.

Hence, for the partial zeta-function $\zeta_\varepsilon(s, h, c)$ defined by (4.70), it follows that

$$\zeta_0(s,h,c) = \begin{cases} 2c^{-s}\zeta(s), & c \mid h \\ c^{-s}(\zeta(s,\{\frac{h}{c}\}) + \zeta(s,1-\{\frac{h}{c}\})), & c \nmid h \end{cases} \tag{4.72}$$

and

$$\zeta_1(s,h,c) = \begin{cases} 0, & c \mid h \\ c^{-s}\left(\zeta(s,\{\frac{h}{c}\}) - \zeta(s,1-\{\frac{h}{c}\})\right), & c \nmid h. \end{cases} \tag{4.73}$$

From well-known properties of the Hurwitz and the Riemann zeta-function, it follows that $\zeta_\varepsilon(s,h,c)$ may be continued over the whole plane and the singularity occurs only when $\varepsilon = 0$, in which case there is a simple pole at $s = 1$ with residue $\frac{2}{c}$.

Appealing to other well-known properties of the Hurwitz zeta-function, i.e. for $0 \leq l \in \mathbb{Z}$,

$$\zeta(-l,x) = -\frac{1}{l+1}\overline{B}_{l+1}(x) \tag{4.74}$$

and

$$\overline{B}_{l+1}(1-x) = (-1)^{l+1}\overline{B}_{l+1}(x) \tag{4.75}$$

(cf. e.g. [Kanemitsu (2007), Chapter 4]), we may find the values of $\zeta_\varepsilon(s,h,c)$ at non-positive integers $-l$.

$$\zeta_0(-l,h,c) = \begin{cases} 2c^l\zeta(-l), & c \mid h \\ -\frac{c^l}{l+1}\overline{B}_{l+1}(\frac{h}{c})(1+(-1)^{l+1}), & c \nmid h \end{cases} \tag{4.76}$$

and

$$\zeta_1(-l,h,c) = \begin{cases} 0, & c \mid h \\ -\frac{c^l}{l+1}\overline{B}_{l+1}(\frac{h}{c})(1+(-1)^l), & c \nmid h. \end{cases} \tag{4.77}$$

We note that (4.76) and (4.77) give advantages over the method of [Goldstein and Torre (1974)]:

$$\zeta_1(-1,h,c) = 0 \quad c \mid h$$

([Goldstein and Torre (1974), p. 12, ll.12-13]),

$$\zeta_0(0,h,c) = 0 \quad c \nmid h$$

([Goldstein and Torre (1974), (16)]), and

$$\zeta_1(0,h,c) = -2\left(\left(\frac{h}{c}\right)\right)$$

whose proof depend on the functional equation.

Although Goldstein-de la Torre [Goldstein and Torre (1974)] state that (4.70) may be proved by the θ-transformation formula (§4.2.1) or by the functional equation (4.79) for the Hurwitz zeta-function, we note that the latter statement lacks an essential ingredient, i.e. Eisenstein's formula ([Iseki (1952), (8.19), p. 171])

$$\sum_{a=1}^{c-1} e^{-2\pi i \frac{h}{c} a} l_s \left(\frac{a}{c}\right) = c^{1-s} \zeta\left(s, \left\{\frac{h}{c}\right\}\right) - \zeta(s), \qquad (4.78)$$

which is a relation between two bases of the vector space of periodic Dirichlet series of period c. Thus, giving a proof of the following theorem first proved by [Goldstein and Torre (1974)] is quite instructive.

For the proof we need the following. The functional equation (e.g. [Kanemitsu (2007), (3.67), p. 75]) in one of its equivalent forms reads $(z \notin \mathbb{Z})$

$$\ell_s(x) = \frac{\Gamma(s)}{(2\pi)^s} \left(e^{-\frac{\pi i s}{2}} \zeta(1-s, \{x\}) - e^{\frac{\pi i s}{2}} \zeta(1-s, 1-\{x\}) \right), \qquad (4.79)$$

whose special case is the functional equation for the Riemann zeta-function (2.90). Another form is (2.104).

Theorem 4.4. *The ramified functional equation for $\zeta_\varepsilon(s, h, c)$ of Hecke type reads*

$$\left(\frac{\pi}{c}\right)^{-\frac{s}{2}} \Gamma\left(\frac{s+\varepsilon}{2}\right) \zeta_\varepsilon(s, h, c)$$

$$= (-i)^\varepsilon c^{-\frac{1}{2}} \left(\frac{\pi}{c}\right)^{-\frac{1-s}{2}} \Gamma\left(\frac{1-s+\varepsilon}{2}\right) \sum_{\lambda \bmod c} e^{2\pi i \lambda h/c} \zeta_\varepsilon(1-s, \lambda, c). \qquad (4.80)$$

Proof. We may suppose that $c \nmid h$. By (4.78)

$$\sum_{a=1}^{c-1} e^{-2\pi i \frac{h}{c} a} l_s\left(\frac{a}{c}\right) = c^{1-s} \zeta\left(s, \left\{\frac{h}{c}\right\}\right) - \zeta(s),$$

which has its counterpart

$$\sum_{a=1}^{c-1} e^{-2\pi i \frac{h}{c} a} l_s\left(1 - \frac{a}{c}\right) = c^{1-s} \zeta\left(s, \left\{\frac{h}{c}\right\}\right) - \zeta(s),$$

(which follows by the change of variable $a \leftrightarrow c - a$). Hence

$$c\zeta_0(s, h, c) = c^{1-s}\left(\zeta\left(s, \left\{\frac{h}{c}\right\}\right) + \zeta\left(s, 1 - \left\{\frac{h}{c}\right\}\right)\right)$$

$$= \sum_{a=1}^{c-1} e^{-2\pi i \frac{h}{c}a}\left(l_s\left(\frac{a}{c}\right) + l_s\left(1 - \frac{a}{c}\right) + 2\zeta(s)\right)$$

$$= \sum_{a=1}^{c-1} e^{-2\pi i \frac{h}{c}a}\frac{i\Gamma(1-s)}{(2\pi)^{1-s}}\left(e^{-\frac{\pi i s}{2}}\zeta\left(1 - s, \left\{\frac{a}{c}\right\}\right)\right.$$

$$- e^{\frac{\pi i s}{2}}\zeta\left(1 - s, 1 - \left\{\frac{a}{c}\right\}\right)$$

$$+ e^{-\frac{\pi i s}{2}}\zeta\left(1 - s, 1 - \left\{\frac{a}{c}\right\}\right) - e^{\frac{\pi i s}{2}}\zeta\left(1 - s, \left\{\frac{a}{c}\right\}\right)\right)$$

$$+ 2^{s+1}\pi^{s-1}\Gamma(1-s)\sin\frac{\pi s}{2}\zeta(1-s),$$

whose right-hand side becomes, by (2.8) and (2.9),

$$= \frac{i\Gamma(1-s)}{(2\pi)^{1-s}}\left\{(e^{-\frac{\pi i s}{2}} - e^{\frac{\pi i s}{2}})\sum_{a=1}^{c-1} e^{-2\pi i \frac{h}{c}a}\zeta_0(1 - s, a, c) + 2\zeta(1-s)\right\}$$

$$= \frac{2\Gamma(1-s)}{(2\pi)^{1-s}}\sin\pi(1 - \frac{s}{2})(\sum_{a=1}^{c-1} e^{-2\pi i \frac{h}{c}a}\zeta_0(1 - s, a, c) + 2\zeta(1-s)).$$

The subsequent procedure is very standard and will be given as an exercise.

This proves the $\varepsilon = 0$ case. $\qquad\square$

Exercise 24. Give the details of the last part of the proof.

Solution. By the duplication formula

$$\Gamma(1-s) = 2^{1-s}\pi^{-\frac{1}{2}}\Gamma\left(\frac{1-s}{2}\right)\Gamma\left(1 - \frac{s}{2}\right) \qquad (4.81)$$

and the reciprocation formula

$$\Gamma\left(1 - \frac{s}{2}\right)\sin\pi\left(1 - \frac{s}{2}\right) = \frac{\pi}{\Gamma\left(\frac{s}{2}\right)}. \qquad (4.82)$$

Adjusting the factors suitably, we deduce that

$$\left(\frac{\pi}{c}\right)^{-\frac{s}{2}}\Gamma\left(\frac{s}{2}\right)\zeta_0(s, h, c)$$

$$= c^{-\frac{1}{2}}\left(\frac{\pi}{c}\right)^{-\frac{1-s}{2}}\Gamma\left(\frac{1-s}{2}\right)\sum_{a=1}^{c-1} e^{-2\pi i \frac{h}{c}a}\zeta_0(1-s, a, c) + c^{-s}\Gamma\left(\frac{1-s}{2}\right)\zeta(1-s)$$

$$= c^{-\frac{1}{2}}\left(\frac{\pi}{c}\right)^{-\frac{1-s}{2}}\Gamma\left(\frac{1-s}{2}\right)\sum_{a \bmod c} e^{2\pi i \frac{h}{c}a}\zeta_0(1-s, a, c),$$

on incorporating the Riemann zeta-function part.

Let c, r be positive integers with r *odd*. For arbitrary integers p, q we define the Lambert series

$$G_{p,q;\varepsilon}^{(r)}(x) = \sideset{}{'}\sum_{\substack{m,n=-\infty \\ (m,n)\equiv(q,p)\,\mathrm{mod}\,c}}^{\infty} \frac{e^{-(2\pi|mn|/c)x}}{|m|^r}\nu_\varepsilon(mn) \qquad (4.83)$$

and consider the completed zeta-function

$$F_{p,q;\varepsilon}^{(r)}(s) = \Gamma(s)\left(\frac{c}{2\pi}\right)^s \zeta_\varepsilon(s,p,c)\zeta_\varepsilon(s+r,q,c). \qquad (4.84)$$

This, in view of (4.80) and the duplication formula for the gamma function, satisfies a ramified functional equation of Hecke type:

$$F_{p,q;\varepsilon}^{(r)}(s) = c^{-1}(-1)^{\frac{r-1}{2}} \sum_{\alpha,\beta\,\mathrm{mod}\,c} e^{2\pi i(\alpha p+\beta q)/c} F_{\beta,\alpha;\varepsilon}^{(r)}(1-r-s). \qquad (4.85)$$

Also, $F_{p,q;\varepsilon}^{(r)}(s)$ is the Mellin transform of $G_{p,q;\varepsilon}^{(r)}(s)$ for $\sigma > 1$, or what is the same thing,

$$G_{p,q;\varepsilon}^{(r)}(x) = \frac{1}{2\pi i}\int_{(\kappa)} F_{p,q;\varepsilon}^{(r)}(s)x^{-s}ds, \qquad (4.86)$$

where (κ) signifies the vertical line $s = \kappa + it$, $-\infty < t < \infty$ with $\kappa > 1$. We choose $\kappa = \frac{3}{2}$ in what follows.

The modular relation for $G_{p,q;\varepsilon}^{(r)}(x)$ now reads

Lemma 4.1. *For* $\mathrm{Re}\,x > 0$,

$$G_{p,q;\varepsilon}^{(r)}(x) = c^{-1}(-1)^{\frac{r-1}{2}} \sum_{\alpha,\beta\,\mathrm{mod}\,c} e^{2\pi i(\alpha p+\beta q)/c} G_{\beta,\alpha;\varepsilon}^{(r)}(x^{-1})$$

$$+ \sum_{s_0\in S} \mathrm{Res}_{s=s_0} F_{p,q;\varepsilon}^{(r)}(s)x^{-s}, \qquad (4.87)$$

where

$$S = \{1, \ -r+1\}\cup\{0, \ -1, \ -2,\cdots,-r\}$$

is the set of all singularities of F *lying between* $-\frac{1}{2}-r < \sigma < \frac{3}{2}$.

Proof. By the Phragmén-Lindelöf principle and the Stirling formula applied to (4.84), we may move the line of integration $\sigma = \kappa = \frac{3}{2}$ in (4.80) to $\sigma = 1-r-\kappa = -\frac{1}{2}-r$ to obtain

$$G_{p,q;\varepsilon}^{(r)}(x) = \frac{1}{2\pi i}\int_{(-r-\frac{1}{2})} F_{p,q;\varepsilon}^{(r)}(s)x^{-s}ds$$

$$+ \sum_{s_0\in S} \mathrm{Res}_{s=s_0} F_{p,q;\varepsilon}^{(r)}(s)x^{-s}.$$

Applying (4.85) to the integral on the right of this equality shows that it is

$$c^{-1}(-1)^{\frac{r-1}{2}} \sum_{\alpha,\beta \bmod c} e^{2\pi i(\alpha p+\beta q)/c} \frac{1}{2\pi i} \int_{(\frac{3}{2})} F_{p,q;\varepsilon}^{(r)}(s) x^{-s} ds$$

on writing s for $1 - r - s$. This completes the proof. $\square$

For an integer a such that $(a, c) = 1$, we denote by a^* the multiplicative inverse of $a \bmod c$, $aa^* \equiv 1 \bmod c$. We may now state the transformation formula. Proof follows on summing the identity in Lemma 4.1.

Theorem 4.5. *Let c and r be positive integers with r being odd. Then for $\operatorname{Re} x > 0$ we have the transformation formula*

$$\sum_{\varepsilon=0,1} \sum_{p,q \bmod c} e^{2\pi i(apq/c)} G_{p,q;\varepsilon}^{(r)}(x)$$

$$= c^{-1}(-1)^{\frac{r-1}{2}} x^{r-1} \sum_{\varepsilon=0,1} \sum_{\alpha,\beta \bmod c} e^{2\pi i(a^*\alpha\beta/c)} G_{\beta,\alpha;\varepsilon}^{(r)}(x^{-1}) + P(x), \quad (4.88)$$

where $P(x) = P(x; r)$ is the residual function given by

$$P(x; r)$$

$$= \frac{2(2\pi i)^r}{r!} s_r(a, c) + \frac{(-1)^{\frac{r-1}{2}} i c x^{r-1}}{\pi} \sum_{q \bmod c} \zeta_1(1, -q, c) \zeta_1(r, -aq, c)$$

$$+ \sum_{k=1}^{\frac{r-1}{2}} (-1)^{k-1} 2^{-2k+1} \pi^{-1} x^{2k-1} \sum_{q \bmod c} \zeta_0(2k, -aq, c) \zeta_0(r - 2k + 1, q, c)$$

$$+ \left\{ 2 + (-1)^{\frac{r-1}{2}} 2 x^{r-1} \right\} \zeta(r) + \left\{ \frac{2}{\pi c^r x} + (-1)^{\frac{r-1}{2}} \frac{2 x^r}{\pi c^r} \right\} \zeta(r + 1) \quad (4.89)$$

for $r \geq 3$ and

$$P(x; 1) = 4\pi i s_1(a, c) + 2 \log x + \frac{\pi}{3c} \frac{1}{x} - \frac{\pi}{3c} x. \quad (4.90)$$

4.1.5   *The RH*

This subsection is a sequel to §3.5 and mainly state results related to the Riemann Hypothesis (RH) defined by (3.50).

Under the RH, Theorem 3.20 is applicable, which is a consequence of Theorem 3.16, and the counterpart of (3.49) reads

$$\sum_{n \leq x} \mu(n) = O(x^{\frac{1}{2}+\varepsilon}) \iff \sum_{n \leq x} \lambda(n) = O(x^{\frac{1}{2}+\varepsilon}), \quad (4.91)$$

which is equivalent to

$$\pi(x) = \text{li}x + O(x^{\frac{1}{2}+\varepsilon}), \tag{4.92}$$

the counterpart of (3.45).

Here we state another theorem connecting (4.28) to RH.

Let $F_x = F_{[x]}$ denote the Farey series of order x:

$$F_x = \left\{ \rho_\nu \ \middle| \ \rho_\nu = \frac{a_\nu}{b_\nu}, \ (a_\nu, b_\nu) = 1, \ 0 \leq a_\nu < b_\nu \leq x \right\}.$$

Then

$$\#F_x = \sum_{n \leq x} \varphi(n) = \Phi(x),$$

where $\varphi(n)$ signifies the Euler function in Example 3.2.

Define the error function

$$E_f(x) := \sum_{\nu=1}^{\Phi(x)} f(\rho_\nu) - \Phi(x) \int_0^1 f(u)\mathrm{d}u.$$

Theorem 4.6. ([Kanemitsu (2000), p. 442, Corollary]) *For every $\varepsilon > 0$,*

$$\text{RH} \Longleftrightarrow E_f(x) = O(x^{\frac{1}{2}+\varepsilon}).$$

More generally, we may formulate the equivalent condition to the RH for

$$f_{\kappa,l}(x) = \sum_{n=1}^{\infty} \frac{1}{n^\kappa} \cos 2\pi n^l x, \quad \Re\kappa > 1, \ l \in \mathbb{N},$$

whence

$$f_{2,2}(x) = f(x)$$

above.

For GRH for Selberg class zeta-functions, cf. Remark 2.2.

In §6.5 we come to the notion of finite and infinite energy signals and the weak Lindöf hypothesis. The **Lindelöf Hypothesis** (LH) asserts that

$$\zeta\left(\frac{1}{2} + it\right) = O(|t|^\varepsilon) \tag{4.93}$$

for every $\varepsilon > 0$. It is apparent that RH implies LH. We have a speculation that PNT is in the realm of infinite power signals while signals in control theory lies in the world of finite power signals.

4.2　Quadratic reciprocity and Riemann's non-differentiable function

Let p be a natural number, b an arbitrary real number and $\mathfrak{z} = h + i\epsilon$ is a complex variable tending to 0 with imaginary part ϵ positive.

By the Euler-Maclaurin summation formula, we obtain

$$\sum_{n=-\infty}^{\infty} e^{(b+pn)^2 i\mathfrak{z}} = \frac{2\sqrt{\pi}}{p}\sqrt{i\mathfrak{z}} - \frac{p}{2}\left(2b + \frac{1}{3}p\right)i\mathfrak{z} + O(\mathfrak{z}^2) \tag{4.94}$$

$$= \frac{2\sqrt{\pi}}{p}e^{-\pi i/4}\sqrt{\mathfrak{z}} + O(\mathfrak{z}),$$

where the branch of $\sqrt{\mathfrak{z}}$ is chosen so that it is positive for $\mathfrak{z} > 0$.

We integrate this along the line segment parallel to the real axis, say over $[\mathfrak{z}', \mathfrak{z}]$ with $\mathfrak{z} - \mathfrak{z}' = h$. Separating the case $(b, n) = (0, 0)$, we have

$$h + \sum_{\substack{n=-\infty \\ (n,b)\neq(0,0)}}^{\infty} \frac{e^{(b+pn)^2 i\mathfrak{z}}}{i(b+pn)^2} - \sum_{\substack{n=-\infty \\ (n,b)\neq(0,0)}}^{\infty} \frac{e^{(b+pn)^2 i(h'+i\epsilon)}}{i(b+pn)^2} = \frac{2\sqrt{\pi}}{p}e^{-\pi i/4}\sqrt{\mathfrak{z}} + O(\mathfrak{z}^2) \tag{4.95}$$

or

$$T(\mathfrak{z}) - T(i\epsilon) = \frac{2\sqrt{\pi}}{p}e^{-\pi i/4}\sqrt{\mathfrak{z}} - h(1 + o(1)) + O(\mathfrak{z}^2), \tag{4.96}$$

where

$$T(\mathfrak{z}) = T(\mathfrak{z}, b) = \sum_{\substack{n=-\infty \\ (n,b)\neq(0,0)}}^{\infty} \frac{e^{(b+pn)^2 i\mathfrak{z}}}{i(b+pn)^2}. \tag{4.97}$$

For $z \in \mathcal{H} \cup \mathbb{R}$ let

$$F(z) = \sum_{n=1}^{\infty} \frac{e^{\pi i n^2 z}}{\pi i n^2} = \frac{1}{2}\sum_{\substack{n=-\infty \\ n\neq 0}}^{\infty} \frac{e^{\pi i n^2 z}}{\pi i n^2}. \tag{4.98}$$

Then by the decomposition into residue classes, we obtain

$$F\left(\frac{2q}{p} + \mathfrak{z}\right) = \frac{1}{2}\sum_{\substack{n=-\infty \\ n\neq 0}}^{\infty} \frac{e^{\pi i n^2\left(\frac{2q}{p}+\mathfrak{z}\right)}}{\pi i n^2} = \frac{1}{2}\sum_{b=0}^{p-1} e^{2\pi i b^2 \frac{q}{p}} \sum_{\substack{n\equiv b \\ (n,b)\neq(0,0)}} \frac{e^{\pi i n^2 \mathfrak{z}}}{\pi i n^2}$$

$$= \frac{1}{2}\sum_{b=0}^{p-1} e^{2\pi i b^2 \frac{q}{p}}\frac{1}{\pi}T(\pi\mathfrak{z}, b). \tag{4.99}$$

Substituting from (4.97), we deduce that

$$F\left(\frac{2q}{p}+\mathfrak{z}\right) = \frac{1}{2}\sum_{b=0}^{p-1} e^{2\pi i b^2 \frac{q}{p}} \frac{1}{\pi}\left(\frac{2\sqrt{\pi}}{p}e^{-\pi i/4}\sqrt{\pi\mathfrak{z}} - \frac{1}{2}h + T(i\epsilon, b)\right) + O(\mathfrak{z}^2)$$

$$= \frac{1}{2}S(p,q)\left(\frac{2\sqrt{\pi}}{p}e^{-\pi i/4}\sqrt{\mathfrak{z}} - \mathfrak{z}\right) + \frac{1}{2\pi}\sum_{b=0}^{p-1} e^{-2\pi i b^2 \frac{q}{p}} T(i\epsilon, b) + O(\mathfrak{z}^2)$$

$$= S(p,q)\left(\frac{1}{p}e^{-\pi i/4}\sqrt{\mathfrak{z}} - \frac{1}{2}h\right) + F\left(\frac{2q}{p}+i\epsilon\right) + O(\mathfrak{z}^2),$$

where $S(p,q)$ indicates the quadratic Gauss sum defined by

$$S(b,a) = \sum_{j=0}^{b-1} e^{2\pi i j^2 \frac{a}{b}} \tag{4.100}$$

for b a natural number. In general, for a non-zero integral values of b, we define

$$S(b,a) = S(|b|, \operatorname{sgn}(b)a). \tag{4.101}$$

We note that $S(|b|, -a) = \overline{S(|b|, a)}$ and $S(ka, kb) = S(a, b)$.

Theorem 4.7. *For any integers $p > 0$, q we have*

$$F\left(\frac{2q}{p}+\mathfrak{z}\right) - F\left(\frac{2q}{p}+i\epsilon\right) = S(p,q)\frac{e^{-\pi i/4}}{p}\sqrt{\mathfrak{z}} - \frac{1}{2}h + O(\mathfrak{z}^2), \tag{4.102}$$

where for a non-zero integer p, the coefficient is to be understood $S(|p|, \operatorname{sgn}(p)q)$.

In (4.2.2), the variable can be $\frac{2q}{p}+\mathfrak{z}$ and $\frac{2q}{p}+\mathfrak{z}'$ and then instead of h we have $\mathfrak{z} - \mathfrak{z}'$. This will be used in deriving (4.121).

Eq. (4.98) in this form is essentially Theorem 1 of Itatsu [Itatsu (1981)] from which differentiability of Riemann's function can be deduced. Indeed, let $\mathfrak{z} = h + i\epsilon$ and let $\epsilon \to 0+$, in which we have to pay attention to the sign $\operatorname{sgn} h$ of h. Then

$$F\left(\frac{2q}{p}+h\right) - F\left(\frac{2q}{p}\right) = S(p,q)\frac{e^{-\pi i/4\,\operatorname{sgn} h}}{p}\sqrt{|h|} - \frac{1}{2}h + O(h^2). \tag{4.103}$$

Hence differentiability follows only in the case $S(p,q) = 0$ with differential coefficient $-\frac{1}{2}$. This will be done in §4.2.2 appealing to Corollary 4.1 and the formula in [Murty (2004), p. 110]

$$S(p,q) = \left(\frac{q}{p}\right) \varepsilon(p) \sqrt{p} \tag{4.104}$$

valid for $(p,q) = 1$. At the same time this is an elaboration of [Murty (2004), (47)] (on the right-hand side of which the factor $\sqrt{\pi}$ is to be deleted). Arguing as in [Murty (2004)] using the theta transformation formula, we may deduce the Landsberg-Schaar identity, from which the quadratic reciprocity may be deduced.

Remark 4.1. In [Duistermaat (1991), p. 4, $\ell\ell$. 1-2] Duistermaat says that "this selfsimilarity formula was just an integrated version of the well-known transformation formula (4.111)." By this [Duistermaat (1991), Theorem 4.2] is meant. [Duistermaat (1991), (3.4)] reads for an irreducible fraction $r = \frac{q}{p}, p > 0$

$$\mu_\gamma(x) = e^{\frac{\pi}{4}m} p^{-\frac{1}{2}} (x-r)^{-\frac{1}{2}} = e^{\frac{\pi}{4}} p^{-1} S(2p,q)(x-r)^{-\frac{1}{2}}. \tag{4.105}$$

Incorporating this in [Duistermaat (1991), (4.1)], we see that [Duistermaat (1991), (4.1)] refers to the case $S(2p,q)$ of Theorem 4.7. Hence by Corollary 4.1, differentiability of Riemann's function can be read off. We note that [Duistermaat (1991), (3.4)] is already proved by Cauchy [Cauchy (1840), pp. 157-159].

Further, on [Duistermaat (1991), p. 9, ℓ 7 from below], [Murty (2004), (47)] is stated in the form

$$\Theta\left(\frac{2q}{2p} + i\epsilon\right) \sim \frac{1}{p\sqrt{\epsilon}} S(2p,q), \quad \epsilon \to 0+. \tag{4.106}$$

Thus, we could say that [Duistermaat (1991)] gives already material to deduce the reciprocity. [Duistermaat (1991), Theorem 3.4] states that

$$\Theta(z) = \begin{cases} \Theta(\gamma z) e^{\frac{\pi i}{4}p} \left(\frac{-q}{p}\right) p^{-\frac{1}{2}} (z-r)^{-\frac{1}{2}} & p \text{ odd} \\ \Theta(\gamma z) e^{\frac{\pi i}{4}(q+1)} \left(\frac{p}{|q|}\right) p^{-\frac{1}{2}} (z-r)^{-\frac{1}{2}} & q \text{ odd} \end{cases}. \tag{4.107}$$

From (4.107), the reciprocity follows. However, it is used in its proof and unfortunately, this does not lead to the proof of reciprocity law.

4.2.1 *Theta-transformation formula*

For $z \in \mathcal{H}$ let

$$\Theta(z) = \sum_{n=-\infty}^{\infty} e^{\pi i n^2 z} = 1 + 2 \sum_{n=1}^{\infty} e^{\pi i n^2 z} \tag{4.108}$$

and for $\operatorname{Re} s > 0$ let

$$\theta(s) = \Theta(iz) = \sum_{n=-\infty}^{\infty} e^{-\pi n^2 z} = 1 + 2\omega(s), \qquad (4.109)$$

where

$$\omega(s) = \sum_{n=1}^{\infty} e^{-\pi n^2 s}. \qquad (4.110)$$

The **theta-transformation formula** reads

$$\Theta(z) = e^{\frac{\pi i}{4}} z^{-\frac{1}{2}} \Theta\left(-\frac{1}{z}\right). \qquad (4.111)$$

$F(z)$ is essentially the integral of $\Theta(z)$:

$$\int_0^z \theta(-iz)\,\mathrm{d}z = \int_0^z \Theta(z)\,\mathrm{d}z$$

$$= z + 2\left(\sum_{n=1}^{\infty} \frac{e^{\pi i n^2 z}}{\pi i n^2} - \sum_{n=1}^{\infty} \frac{e^{\pi i n^2 z}}{\pi i n^2}\right) = z + 2(F(z) - F(0)).$$

$$(4.112)$$

In particular, for $z = x + u + i\epsilon \in \mathbb{C}$, $\epsilon > 0$, u varies between 0 and h, this reads

$$\int_x^{x+h} \theta(\epsilon - iu)\,\mathrm{d}u = \int_{x+i\epsilon}^{x+h+i\epsilon} \Theta(z)\,\mathrm{d}z = h + 2(F(x+h+i\epsilon) - F(x+i\epsilon)).$$

$$(4.113)$$

The theta-transformation formula (4.111) reads for $y > 0$

$$\theta(y - iu) = e^{\frac{\pi}{4}i} \frac{1}{\sqrt{u+iy}} \theta\left(\frac{i}{u+iy}\right) = e^{\frac{\pi}{4}i} \frac{1}{\sqrt{u+iy}} \sum_{n=-\infty}^{\infty} e^{\frac{i\pi n^2}{u+iy}}. \quad (4.114)$$

By the change of variable

$$\frac{i}{u+i\epsilon} = \frac{i}{x+v+i\epsilon} = \tau + \frac{1}{x}i, \qquad (4.115)$$

i.e. $\tau = \frac{\epsilon - iv}{x(x+v+i\epsilon)} \sim \frac{\epsilon - iv}{x^2}$ the integral in (4.113) becomes

$$\int_x^{x+h} \theta(\epsilon - iu)\,\mathrm{d}u = -ie^{\pi/4} \int_{\frac{i}{x+h+i\epsilon} - \frac{i}{x}}^{\frac{i}{x+i\epsilon} - \frac{i}{x}} \frac{1}{\left(\tau + \frac{1}{x}i\right)^{\frac{3}{2}}} \theta\left(\tau + \frac{1}{x}i\right)\,\mathrm{d}\tau. \quad (4.116)$$

To apply integration by parts, the following is useful which is equivalent to (4.113).

$$\int \theta(\tau + \frac{i}{x})\,\mathrm{d}u = \tau - 2i(F(-\frac{1}{x} + i\tau) + C. \qquad (4.117)$$

$$\int_{\frac{i}{x+h+i\epsilon}-\frac{i}{x}}^{\frac{i}{x+i\epsilon}-\frac{i}{x}} \frac{1}{\left(\tau+\frac{1}{x}i\right)^{\frac{3}{2}}} \theta\left(\tau+\frac{1}{x}i\right) \, d\tau$$

$$= \left[\left(\tau+\frac{i}{x}\right)^{3/2}\left(\tau-2iF\left(-\frac{1}{x}+i\tau\right)\right)\right]_{\frac{i}{x+i\epsilon}-\frac{i}{x}}^{\frac{i}{x+h+i\epsilon}-\frac{i}{x}} - \frac{3}{2}\int \tag{4.118}$$

$$= \left(\frac{x+h+i\epsilon}{i}\right)^{3/2}\left(\frac{i}{x+h+i\epsilon}-\frac{i}{x}-2iF\left(-\frac{1}{x+h+i\epsilon}\right)\right)$$

$$- \left(\frac{x+i\epsilon}{i}\right)^{3/2}\left(\frac{i}{x+i\epsilon}-\frac{i}{x}-2iF\left(-\frac{1}{x+i\epsilon}\right)\right) + O(h).$$

Noting that

$$\frac{-1}{x+h+i\epsilon} = -\frac{1}{x}+\frac{1}{x^2}(\mathfrak{z}(1+o(1))) \tag{4.119}$$

we find that the main term in (4.118) is

$$- 2e^{\frac{\pi}{4}i}\left((x+h+i\epsilon)^{3/2}F\left(-\frac{1}{x+h+i\epsilon}\right)-(x+i\epsilon)^{3/2}F\left(-\frac{1}{x+i\epsilon}\right)\right) \tag{4.120}$$

$$= 2e^{\frac{\pi}{4}i}(x+i\epsilon)^{3/2}\left(F\left(-\frac{1}{x}+\frac{1}{x^2}\mathfrak{z}'\right)-F\left(-\frac{1}{x}+\frac{1}{x^2}\epsilon'\right)\right) + O(h),$$

where $\mathfrak{z}' = \mathfrak{z}(1+o(1))$ and $\epsilon' = \epsilon(1+o(1))$.

Now we specify $x = \frac{2q}{p}$, so that $\frac{1}{x} = \frac{p}{2q}$ and apply Theorem 4.7. (4.120) becomes under this specification,

$$= 2e^{\frac{\pi}{4}i}\left(\frac{2q}{p}+i\epsilon\right)^{3/2}\left(F\left(-\frac{2p}{4q}+\left(\frac{p}{2q}\right)^2\mathfrak{z}'\right)-F\left(-\frac{2p}{4q}+\left(\frac{p}{2q}\right)^2\epsilon'\right)\right) \tag{4.121}$$

$$+ O(h)$$

$$= 2e^{\frac{\pi}{4}i}\left(\frac{2q}{p}+i\epsilon\right)^{3/2}S(4q,-p)e^{-\frac{\pi}{4}i}\frac{1}{4|q|}\left|\frac{p}{2q}\right|\sqrt{\mathfrak{z}'} + O(h)$$

$$= \left(\frac{p}{2|q|}\right)^{1/2}\frac{1}{p}S(4|q|,-\text{sgn}(q)p)\sqrt{\mathfrak{z}'} + O(h),$$

on letting $\epsilon \to 0$.

We must have, correspondingly to [Murty (2004), (52)],

Theorem 4.8. *For $p \in \mathbb{N}, 0 \neq q \in \mathbb{Z}$, we have the reciprocity law*

$$S(p,q) = e^{\frac{\pi}{4} \operatorname{sgn}(q)i} \left(\frac{p}{2|q|} \right)^{1/2} S(4|q|, -\operatorname{sgn}(q)p). \quad (4.122)$$

Corollary 4.1. *Let $x = \frac{q}{p}$ be of the form $\frac{2A+1}{2B+1}$, i.e. p, q both being odd. Then*

$$R(2A+1, 2B+1) = S(2p,q) = 0, \quad (4.123)$$

where R is the coefficient in (4.127).

Proof.

$$\begin{aligned}
S(2p,q) &= e^{\frac{\pi}{4}i} \left(\frac{p}{2|q|} \right)^{1/2} S(4|q|, 2\operatorname{sgn}(q)p) \\
&= e^{\frac{\pi}{2}i} \left(\frac{p}{2|q|} \right)^{1/2} \left(\frac{4|q|}{2|2p|} \right)^{1/2} S(4 \cdot 2p, 2\operatorname{sgn}(q)|q|) \quad (4.124) \\
&= e^{\frac{\pi}{2}i} \frac{p}{|q|} \sqrt{\operatorname{sgn}(q)} S(2p, \operatorname{sgn}(q)|q|).
\end{aligned}$$

Noting that $\operatorname{sgn}(q)|q| = q$, we conclude (4.123). $\qquad \square$

Supposing $p, q > 0$ coprime integers, (4.122) leads to the Landsberg-Schaar identity [Murty (2004), (5)]

$$\frac{1}{\sqrt{p}} \sum_{j=0}^{p-1} e^{2\pi i j^2 \frac{q}{p}} = \frac{e^{\frac{\pi}{4}i}}{\sqrt{2q}} \sum_{j=0}^{2q-1} e^{2\pi i j^2 \frac{p}{2q}}. \quad (4.125)$$

Lemma 4.2. *For a natural number p,*

$$S(p,q) = \varepsilon(p) \left(\frac{q}{p} \right) \sqrt{p}, \quad (4.126)$$

where $\left(\frac{q}{p} \right)$ indicates the Jacobi symbol and $\varepsilon(p) = \begin{cases} 1 & p \equiv 1 \bmod 4 \\ i & p \equiv 3 \bmod 4 \end{cases}$.

We are now ready to state a seemingly more general version of Theorem 4.7, which implies differentiability of Riemann's function at the rational point $\frac{2A+1}{2B+1}$ on putting $\mathfrak{z} = h + i\epsilon$ and $\epsilon \to +0$. Similar and more elaborate result is obtained in [Duistermaat (1991), Theorem 4.2].

Corollary 4.2.

$$F\left(\frac{q}{p}+\mathfrak{z}\right)-F\left(\frac{q}{p}+i\epsilon\right)=R(p,q)\frac{e^{-\pi i/4}}{p}\sqrt{\mathfrak{z}}-\frac{1}{2}h+O(\mathfrak{z}^2), \qquad (4.127)$$

where

$$R(p,2q)=S(p,q)=\varepsilon(p)\left(\frac{q}{p}\right)\sqrt{p},$$

$$R(2p,q)=S(4p,q)=e^{\frac{\pi}{4}i}\sqrt{2p}\left(\frac{-p}{q}\right) \qquad (4.128)$$

$$R(2B+1,2A+1)=0.$$

Proof. By Corollary 4.1, only the case $R(2p,q)$ case needs to be considered. By (4.122),

$$R(2p,q)=S(4p,q)=e^{\frac{\pi}{4}i}\left(\frac{4p}{2|q|}\right)^{1/2}S(4|q|,-4\operatorname{sgn}(q)p)$$

$$=e^{\frac{\pi}{4}i}\left(\frac{2p}{|q|}\right)^{1/2}S(|q|,-\operatorname{sgn}(q)p)$$

$$=e^{\frac{\pi}{4}i}\left(\frac{2p}{|q|}\right)^{1/2}\sqrt{|q|}\varepsilon(|q|)\left(\frac{-\operatorname{sgn}(q)p}{|q|}\right) \qquad (4.129)$$

$$=e^{\frac{\pi}{4}i}\sqrt{2p}\left(\frac{-p}{q}\right).$$

$$\square$$

4.2.2   Remarks

It seems that Itatsu's main formula is not correct.

$$\int_0^x \theta(y-iu)\,du=e^{\frac{\pi}{4}i}\left[2\sqrt{u+iy}+2(u+iy)^{3/2}F\left(\frac{-1}{u+iy}\right)\right]_0^x$$
$$\qquad -\frac{3}{2}e^{\frac{\pi}{4}i}\int_0^x(u+iy)^{1/2}F\left(\frac{-1}{u+iy}\right)du. \qquad (4.130)$$

For he seems to have used the wrong formula

$$\int\left(2\theta\left(\frac{i}{u+iy}\right)+1\right)du=2(u+iy)^2F\left(\frac{-1}{u+iy}\right)+u, \qquad (4.131)$$

which should be

$$\int\left(2\theta\left(\frac{i}{u+iy}\right)+1\right)du=2(u+iy)^2F\left(\frac{-1}{u+iy}\right)$$
$$+u-2\int(u+iy)^3F\left(\frac{-1}{u+iy}\right)du. \qquad (4.132)$$

Using (4.130), we obtain

$$\int_0^x \theta(y - iu)\, du = e^{\frac{\pi}{4}i}\left[\frac{1}{\sqrt{u + iy}}\left(2(u + iy)^2 F\left(\frac{-1}{u + iy}\right) + u\right)\right]_0^x$$

$$(4.133)$$

$$- e^{\frac{\pi}{4}i}\int_0^x \frac{-1}{2}(u + iy)^{-3/2}\left(2(u + iy)^2 F\left(\frac{-1}{u + iy}\right) + u\right) du.$$

The last integral on the right of (4.133) becomes

$$e^{\frac{\pi}{4}i}\int_0^x (u + iy)^{1/2} F\left(\frac{-1}{u + iy}\right) du + \frac{1}{2}e^{\frac{\pi}{4}i}\int_0^x (u + iy)^{-3/2} u\, du$$

$$= \cdots + \left[-\frac{u}{\sqrt{u + iy}} + 2\sqrt{u + iy}\right]_0^x.$$

$$(4.134)$$

Substituting (4.134) in (4.133), we obtain

$$\int_0^x \theta(y - iu)\, du = \left[2\sqrt{u + iy} + 2(u + iy)^{3/2} F\left(\frac{-1}{u + iy}\right)\right]_0^x \qquad (4.135)$$

$$+ e^{\frac{\pi}{4}i}\int_0^x (u + iy)^{1/2} F\left(\frac{-1}{u + iy}\right) du,$$

which is not totally in conformity with (4.130).

Remark 4.2. We make a historical remark on Riemann's function. [Butzer and Stark (1986)] contains an almost complete list of references up to 1986. One addition is a correction to [Smith (1972)] in 1983. After this, the review of [Gerver (2003)] contains an almost complete list after [Butzer and Stark (1986)] except for [Meyer (1994)] (esp. 619) and [Ulrich (1997)]. Among the papers listed in the review of [Gerver (2003)], we mention [Holschneider (1991)] and [Jaffard (1996)] for consideration from the point of wavelets and [Duistermaat (1991)] for self-similarity.

Hecke

Chapter 5

Hilbert space and number theory

Abstract: In this chapter, we shall present applications of Hilbert space theory to number theory. To enhance easy understanding of the contents, we shall provide basics of the theory in §5.1 depending mostly on [Yoshida (1967)]. In §5.2 we keep the contents in §3.4, **I** of Romanoff's (and Wintner's [Wintner (1944)]) theory in a slightly expanded form. §5.3 is the core of the chapter, presenting an accessible and partial introduction to automorphic L-functions by way of the **Kuznetsov trace formula**. There are a few different statements of the formula. In literature [Baker (2003)], [Miatello (1990)], [Venkov (1990)], the form stated as the Kuznetsov sum formula is the reversed form according to Iwaniec. Thus comparing these different versions in a more unified way will help the reader to comprehend the situation and at the same time it will pave a road to the coming book by Liu [Liu]. The most comprehensive one for a Fuchsian group of the first kind is given in Iwaniec [Iwaniec (1995)]. In Motohashi [Motohashi (1997)], the case of the full modular group is treated and the sum formula is the reversed form. In Iwaniec [Iwaniec (1995)], the reversed Kuznetsov formula is stated in the form of two complementary theorems, Theorem 5.27 and Theorem 5.28 below, the former the Titchmarsh series and the latter Neumann series. The full version of the Kuznetsov sum formula is Theorem 5.29 below which follows immediately from these two theorems, where (1) functional equation for the weighted Kloostermann zeta for the former and the inversion of the Hankel, Titchmarsh transforms and (2) spectral expansion of the Kloostermann zeta.

Book	name	name for the reversed	proof
Baker [Baker (2003)]	trace formula	-	-
Iwa [Iwaniec (1995)]	Bruggemnn-Kuznetsov	Kuznetsov reversed	(1)
Moto [Motohashi (1997)]	trace formula	sum formula	(2)
Venk [Motohashi (1997)]	trace formula	-	-

Table 5.1. Kuznetsov's formula

We shall show that the statement of Motohashi is another version of Theorem 5.29 and that it has a different outlook since the Neumann series part remains inexplicit. Since in Iwaniec, Theorem 5.29 is deduced by reversing Theorem 5.24 and we slightly generalize the method of Selberg adopted by Motohashi [Motohashi (1997)] of **equating the two different expressions for the inner product of two Poincaré series** to prove Theorem 5.29. The guiding principle is the philosophy of Iwaneic to the effect that Kuznetsov sum formula is a sort of the Poisson sum formula, which is equivalent to the functional equation. Indeed, the Bruggeman-Kuznetsov theorem, Theorem 5.29 is most naturally deduced from the functional equation for the Kloosterman sum zeta function with Bessel function weight. We try to elucidate this zeta-symmetry aspect of the Kuznetsov sum formula in Theorem 5.29 as the Sears-Titchmarsh inversion in the light of the Hankel inversion. *G*-functions give rise to the clarification of the proofs and statement of the results.

5.1   Hilbert spaces

There is an enormous amount of excellent books on the theory of Hilbert spaces and the interested reader can consult them. What is given in this section is a very brief summary of the theory which is meant to give a feeling to beginners so as to go on reading topics using the theory. For an excellent exposition of the theory of Hilbert spaces, cf. e.g. [Yoshida (1967)] and [Halmos (1950)]. Our exposition follows closely the former book with slight deviations and modifications.

A (complex) **pre-Hilbert space** is a vector space over the complex numbers $\mathbb{C}$ with inner product defined below.

A (complex) **Hilbert space** H is a pre-Hilbert space such that it is **complete**, i.e. every Cauchy sequence in H ((5.9) below) is convergent with respect to the distance arising from the inner product.

The **inner product** of $a, b \in H$ is denoted (a, b) and it satisfies for $a, b, c \in H$ and $\lambda \in \mathbb{C}$ the following
(i) $(a, a) \geq 0$ and $(a, a) = 0 \iff a = 0$.
(ii) $(b, a) = \overline{(a, b)}$ the bar meaning the complex conjugate.
(iii) (linearity in the first factor) $(a + b, c) = (a, b) + (a, c)$, $(\lambda a, b) = \lambda(a, b)$.
It follows that $(a, b + c) = (a, b) + (a, c)$ and $(a, \lambda b) = \bar{\lambda}(a, b)$.

Exercise 25. Let

$$\|a\| = \sqrt{(a, a)}. \tag{5.1}$$

Then prove that this satisfies the properties of **norm**:
(i) $\|a\| \geq 0$ and $\|a\| = 0 \iff a = 0$.
(ii) $\|\lambda a\| = |\lambda|\|a\|$.
(iii) (triangular inequality) $\|a + b\| \leq \|a\| + \|b\|$.

Solution. To prove (iii) we first need to prove the **Cauchy-Schwarz inequality**

$$|(a, b) \leq \|a\|\|b\|. \tag{5.2}$$

If either of a or b is 0, (5.2) is clear. Supposing $a \neq 0$ we consider the quadratic expression in t:

$$f(t) = \|ta + b\|^2 = \|a\|^2 t^2 + 2\operatorname{Re}(a, b)t + \|b\|^2. \tag{5.3}$$

Since this is non-negative, its discriminant is non-negative, implying (5.2). Equality holds when $a + tb = 0$, i.e. when a and b are parallel.

Now (5.3) with $t = 1$ reads

$$\|a + b\|^2 = \|a\|^2 + 2\operatorname{Re}(a, b) + \|b\|\|^2 \leq \|a\|^2 + 2\|a\|\|b\| + \|b\|^2 = (\|\|a\| + \|b\|)^2, \tag{5.4}$$

whence (iii).

Exercise 26. Let

$$d(a, b) = \|a - b\|. \tag{5.5}$$

Then prove that d satisfies the axioms of distance
(i) $d(a, b) \geq 0$ and $d(a, b) = 0 \iff a = b$.
(ii) $d(b, a) = d(a, b)$.
(iii) (triangular inequality) $d(a, b) \leq d(a, c) + d(c, b)$.

Definition 5.1. We may speak of topological properties of a Hilbert space as a metric space with metric d.

If the sequence $\{a_n\} \subset H$ satisfies

$$\lim_{n \to \infty} \|a_n - a\| = 0, \tag{5.6}$$

we write $\lim_{n \to \infty} a_n = a$ (strong) or $a_n \to a$ (strong) and say that $\{a_n\}$ converges (or is convergent) **strongly** to a or simply converges when it is clear that convergence is with respect to norm. On the other hand, a sequence $\{y_n\}$ is said to be **weakly** convergent if

$$\lim_{n \to \infty} (x, y_n) = (x, y) \tag{5.7}$$

for all $x \in H$ and

$$\|y\| \leq \liminf_{n \to \infty} \|y_n\| \leq \limsup_{n \to \infty} \|y_n\| < \infty, \tag{5.8}$$

denoted $y_n \to y$ (weak).

Exercise 27. Prove that a closed subspace of a Hilbert space is a Hilbert space.

Solution. It suffices to prove completeness. Given a Cauchy sequence $\{a_n\}$ in a subspace K, it is convergent to $a \in H$. Since a is an accumulation point of $\{a_n\}$ and so it belongs to K, i.e. completeness follows.

Exercise 28. Prove that a convergent sequence satisfies the Cauchy condition

$$\lim_{m,n \to \infty} \|a_m - a_n\| = 0 \tag{5.9}$$

and it is bounded.

Solution. (5.9) follows from

$$\|a_m - a_n\| \leq \|a_m - a\| + \|a - a_n\|$$

and boundedness follows from

$$\|a_n\| \leq \|a_n - a + a\| \leq \|a_n - a\| + \|a\|.$$

Proposition 5.1. *The inner product* (x, y) *is a continuous function in* x, y.

Proof. Suppose $x_n \to x$ and $y_n \to y$. Then from the triangular inequality

$$\|(x_n, y_n) - (x, y)\| \leq \|(x_n - x, y_n) + (x, y_n - y)\| \leq \|x_n - x\| \|y_n\| + \|y\| \|y_n - y\|.$$

By boundedness of $\{y_n\}$, the right-hand side of the above inequality can be made as small as we please.   □

This proposition will be applied very often in the sequel without specific mentioning.

Example 5.1. A (complex-valued Lebesuge) measurable function $f(t)$ defined on an interval $(a, b) \subset \mathbb{R}$, finite or infinite, such that $|f(t)|^2$ is integrable in the Lebesgue sense is called a **square integrable function**. The set of all square integrable functions on (a, b) forms a Hilbert space $L^2(a, b)$ with addition, scalar multiplication, and inner product defined respectively by

$$(f + g)(t) = f(t) + g(t), \quad (\lambda f)(t) = \lambda f(t), \quad (f, g) = \int_a^b f(t) \bar{g}(t) \, dt \tag{5.10}$$

for $f, g \in L^2(a, b)$ and $\lambda \in \mathbb{C}$, where two functions f, g are equal if they coincide **almost everywhere** (a.e.), i.e. their values coincide except for measure zero sets. Hence the elements in $L^2(a, b)$ are equivalence classes.

That $f + g \in L^2(a, b)$ follows from $\|f + g\|^2 \leq 4(\|f\|^2 + \|g\|^2)$ and the existence of the inner product integral follows from $\|f\bar{g}\| \leq \frac{1}{2}(\|f\|^2 + \|g\|^2)$. Completeness may be proved in a more general situation, which we state as Theorem 5.1.

Definition 5.2. For $p \geq 1$, let L^p denote the linear space of all pth **power integrable functions** with p-**norm** defined by

$$\|f\|_p = \left(\int_0^\infty \|f(t)\|^p \, dt \right)^{\frac{1}{p}}, \tag{5.11}$$

where $\|f(t)\|$ is any Euclidean norm in Definition 6.2.

A complete space with respect to norm is called a **Banach space**.

Theorem 5.1. *The space L^p of all pth power integrable functions forms a Banach space, and a fortiori, L^2 is a complete space w.r.to the norm induced by inner product, i.e. a Hilbert space.*

Suppose $\{f_n\}$ is a Cauchy sequence in L^p. We may choose ν_1 so that for $n > \nu_1$ we have

$$\|f_n - f_{\nu_1}\|_p < 2^{-1}$$

and then choose $n\nu_2 > \nu_1$ so that for $n > n\nu_2$ we have

$$\|f_n - f_{\nu_2}\|_p < 2^{-2}$$

and so on. Thus we may find a strictly increasing subsequence $\{\nu_k\}$ such that for $n > \nu_k$ we have

$$\|f_n - f_{\nu_k}\|_p < 2^{-k}$$

and in particular

$$\|f_{\nu_{k+1}} - f_{\nu_k}\|_p < 2^{-k}, \tag{5.12}$$

whence

$$\sum_{k=1}^\infty \|f_{\nu_{k+1}} - f_{\nu_k}\|_p \leq 1. \tag{5.13}$$

We write f_{ν_m} as a telescoping series $(l < m)$

$$f_{\nu_m} = f_{\nu_l} + \sum_{k=l}^{m-1} (f_{\nu_{k+1}} - f_{\nu_k}) \tag{5.14}$$

and consider the majorant (with $l = 1$)

$$g_{\nu_m} = |f_{\nu_l}| + \sum_{k=1}^{m-1} |f_{\nu_{k+1}} - f_{\nu_k}| (\geq |f_{\nu_m}|) \tag{5.15}$$

which is a monotone increasing sequence $\in L^p$. By the triangular inequality and (5.13)

$$\|g_{\nu_m}\|_p \leq \|f_{\nu_l}\|_p + \sum_{k=1}^{m-1} \|f_{\nu_{k+1}} - f_{\nu_k}\|_p \leq \|f_{\nu_l}\|_p + 1 \qquad (5.16)$$

a.e., and *a fortiori*, bounded a.e., so that the sequence $\{g_{\nu_m}\}$ is convergent to $g(x)$, say. Hence f_{n_m} in (5.14) is absolutely convergent as $m \to \infty$ to $f_\infty(x)$ a.e. By one of Lebesgue's theorems on termwise integration and (5.16), we have

$$\|g(x)\|_p^p = \int \lim_{m \to \infty} g_{\nu_m}(x)^p \, dx = \lim_{m \to \infty} \int g_{\nu_m}(x)^p \, dx = \lim_{m \to \infty} \|g_{\nu_m}\|_p^p$$

$$\qquad (5.17)$$

$$\leq (\|f_{\nu_l}\|_p + 1)^p.$$

Hence $g \in L^p$. Since $|f_\infty| \leq g$, $f_\infty \in L^p$. Now letting $m \to \infty$ in (5.14), we have

$$f_\infty - f_{\nu_l} = \sum_{k=l}^{\infty} (f_{\nu_{k+1}} - f_{\nu_k}). \qquad (5.18)$$

Hence

$$\|f_\infty - f_{\nu_k}\|_p \leq \sum_{j=k}^{\infty} \|f_{\nu_{j+1}} - f_{\nu_j}\|_p \to 0$$

as $k \to \infty$, whence

$$\lim_{k \to \infty} f_{\nu_k} = f_\infty.$$

From this we may conclude that

$$\limsup_{k \to \infty} \|f_\infty - f_n\|_p \leq \lim_{k \to \infty} \|f_\infty - f_{\nu_k}\|_p + \lim_{k \to \infty} \|f_{\nu_k} - f_n\|_p \to 0, \quad (5.19)$$

i.e. the sequence $\{f_n\}$ is convergent.

Example 5.2. Let l^2 denote the set of all square-summable sequences, i.e. complex sequences $x = \{x_n\}$ such that

$$\sum_{n=1}^{\infty} |x_n|^2 < \infty. \qquad (5.20)$$

Then l^2 forms a Hilbert space with addition, scalar multiplication and inner product defined respectively by

$$x + y = \{x_n + y_n\}, \quad \lambda x = \{\lambda x_n\}, \quad (x, y) = \sum_{n=1}^{\infty} x_n \bar{y}_n \qquad (5.21)$$

for $x = \{x_n\}, y = \{y_n\} \in l^2$.

Example 5.3. The Cartesian product $H \times H = \{\{x, y\} | x, y \in H\}$ of a Hilbert space H is again a Hilbert space with respect to componentwise addition, scalar multiplication and the inner product defined by

$$(\{x_1, y_1\}, \{x_2, y_2\}) = (x_1, y_1) + (x_2, y_2). \tag{5.22}$$

$H \times H$ is called the **direct product** of H.

Definition 5.3. A Hilbert space H is called **separable** if there exists an at most countable set $A = \{a_k\}$ which is **dense** in H, i.e. $\bar{A} = H$, where $\bar{A}$ means the **closure** of a set A. I.e. each element may be expressed as the limit of a subsequence of A.

$x, y \in H$ are called orthogonal if $(x, y) = 0$. A system $\{x_\lambda\}$ in a Hilbert space is called an orthonormal system (**ONS**) if $(x_\lambda, x_\mu) = \delta_{\lambda\mu}$ with *delta* the Kronecker delta. An ONS $\{x_\lambda | \lambda \in \Lambda\}$ is called **complete** if $(f, x_\lambda) = 0$ for all $\lambda \in \Lambda$ implies $f = 0$.

Theorem 5.2. *In a separable Hilbert space there exists an at most countable complete ONS and we may speak of Fourier analysis: Let $\{\varphi_n\}$ be an ONS. Then for any $f \in H$ we have the Fourier expansion*

$$f = \sum_{n=1}^{\infty} (f, \varphi_n) \varphi_n \tag{5.23}$$

and the **Bessel inequality** *holds*

$$\|f\|^2 \geq \sum_{n=1}^{\infty} \|(f, \varphi_n)\|^2. \tag{5.24}$$

If $\{\varphi_n\}$ is a complete ONS, then we have the **Parseval identity**

$$\|f\|^2 = \sum_{n=1}^{\infty} \|(f, \varphi_n)\|^2. \tag{5.25}$$

The Parseval identity holds true if and only if the system be complete.

Definition 5.4. Let H be a Hilbert space and let D be its subspace. A linear mapping T from D to another Hilbert space H_1 is called an **additive operator** or a linear operator and $D = D(T)$ and $W = \{y | y = Tx, x \in D\} = W(T)$ are called the domain and region (Wertevorrat), respectively. I.e. for $x, y \in D$ and $\lambda, \mu \in \mathbb{C}$ we have

$$T(\lambda x + \mu y) = \lambda Tx + \mu Ty. \tag{5.26}$$

If $H_1 = \mathbb{C}$, an additive operator is often called a **linear functional**. An additive operator is continuous if $x_n, x \in D(T)$ and $x_n \to x$, then $Tx_n \to Tx$. If an operator is a one-to-one and onto mapping from $D(T)$ to $W(T)$, the inverse mapping T^{-1} exists such that

$$D(T^{-1}) = W(T), \quad W(T^{-1}) = D(T)$$

which is an additive operator called the **inverse** operator of T.

Exercise 29. Prove the following. For an additive operator to be continuous it is necessary and sufficient that there exists a constant $c > 0$ such that

$$\|Tx\| \le c\|x\| \tag{5.27}$$

for all $x \in D(T)$. If $D(T) = H$, then continuity of T is equivalent to the boundedness of $\|Tx\|$ over the unit ball $\|x\| \le 1$.

Solution. By linearity, $T0 = 0$. To prove necessity, suppose there exists a sequence $\{x_n\}$ such that $\|Tx_n\| > n\|x_n\|$. Then $x_n \ne 0$ and we may define a new sequence $y_n = \frac{1}{\sqrt{n}\|x_n\|}x_n$. Then $\|y_n\| = \frac{1}{\sqrt{n}} \to 0$ while $Ty_n = \frac{1}{\sqrt{n}\|x_n\|}\|Tx_n\| > \sqrt{n}$, which contradicts continuity of T at 0. Sufficiency follows from $\|Tx_n - Tx\| = \|T(x_n - x)\| \le c\|x_n - x\|$.

From Exercise 29 we make

Definition 5.5. We call a continuous additive operator with $D(T) = H$ a **bounded operator** and define its norm by

$$\|T\| = \inf\left\{ c \,\middle|\, c \ge \frac{\|Tx\|}{\|x\|}, x \ne 0 \right\} \tag{5.28}$$

in the light of (5.27). Addition, scalar multiplication of operators are defined in a similar way as with linear mappings. The multiplication is defined as a composition of operators.

Remark 5.1. From (5.28), we have

$$\|Tx\| \le \|T\|\|x\| \tag{5.29}$$

for all $x \in H$, which we will use freely without further notice.

Exercise 30. Prove that the norm $\|T\|$ can be also defined as

$$\|T\| = \sup_{\|x\| \le 1} \|Tx\|. \tag{5.30}$$

Cf. the last passage in Exercise 29.

Definition 5.6. A linear subspace of a Hilbert space is called a **closed subspace** if it is closed with respect to norm convergence. The set of all orthogonal elements to a subset A is denoted $A^\perp$ called the **orthogonal complement**:

$$A^\perp = \{y : y \perp x, x \in A\}. \tag{5.31}$$

Exercise 31. Show that $A^\perp$ is a closed subspace.

Theorem 5.3. *Given a closed subspace A of a Hilbert space H, every $x \in H$ is decomposed uniquely as*

$$x = y + z, \quad y \in A, z \in A^\perp. \tag{5.32}$$

Definition 5.7. The component y in (5.32) is called the **projector** of x on A and denoted $y = P(A)x$. A bounded additive operator P is called a **projector** if it satisfies
(i) (idempotency) $P^2 = P$.
(ii) (symmetry) $(Px, y) = (x, Py)$.

Theorem 5.4. *A projector P of a Hilbert space is given as the projector $P(A)$ with a closed subspace A.*

Proof. Suppose $W(P) = A$ and let us prove that A is a closed subspace and $P = P(A)$. Note that $x \in A$ and $x = Px$ are equivalent. For $x \in A$ is given as $x = Py$, $y \in A$. Hence $x = Py = P^2 y = Px$. That A is closed follows from the following. If $A \supset \{x_n\} \to y$, then P is continuous, P being bounded (Exercise 29), whence $y \leftarrow \{x_n\} = \{Px_n\} \to Py$ and $y = Py \in A$.

Note that (5.32) may be expressed as

$$P(A)x + P(A^\perp)x = x \tag{5.33}$$

and $P(A)x \perp P(A^\perp)y$ and that $P(A)P(A^\perp)x = 0$ for any $x \in H$.

Now we prove $P = P(A)$, where $W(P) = A$. If $x \in A$, then $Px = x = P(A)x$ while if $x \in A^\perp$, then $P(A)x = 0$ and $\|Px\|^2 = (Px, Px) = (x, PPx) = (x, Px) = 0$, the last equality because $Px \in A$ being orthogonal to x, whence $Px = 0$. Linearity of P and $P(A)$ together with (5.34) proves their equality.

Conversely, we show that $P(A)$ satisfies conditions in Definition 5.7. Boundedness follows from

$$\|P(A)x\|^2 \leq \|x\|^2. \tag{5.34}$$

$\square$

Exercise 32. Prove that $P(A)$ satisfies conditions in Definition 5.7.

Solution. (i) is true by definition. To prove idempotency and (5.34) we consider

$$(x, y) = (P(A)x + P(A^\perp)x, P(A)y + P(A^\perp)y) \qquad (5.35)$$
$$= (P(A)x, P(A)y) + (P(A^\perp)x, P(A^\perp)y).$$

With $y = x$, this leads to

$$\|x\|^2 = \|P(A)x\|^2 + \|P(A^\perp)x\|^2,$$

whence (5.34) follows. With y replaced by $P(A)y$, (5.34) reads

$$(x, P(A)y) = (P(A)x, P(A)^2 y) + (P(A^\perp)x, P(A^\perp)P(A)y) = (P(A)x, P(A)y) \qquad (5.36)$$

$$= (P(A)x, P(A)y + P(A^\perp)y) = (P(A)x, y),$$

i.e (ii).

Remark 5.2. From Definition 5.7, we obtain

$$(Px, x) = (PPx, x). = (Px, Px) = \|Px\|^2. \qquad (5.37)$$

This and the inequality (5.34) for a projector will be used often in what follows without notice.

Theorem 5.5. *Every bounded sequence in a Hilbert space contains a weakly convergent subsequence.*

Proof. Suppose $\{t_n\} \subset H$ is bounded, $\|t_n\| \leq c$, $c > 0$, say. Let A denote the set of all linear combinations $\sum_{k=1}^m \alpha_k t_k$ together with all of their accumulation points (in the sense of strong convergence), where $\alpha_k \in \mathbb{C}$ and $m \in \mathbb{N}$ are arbitrary. Then A is a Hilbert space as a closed subspace of H. Those linear combinations $A_1 \subset A$ with coefficients α_k whose real and imaginary parts are rational numbers from a countable dense subset. Hence A is separable. Denoting this set A_1 as a sequence $\{y_m\}$, we consider the double sequence (y_m, t_n) which is bounded for each m since $|(y_m, t_n)| \leq c\|y_m\|$. Hence by Lemma 5.1 below we may choose a subsequence (y_m, t_{ν_n}) such that

$$\lim_{n \to \infty} (y_m, t_{\nu_n}) = (y_m, t) \qquad (5.38)$$

for each y_m. Since $\{y_m\}$ is dense, it follows that given $\varepsilon > 0$, for any $y \in A$ there exists an $m_0 \in \mathbb{N}$ such that $|y - t_{m_0}| < \varepsilon$. Hence it follows that

$$|(y, t_{\nu_k}) - (y, t_{\nu_l}|$$ (5.39)
$$\leq \|(y, t_{\nu_k}) - (y_{m_0}, t_{\nu_k})\| + \|(y_{m_0}, t_{\nu_k}) - (y_{m_0}, t_{\nu_l})\| + \|(y_{m_0}, t_{\nu_l}) - (y, t_{\nu_l})\|$$
$$\leq c\|y - y_{m_0}\| + \|(y_{m_0}, t_{\nu_k}) - (y_{m_0}, t_{\nu_l})\| + c\|y - y_{m_0}\|.$$

The mid term on the far right-hand side is $< \varepsilon$ by (5.38). Hence (y, t_{ν_k}) is a Cauchy sequence and so it converges. Now for any $z \in H$ let $P(A)z = y$, where $P(A)$ is the projection. Then since $t_{\nu_k} \in A$, we have

$$(z, t_{\nu_k}) = (z, P(A)t_{\nu_k}) = (P(A)z, t_{\nu_k}) = (y, t_{\nu_k}).$$ (5.40)

Hence $t_{\nu_k} \to t$ (weak). $\qquad\square$

Lemma 5.1. (Cantor diagonal argument) *Suppose the double sequence $\{a_{mn}\}$ is bounded. Then one can choose a strictly increasing subsequence $\{\nu_n\} \subset \mathbb{N}$ such that $\{a_{m\nu_n}\}$ is convergent for each m.*

Proof. By the Bolzano-Weierstrass theorem, one can choose a subsequence $\{\nu_{1n}\}$ of $\{n\}$ so that $\{a_{m\nu_{1n}}\}$ is convergent for each m. Inductively one can choose a subsequence $\{\nu_{mn}\}$ of $\{\nu_{m-1,n}\}$ such that $\{a_{m\nu_{mn}}\}$ is convergent for each m. Now put $\nu_n = \nu_{nn}$. Then save for a finite number of terms, $\{\nu_n\}$ is a subsequence of $\{\nu_{mn}\}$ and so $\{a_{m\nu_n}\}$ is convergent for each m. $\quad\square$

The following theorem of F. Riesz plays a fundamental role in the theory of Hilbert spaces asserting that a continuous linear functional on a Hilbert space is given in the form of an inner product.

Theorem 5.6. (F. Riesz) *Let H be a Hilbert space and f a bounded linear functional such that $D(f) = H$. Then there exists a y_f such that*

$$f(x) = (x, y_f).$$ (5.41)

Conversely, $f(x) = (x, y)$ is a bounded linear functional for any $y \in H$.

Definition 5.8. Given a bounded operator T on H such that $D(T) = W(T) = H$, the unique bounded operator T^* such that

$$(Tx, y) = (x, T^*y)$$ (5.42)

for all $x, y \in H$ is called the **conjugate operator** of T.

An additive operator such that $D(T) = W(T) = H$ and satisfies the isometry condition is called a **unitary** operator:

$$\|Tx\| = \|T\|.$$ (5.43)

Theorem 5.7. *The conjugate operator uniquely exists and satisfies*

$$(T^*)^* = T \tag{5.44}$$

and

$$\|T^*\| = \|T\|. \tag{5.45}$$

Proof. $F_y(x) := (Tx, y)$ is a continuous additive functional. For by familiar inequalities,

$$\|F_y(x)\| = \|(Tx, y)\| \le \|Tx\| \|y\| \le \|T\| \|y\| \|x\| \tag{5.46}$$

whence continuity follows. By Theorem 5.6, there exists a unique y^* satisfying (5.43). Putting $T^*y = y^*$, we see that T^* is an additive operator and that it is a continuous functional. The latter follows from

$$\|(x, T^*y)\| = \|(Tx, y)\| \le \|Tx\| \|y\| \le \|T\| \|x\| \|y\| \tag{5.47}$$

which is a consequence of (5.46).   □

Exercise 33. Prove (5.45).

Solution. Putting $x = T^*y$ in (5.47), we derive that

$$\|T^*y\|^2 \le \|T\| \|T^*y\| \|y\|,$$

whence $\|T^*\| = \|T\|$. By (5.44), the inequality is deduced, proving the assertion.

Definition 5.9. For an additive operator T on H, the subset of the direct product in Example 5.3 is called its **graph**

$$G(T) = \{\{x, Tx\} | x \in D(T)\}. \tag{5.48}$$

An additive operator T' is called an **extension** of an additive operator T and denoted $T' \supset T$ if $D(T') \supset D(T)$ and $T^*|_{D(T)} = T'$, i.e. $Tx = T'x$ for each $x \in D(T)$. $T' \supset T$ is equivalent to $G(T') \supset G(T)$.

The **skew commutator** V is defined as the additive operator $V : H \times H \to H \times H$ satisfying

$$V\{x, y\} = \{-y, x\}. \tag{5.49}$$

Theorem 5.8. *Let T be an additive operator of H. That $VG(T)^{\perp}$ is a graph of an additive operator T^* of H is equivalent to $\overline{D(T)} = H$.*

For an additive operator satisfying the condition in Theorem 5.8, the operator T^* exists such that $VG(T)^\perp = G(T^*)$, called the **conjugate** operator. Since T^* maps $x \in D(T)$ to y^* such that

$$(Tx, y) = (x, y^*),$$

we see that the definition of the conjugate operator in Definition 5.8 remains true. Hereafter we mean the conjugate operator in this wider sense.

Definition 5.10. An additive operator T is called a **symmetric** operator if there exists the conjugate T^* such that $T^* \supset T$.

An additive operator T is said to be **self-adjoint** or hermitian (cf. §2.4.1) if $T = T^*$ holds true.

A symmetric operator T is said to be **lower semi-bounded** if there exists a positive constant c such that for all $x \in D(T)$, the lower bound

$$(Tx, x) \geq c\|x\|^2 \tag{5.50}$$

is true. If the reverse inequality holds, it is called upper semi-bounded.

Lemma 5.2. *A symmetric operator T on H with $D(T) = H$ is a bounded self-adjoint operator. A bounded symmetric operator is self-adjoint.*

Proof. Since $T = T^*|_{D(T)} = T^*|_H = T^*$, T is self-adjoint. Boundedness follows from the fact that a closed operator T on H with $D(T) = H$ is bounded and that $T^* = T$ is closed.

Regarding the second passage, recall from Definition 5.5 that for a bounded operator T, $D(T) = H$. Hence together with $T \subset T^*$, the preceding argument applies. $\square$

Theorem 5.9. *A symmetric operator T on H with $W(T) = H$ is self-adjoint.*

Proof. Note that such an operator is one-to-one. For suppose $Tx = 0$. Then for every $y \in D(T)$, we have $0 = (Tx, y) = (x, Ty)$. Since $W(T) \; H$, there is a $y \in H$ such that $Ty = x$, whence $\|x\|^2 = 0$, $x = 0$. Hence it has the inverse operator. Since $(T^{-1}x, y) = (T^{-1}x, TT^{-1}y) = (TT^{-1}x, T^{-1}y) = (x, T^{-1}y)$, T^{-1} is symmetric and $D(T^{-1}) = W(T) = H$. By Lemma 5.2, T^{-1} is self-adjoint. Hence by Exercise 34, $(T^{-1})^{-1} = T$ is self-adjoint. $\square$

Exercise 34. Show that if a self-adjoint operator T on H has the inverse operator T^{-1}, then T^{-1} is also self-adjoint.

For a symmetric operator with domain $D(T)$, we have

$$T^*x = Tx, \quad x \in D(T)$$

and *a fortiori*, by (5.43)

$$(Tx, y) = (x, T^*y) = (x, Ty) \quad \text{for} \quad x, y \in D(T). \tag{5.51}$$

Theorem 5.10. (K. Friedrichs) *A lower [upper] semi-bounded symmetric operator T can be extended to a self-adjoint operator.*

Proof. As in §2.4, **I**, every metric space has its completion and *a fortiori* every pre-Hilbert space H_0 has its completion, i.e. there exists a Hilbert space $\tilde{H}$ which has a dense subspace $\tilde{H}_0$ such that H_0 is isomorphic to $\tilde{H}_0$ as pre-Hilbert spaces.

We introduce a new inner product $(x, y)_0$ in $D(T)$ by (recall (5.43))

$$(x, y)_0 = (Tx, x) \tag{5.52}$$

and let $\widetilde{D(T)}$ be the completion of $D(T)$. For every Cauchy sequence $\widetilde{D(Y)} \ni \tilde{x} = \{x_n\} \subset D(T)$ we have by (5.50)

$$\|x_n - x_m\|^2 \le c(T(x_n - x_m), x_n - x_m) = c\|x_n - x_m\|_0^2 \to 0 \tag{5.53}$$

as $m, n \to \infty$. Hence it is a Cauchy sequence in H and there is an $x \in H$ such that

$$\lim_{n \to \infty} \|x_n - x\| = 0. \tag{5.54}$$

The correspondence

$$\widetilde{D(Y)} \to H; \quad \tilde{x} \mapsto x \tag{5.55}$$

is one to one. For if $\{y_n\} \subset D(T)$ is another sequence converging to x, then letting $z_n = x_n - y_n$ and using the continuity of the inner product in $\tilde{T}(D)$, we have

$$\lim_{n \to \infty} \|z_n\|^2 = \lim_{n \to \infty} (z_n, z_n)_0 = \lim_{m, n \to \infty} (z_m, z_n)_0 = \lim_{m, n \to \infty} (Tz_m, z_n) \tag{5.56}$$

$$= \lim_{m \to \infty} (Tz_m, \lim_{n \to \infty} z_n) = \lim_{m \to \infty} (Tz_m, 0) = 0.$$

Let

$$D_0 = D(T^*) \cap \widetilde{D(T)}. \tag{5.57}$$

Since $D(T) \subset D(T^*)$, it follows that

$$D(T) \subset D_0 \subset D(T^*).$$

Let

$$T_0 = T^*|_{D_0}. \tag{5.58}$$

Since $T^*x = Tx$ for $x \in D(T)$, T_0 is an extension of T. It suffices to prove that it is symmetric. If $x, y \in D_0$, then $x, y \in \widetilde{D(T)}$, whence there are sequences $\{x_n\}, \{y_n\}$ such that $\lim_{n\to\infty} x_n = x$, $\lim_{n\to\infty} y_n = y$. By continuity of the inner product, the limit $\lim_{m,n\to\infty}(x_m, y_n)_0$ exists, so that repeated limits exist.

On one hand, the transformation is the same as in (5.56):

$$\lim_{m,n\to\infty} (x_m, y_n)_0 = \lim_{m,n\to\infty} (Tx_m, y_n) = \lim_{m\to\infty} (Tx_m, \lim_{n\to\infty} y_n) = \lim_{m\to\infty} (Tx_m, y)$$

$$\tag{5.59}$$

$$= \lim_{m\to\infty} (x_m, T_0 y) = (\lim_{m\to\infty} x_m, T_0 y) = (x, T_0 y)$$

and on the other hand,

$$\lim_{m,n\to\infty} (x_m, y_n)_0 = \lim_{m,n\to\infty} (x_m, T^* y_n) = \lim_{n\to\infty} (\lim_{n\to\infty} x_m, T y_n) = \lim_{n\to\infty} (x, T y_n)$$

$$\tag{5.60}$$

$$= \lim_{n\to\infty} (T_0 x, y_n) = (T_0 x, y),$$

so that T_0 is symmetric. We show that $W(T_0)$ H. For every $y \in H$. let $f(x) = f_f(x)$ (x, y) be the functional on $D(T)$. By the Cauchy-Schwarz inequality and (5.50)

$$|f(x)| = |(x, y)| \le \|x\|\|y\| \le \frac{1}{c}\|x\|_0\|y\|,$$

and so $f(x)$ is a continuous functional w.r.t $\|\cdot\|_0$ defined on the dense subset $D(T)$ of $\widetilde{D(T)}$. Hence it can be extended to a continuous functional on $\widetilde{D(T)}$ w.r.t. $\|\cdot\|_0$ extended to it. Hence by Theorem 5.6, there exists a $y_f \in \widetilde{D(T)}$ such that

$$f(x) = (x, y) \ (x, y_f)_0 = (Hx, y_f) = (x, T_0 y_f). \tag{5.61}$$

Hence $y_f \in D(T^*)$ and $T_0 y_f = y$, so that $W(T_0) = H$ and Theorem 5.9 applies. $\qquad\square$

5.1.1   *Self-adjoint operators and Stieltjes integration*

The Stieltjes integration appears in the spectral decomposition of self-adjoint operators on a Hilbert space H over the complex number field. This section corresponds to §3.6.

Lemma 5.3. *If a sequence $\{E(\lambda)\}$ of projectors with parameters $\lambda \in \mathbb{R}$ satisfies the relation*

$$E(\lambda)E(\mu) = E(\min\{\lambda, \mu\})(= E(\mu)E(\lambda)), \qquad (5.62)$$

then for a strictly monotone increasing sequence λ_n converging to λ (written $\lambda_n \uparrow \lambda$), the projector $E(\lambda - 0)$ is determined such that $\lim_{n \to \infty} E(\lambda_n)x = E(\lambda - 0)x$ (strong) for every $x \in H$ and it does not depend on the choice of the sequence. Similarly, for a strictly monotone decreasing sequence λ_n converging to λ (written $\lambda_n \downarrow \lambda$), we may define the projector $E(\lambda + 0)$ and interpreting λ to mean ∞, $E(\infty)$ and $E(-\infty)$ are defined.

Proof. We write for $\beta > \alpha$

$$E(\alpha, \beta] = E(\beta) - E(\alpha) \qquad (5.63)$$

which is a projector. The equality (5.65) below will turn out to be useful. For $\lambda_n \uparrow \lambda$, by **telescoping series technique** we have for $m < n$,

$$E(\lambda_m, \lambda_b] = E(\lambda_n) - E(\lambda_m) = \sum_{j=m}^{n-1} E(\lambda_j, \lambda_{j+1}].$$

Hence

$$\|E(\lambda_m, \lambda_n]x\|^2 = \sum_{j=m}^{n-1} \|E(\lambda_j, \lambda_{j+1}]x\|^2 \qquad (5.64)$$

by (5.67). By (5.34), the left-hand side is upward bounded by $\|x\|^2$. It follows therefore that the right-hand side is a partial sum of monotone increasing sequence which is upward bounded and therefore is convergent. Hence

$$\lim_{m,n \to \infty} \|E(\lambda_m, \lambda_n]x\|^2 = \lim_{m,n \to \infty} \sum_{j=m}^{n-1} \|E(\lambda_j, \lambda_{j+1}]x\|^2 = 0,$$

whence the Cauchy criterion for $\|E(\lambda_n)x\|^2$ holds true, and so $\lim_{n \to \infty} E(\lambda_n)x = E(\lambda - 0)x$ (strong). Uniqueness follows from the observation that given another sequence $\mu_n \uparrow \lambda$, we assemble $\{\lambda_n\}$ and $\{\mu_n\}$ and arrange it in increasing order $\{\nu_n\}$, which is called a **refinement of the division**. Then for this new sequence, we know that $\lim_{n \to \infty} E(\nu_n)x = E_1(\lambda - 0)x$, say. But subsequences must converge to the same limit, so that $\lim_{n \to \infty} E(\lambda_n)x = E_1(\lambda - 0)x = \lim_{n \to \infty} E(\mu_n)x$. $\square$

Exercise 35. (i) Suppose $i \leq j$. Then prove that

$$(E(\lambda_i, \lambda_{i+1}]x, E(\lambda_j, \lambda_{j+1}]y) = \begin{cases} 0 & i < j \\ (E(\lambda_i, \lambda_{i+1}]x, y) & i = j. \end{cases} \tag{5.65}$$

(ii) For $a < b$ prove that

$$E(\lambda)E(a, b] = \begin{cases} E(b) - E(a) = E(a, b] & a < b \leq \lambda \\ E(\lambda) - E(a) = E(a, \lambda] & a < \lambda < b \; . \\ 0 & \lambda \leq a < b \end{cases} \tag{5.66}$$

Solution. Recall the properties of a projector in Definition 5.7. Substituting (5.63) and expanding out, we have four terms including $-(E(\lambda_{i+1})x, E(\lambda_j)y)$ which becomes $-(E(\lambda_{i+1})E(\lambda_j)x, y)$ by property (ii). Then by (5.62), this amounts to

$$\begin{cases} -(E(\lambda_{i+1})x, y) & i < j \\ -(E(\lambda_j)x, y) & i = j. \end{cases}$$

Since other 3 terms are simply calculated, this proves (5.65). For $i \neq j$, (5.65) entails the orthogonality:

$$E(\lambda_i, \lambda_{i+1}] \cdot E(\lambda_j, \lambda_{j+1}] = 0. \tag{5.67}$$

Definition 5.11. A sequence $\{E(\lambda)\}$ of projectors is called a **resolution of the identity** if it satisfies (5.62) and the following conditions

$$E(-\infty) = 0, \; E(\infty) = I, \tag{5.68}$$
$$E(\lambda) = E(\lambda+).$$

Lemma 5.4. *Let* $\{E(\lambda)\}$ *be a resolution of the identity. Then for any* $x, y \in H$, *the real and imaginary parts of* $(E(\lambda)x, y)$ *are of bounded variation as a function in* λ.

Proof. As in §3.6, for any division $\Delta : \lambda_1 < \cdots < \lambda_m$, we show that the sum of differences is bounded. We have by (5.65)

$$\sum_{j=1}^{m-1} |(E(\lambda_{j+1})x, y) - (E(\lambda_j)x, y)| = \sum_{j=1}^{m-1} |(E(\lambda_j, \lambda_{j+1}]x, y)| \tag{5.69}$$

$$= \sum_{j=1}^{m-1} |(E(\lambda_j, \lambda_{j+1}]x, E(\lambda_j, \lambda_{j+1}]y)| \le \sum_{j=1}^{m-1} \|(E(\lambda_j, \lambda_{j+1}]x\| \cdot \|E(\lambda_j, \lambda_{j+1}]y\|$$

$$\le \{\sum_{j=1}^{m-1} \|(E(\lambda_j, \lambda_{j+1}]x\|^2\}^{1/2} \{\sum_{j=1}^{m-1} \|(E(\lambda_j, \lambda_{j+1}]y\|^2\}^{1/2}$$

$$= \{\|(E(\lambda_1, \lambda_m]x\|^2\}^{1/2} \cdot \{\|(E(\lambda_1, \lambda_m]y\|^2\}^{1/2} \le \|x\|\|y\|$$

by the Cauchy-Schwarz inequality applied twice and (5.64) at the last step.
□

Given a system of projectors $\{E(\lambda)\}$ with parameters λ as above and a complex-valued continuous function $\phi(\lambda)$, we define for $x \in H$ the quantity $\in H$

$$\int_a^b \phi(\lambda)\, dE(\lambda)x \tag{5.70}$$

as in Definition 3.7. I.e. in (3.53) we choose $f = \phi$, $g = E$. From each $I_k = [\lambda_k, \lambda_{k+1}]$ choose $\forall \xi_k \in I_k$ and let $\Xi = (\xi_1, \ldots, \xi_m)$.

Theorem 5.11. *The Riemann-Stieltjes sum*

$$S(\phi) = S(\Delta, \phi) = S(\Delta, \phi, \Xi) := \sum_{k=1}^{m-1} \phi(\xi_k) E(\lambda_k, \lambda_{k+1}]x \tag{5.71}$$

converges strongly in H as $|\Delta| \to 0$, irrespectively of the choice of Ξ and the division Δ. The limit is denoted by (5.70).

Proof. Since $\phi(\lambda)$ is continuous on the compact set $[a, b]$, it is uniformly continuous on it and given $\varepsilon > 0$ we may find $\delta = \delta(\varepsilon) > 0$ such that $\phi(\lambda_1) - \phi(\lambda_2)| < \varepsilon$ as long as $|\lambda_1 - \lambda_2| < \delta$. In addition to the division Δ let

$$\Delta_1 : a = \mu_1 < \cdots < \mu_n = b$$

and assume that the norm $|\lambda_{k+1} - \lambda_k| < \varepsilon$ and $|\mu_{j+1} - \mu_j| < \varepsilon$.

We choose any $\eta_j \in J_j = [\mu_j, \mu_{j+1}]$ and form the sum

$$S_1(\phi) = S(\Delta_1, \phi) = S(\Delta_1, \phi, \eta) := \sum_{j=1}^{n-1} \phi(\eta_j) E(\mu_j, \mu_{j+1}]x \tag{5.72}$$

and prove that $\|S - S_1\| \to 0$ as $|\Delta| \to 0$. As in the proof of Lemma 5.4, we form a refinement of the divisions by assembling the points $\{\lambda_k\}$ and $\{\mu_j\}$, calling it $\{\nu_l\}$:

$$\Delta_\nu : a = \nu_1 < \cdots < \nu_\ell = b, \quad \ell \le m + n, \quad \nu_{l+1} - \nu_l < \varepsilon.$$

By the setting, the difference of values of ϕ is $< \varepsilon$. Hence

$$S - S_1 = \sum_{l=1}^{\ell} \varepsilon_l E(\nu_l, \nu_{l+1}]x, \quad |\varepsilon_l| < 2\varepsilon. \tag{5.73}$$

Hence using (5.64), we conclude that

$$\|S - S_1\|^2 \le \varepsilon_l^2 \|E(\nu_1, \nu_\ell]x\|^2 \le 4\varepsilon\|x\|^2, \tag{5.74}$$

whence the Cauchy criterion applies. $\qquad\square$

Exercise 36. Let

$$y = \int_a^b \phi(\lambda)\, \mathrm{d}E(\lambda)x. \tag{5.75}$$

Then prove that

$$E(a, b]y = y \tag{5.76}$$

and that

$$\|y\|^2 = \int_a^b |\phi(\lambda)|^2\, \mathrm{d}\|E(\lambda)x\|^2. \tag{5.77}$$

Solution. We apply $E(a, b]$ to (5.72) and apply (5.66) to the resulting products of projectors. Since it just gives rise to a refinement, we have (5.76). To prove (5.77), we take the square of the norm of (5.72). Applying (5.65) to it, we see that what remains is the sum of $\phi(\xi_j)\phi(\eta_j)(E(\lambda_j, \lambda_{j+1}x, x)$. In the limit this amounts to (5.77).

Lemma 5.5. *For a complex-valued continuous function $\phi(\lambda)$ and $x \in H$ the following three conditions are equivalent:*

$$\int_{-\infty}^{\infty} \phi(\lambda)\, \mathrm{d}E(\lambda)x \tag{5.78}$$

exists.

$$\int_a^b |\phi(\lambda)^2\, \mathrm{d}\|E(\lambda)x\|^2 < \infty. \tag{5.79}$$

$$F(y) = \int_{-\infty}^{\infty} \phi(\lambda)\, \mathrm{d}(E(\lambda)y, x) \tag{5.80}$$

is a bounded linear functional.

Theorem 5.12. *Let $\phi(x)$ be a real-valued continuous function. Then*

$$(Tx, y) = \int_{-\infty}^{\infty} \phi(\lambda)\, d(E(\lambda)x, y) \tag{5.81}$$

defines the self-adjoint operator T with $D(T) = H$ satisfying $TE(\lambda) \supset E(\lambda)T$, In particular for $\phi(\lambda) = \lambda$

$$(Tx, y) = \int_{-\infty}^{\infty} \lambda\, d(E(\lambda)x, y) \tag{5.82}$$

abbreviated as

$$T = \int_{-\infty}^{\infty} \lambda\, dE(\lambda) \tag{5.83}$$

called the **spectral decomposition** *of T.*

5.1.2 *Spectral decomposition of self-adjoint operators*

Definition 5.12. Let T be an additive operator with $\overline{D(T)} = H$. For a complex number λ let

$$T_\lambda = T - \lambda I \tag{5.84}$$

and consider its inverse. λ belongs to the **spectra** $S(T)$ of T if there is no continuous inverse operator T_λ^{-1} such that $\overline{D(T_\lambda^{-1})} = H$. $S(T)$ is divided into three subsets:

$$S(T) = P(T) \cup C(T) \cup R(T), \tag{5.85}$$

where $P(T)$ is the **point spectra**, $C(T)$ is the **continuous spectra** and $R(T)$ is the **residual spectra,** defined respectively as follows.

(i) $\lambda \in P(T)$ means $D(T_\lambda^{-1}) = \emptyset$, i.e. the inverse operator T_λ^{-1} does not exist.

(ii) $\lambda \in P(T)$ means $\overline{D(T_\lambda^{-1})} = H$ but T_λ^{-1} is not continuous.

(iii) $\lambda \in P(T)$ means T_λ^{-1} exists but $\overline{D(T_\lambda^{-1})} \neq H$.

Theorem 5.13. *That $\lambda \in P(T)$ is equivalent to the fact that the equation $Tx = \lambda x$ has a non-trivial solution. If this is the case, λ is called the* **eigenvalue** *of T and $x \neq 0$ the* **eigenvector** *of T.*

Definition 5.13. The set of all eigenvectors and the zero vector forms a subspace $E_T(\lambda)$ of H called the eigenspace of T.

Theorem 5.14. *For a self-adjoint operator* $T = \int \lambda \, dE(\lambda)$ *we have*

(i) $S(T) \subset \mathbb{R}$.

(ii) $\lambda \notin S(T)$ *is equivalent to* $E(\lambda_1) \neq E(\lambda_2)$ *for some open interval* (λ_1, λ_2) *containing* λ.

(iii) $\lambda \in P(T)$ *is equivalent to* $E(\lambda) \neq E(\lambda - 0)$ *and* $E_Y(\lambda) = W(E(\lambda) - E(\lambda - 0))$.

(iv) $\lambda \in C(T)$ *is equivalent to* $E(\lambda) = E(\lambda - 0)$ *and for any open interval* (λ_1, λ_2) *containing* λ *we have* $E(\lambda_1) \neq E(\lambda_2)$.

(v) $R(T) = \emptyset$.

5.1.3 Completely continuous integral operators

Definition 5.14. An additive bounded operator mapping a weakly convergent sequence to a strongly convergent sequence is called **completely continuous**.

Theorem 5.15. *Let* $-\infty \leq a < b \leq \infty$ *and let* $H = L^2(a,b)$. *Suppose E. Schmidt's condition*

$$\int_a^b \int_a^b |k(t,u)| \, dt du < \infty \tag{5.86}$$

is satisfied. Then the operator $K : H \to \mathbb{C}$ *defined by*

$$Kx(u) = X(u) = \int_a^b k(t,u) x(t) \, dt \tag{5.87}$$

is completely continuous.

Proof. By the Cauchy-Schwarz inequality

$$\int_a^b |X(u)|^2 du = \int_a^b \left| \int_a^b k(t,u) x(t) \, dt \right|^2 du$$
$$\leq \int_a^b \left(\int_a^b |k(t,u)|^2 \, dt \right) \left(\int_a^b |x(t)|^2 \, dt \right) du < \infty.$$

Hence the right-hand side of (5.87) exists and $X(u)$ is finite for a.a. $u \in (a.b)$, so that K is a bounded operator.

Also by Schmidt's condition (5.86) and Fubini's theorem, it follows that $k(u) = k(t, u) \in L^2(a, b)$ as a function in u for a.a. t. Suppose $x_n \to x$ (weak). Then

$$\lim_{n \to \infty} X_n(u) = \lim_{n \to \infty} (k, x_n) = (k, x) = \int_a^b k(t, u)x(t)\, dt = X(u),$$

so that $X_n \to X$ (weak) and

$$|X_n(u)|^2 \le \left(\int_a^b |k(t, u)|^2\, dt \right) \left(\int_a^b |x(t)|^2\, dt \right) < \infty$$

by the Cauchy-Schwarz inequality and (5.8). Hence by the theorem on termwise integration, we have

$$\|X_n\|^2 \to \|X\|.$$

Hence, recalling (5.4)

$$\|X_n - X\|^2 = (X_n - X, X_n - X) = \|X_n\|^2 - (X_n, X) - (X, X_n) + \|X\|^2 \to 0,$$

whence $X_n \to X$ (strong). Hence K maps a weakly convergent sequence to a strongly convergent one. Then $W(E(\lambda_1, \lambda_2])$ is finite-dimensional, where $E(\lambda_1, \lambda_2]$ is defined by (5.63). For otherwise, one may construct an orthonormal sequence which by Theorem 5.5 contains a weakly convergent subsequence. $\qquad \square$

Theorem 5.16. *If T is a completely continuous symmetric operator, then (i) it has at most countably many eigenvalues λ_j and they have no accumulation point save for 0, so that they can be arranged as $|\lambda_1| > |\lambda_2| > \cdots$ (ii) the eigenspace belonging to the eigenvalue λ_j is finite dimensional.*

Proof. By Lemma 5.2, T is self-adjoint and so by Theorem 5.12 it is expressed as

$$T = \int \lambda\, dE(\lambda). \tag{5.88}$$

If a closed interval $[\lambda_1, \lambda_2]$ does not contain 0, then we may prove that the region $W = WE(\lambda_1, \lambda_2]$ is finite dimensional. For otherwise, by Hilbert-Schmidt method, one may construct a countable ONS $\{x_n\}$ of W. Hence by Theorem 5.5, we may find a convergent subsequence $\{x_{\nu_k}\}$. The Bessel inequality (5.24) for the ONS $\{x_n\}$ implies that $\lim_{k \to \infty} \{x_{\nu_k}\} = 0$ (weak). Since $x_{\nu_k} \in W$ it follows that

$$E(\lambda_1, \lambda_2]x_{\nu_k} = x_{\nu_k}. \tag{5.89}$$

Hence by (5.88), (5.89) and (5.68) successively

$$\|Tx_{\nu_k}\|^2 = \int \lambda^2 \, \mathrm{d}\|E(\lambda)x_{\nu_k}\|^2 = \int \lambda \, \mathrm{d}\|E(\lambda)E(\lambda_1,\lambda_2]x_{\nu_k}\|^2 \qquad (5.90)$$

$$= \int_{\lambda_1}^{\lambda_2} \lambda \, \mathrm{d}\|E(\lambda) - E(\lambda_1)x_{\nu_k}\|^2$$

$$= \int_{\lambda_1}^{\lambda_2} \lambda \, \mathrm{d}\|E(\lambda_1), E(\lambda)x_{\nu_k}\|^2.$$

The last integral is

$$\geq \|^2 \min\{|\lambda_1|^2, |\lambda_1|^2 \|(E(\lambda_1), E(\lambda_2)]x_{\nu_k}\|^2\} = \|x_{\nu_k}\|^2 \min\{|\lambda_1|^2, |\lambda_1|^2 > 0$$

by (5.89). Hence $\lim_{k\to\infty} T\{x_{\nu_k}\}$ 0 (strong) cannot hold, contrary to the complete continuity of T.

Hence if a closed interval $[\lambda_1, \lambda_2]$ does not contain 0, then there exist at most finitely many eigenvalues in that interval and the eigenspaces belonging to them are finite dimensional, proving (ii).

It also follows that the eigenvalues lying in intervals

$$[n, n+1], [-(n+1), -n], \left[\frac{1}{n}, \frac{1}{n+1}\right], \left[-\frac{1}{n+1}, \frac{1}{n}\right]$$

are at most finite and so in total they are countably many and that they do not have an accumulation point other than 0. It can be proved that neither ∞ nor $-\infty$ are accumulation points. $\qquad \square$

Example 5.4. [Newman (1979)] Let $\xi > 1$, $V > 0$ and only in this example, we let μ denote any non-negative measure on $(1, \xi)$ such that $\mu(\xi) - \mu(1) = V$. Let $L^2 = L^2((1, \xi), \mu)$ denote the Hilbert space of square integrable functions on $(1, \xi)$ w.r.t. the measure μ. Let $\{x\}$ denote the fractional part of the real number x and define the kernel

$$k(x, y) = \int_1^\infty \left\{\frac{t}{x}\right\} \left\{\frac{t}{y}\right\} \frac{\mathrm{d}t}{t^2}. \qquad (5.91)$$

Define the operator $K : L^2 \to L^2$ by

$$Kf = \int_1^\xi k(x, y)f(x) \, \mathrm{d}\mu(x). \qquad (5.92)$$

Since $k(x, y)$ is a bounded function satisfying the condition in Theorem 5.15, it follows that K is self-adjoint and completely continuous. Then we denote the eigenvalues of K according to Theorem 5.16 by $\{\lambda_n\}$ and the eigenvectors belonging to them by $\{\varphi_n\}$, which we may assume to form an ONS:

$$(\varphi_m, \varphi_n) = \int_1^\xi \varphi_m(x)\varphi_n(x \, d\mu(x) = \delta_{mn}. \tag{5.93}$$

Exercise 37. Let

$$\tilde{\alpha}_n = \tilde{\alpha}_n(u) = \int_1^\xi u^{-1}\left\{\frac{u}{x}\right\}\varphi_n(x)\,d\mu(x) \in L^2(1,\infty). \tag{5.94}$$

Then prove that

$$(\tilde{\alpha}_m, \tilde{\alpha}_n) = \lambda_m(\varphi_m, \varphi_n), \tag{5.95}$$

whence that $\lambda_m > 0$.

Solution. Substituting (5.94) in

$$(\tilde{\alpha}_m, \tilde{\alpha}_n) = \int_1^\infty \tilde{\alpha}_m \tilde{\alpha}_n \, dx, \tag{5.96}$$

where dx indicates the Lebesgue measure, we obtain after changing the order of integration which is permissible in view of absolute convergence of three integrals

$$(\tilde{\alpha}_m, \tilde{\alpha}_n) = \int_1^\xi \varphi_n(x)\,d\mu(x)\int_1^\xi k(x,y)\varphi_m(y)\,d\mu(y). \tag{5.97}$$

But the inner integral is by (5.92) $K\varphi_m$, so that (5.97) amounts to

$$(\tilde{\alpha}_m, \tilde{\alpha}_n) = \int_1^\xi K\varphi_m(x)\varphi_n(x)\,d\mu(x) = (K\varphi_m, \varphi_n) = \lambda_m(K\varphi_m, \varphi_n) \tag{5.98}$$

by definition, and (5.95) follows.

If we put $\lambda_m = \lambda_n$ in (5.95), then

$$\lambda_n(\varphi_n, \varphi_n) = (\tilde{\alpha}_n, \tilde{\alpha}_n) = \int_1^\infty \tilde{\alpha}_n(u)^2\,du > 0, \tag{5.99}$$

whence $\lambda_n > 0$. If we put

$$\alpha_n(u) = \lambda_n^{-1/2}\tilde{\alpha}_n(u) = \lambda_n^{-1/2}\int_1^\xi u^{-1}\left\{\frac{u}{x}\right\}\varphi_n(x)\,d\mu(x), \tag{5.100}$$

then we have an ONS $\{\alpha_n\}$.

Now Example 5.4 and Exercise 37 allow us to formulate the following theorem which we are ready to prove.

Theorem 5.17. (Newman-Ryavec-Shure) *Let* $\lambda_1 \geq \lambda_2 \geq \ldots$ *be the eigenvalues of the operator* K *in* (5.92) *and let* $\{\varphi_n\}$ *be an ONS of corresponding eigenvectors. Suppose* $\rho = \beta + i\gamma$ *be a zero of* $\zeta(s)$ *with* $\frac{1}{2} < \beta < 1$. *Then*

$$\Phi(\mu) \leq \frac{|\rho - 1|^2}{2\beta - 1}, \tag{5.101}$$

where

$$\Phi(\mu) = \sum_{n=1}^{\infty} \lambda_n^{-1} \left| \int_1^{\xi} x^{-1} \varphi_n(x) \, d\mu(x) \right|^2. \tag{5.102}$$

In particular, if $\Phi = \sup \Phi(\mu) = \infty$, *where the* sup *is over all* $\xi > 1, V > 0$, *and* μ, *then RH would follow.*

Proof. First we recall (3.76) in the form

$$\zeta(s) = -s \int_0^{\infty} \{u\} u^{-s-1} \, du \tag{5.103}$$

valid for $0 < \sigma < 1$. Putting $u = \frac{v}{x}$ with $x \geq 1$, this leads to

$$-\frac{\zeta(s)}{s} x^{-s} = \int_0^{\infty} \left\{\frac{v}{x}\right\} v^{-s-1} \, dv = -\frac{1}{x(s-1)} + \int_1^{\infty} \left\{\frac{v}{x}\right\} v^{-s-1} \, dv \tag{5.104}$$

on evaluating the integral over $(0, 1)$.

Suppose $\rho = \beta + i\gamma$ indicates a non-trivial zero of the Riemann zeta-function, $\frac{1}{2} < \beta < 1$, then (5.104) reads with $s = \rho$

$$\frac{1}{x(\rho - 1)} = \int_1^{\infty} \left\{\frac{v}{x}\right\} v^{-\rho-1} \, dv. \tag{5.105}$$

Multiplying this by $\lambda_n^{-1/2} \varphi_n(u)$, integrating over $[1, \xi]$ with respect to μ and recalling (5.103), we deduce that

$$\frac{1}{\rho - 1} \lambda_n^{-1/2} \int_1^{\xi} x^{-1} \varphi_n(x) \, d\mu(x) = \int_1^{\xi} u^{-\rho} a_n(u) \, du. \tag{5.106}$$

Taking the inner product of both sides of (5.24) and summing over n, we have on applying the Bessel inequality (5.24)

$$\frac{1}{|\rho - 1|^2} \Phi(\mu) = \sum_{n=1}^{\infty} \left| \int_1^{\xi} u^{-\rho} a_n(u) \, du \right|^2 \leq \int_1^{\infty} u^{-\rho} u^{-\bar{\rho}} \, du = \frac{1}{2\beta - 1}. \tag{5.107}$$

$\square$

5.2   Hilbert space and number theory

This section incorporates some new look at arithmetical functions discussed in Chapter 3 and we take it for granted here. On the ring R of all arithmetical functions, we introduce some additional operations.

First

$$Sf = I * f,$$

where I is the constant function in §3.2, so that by Corollary 3.3, $S^{-1}f = \mu * I$:

$$Sf(n) = \sum_{d|n} f(d), \tag{5.108}$$

$$S^{-1}f(n) = \sum_{d|n} \mu\left(\frac{n}{d}\right) f(d). \tag{5.109}$$

Also $Ef = e * f$, where e is the unity in R (Theorem 3.1):

$$Ef(n) = f(n). \tag{5.110}$$

For a given integer N,

$$Af(n) = f((n, N)), \tag{5.111}$$

and

$$Bf(n) = \varepsilon\left(\frac{N}{n}\right) f(n), \tag{5.112}$$

where $\varepsilon(n) = 1$ if $n \in \mathbb{Z}$ and 0 otherwise. It is the characteristic function of $\mathbb{Z}$ in $\mathbb{R}$.

Exercise 38. Prove that Corollary 3.3 reads

$$SS^{-1} = S^{-1}S = E. \tag{5.113}$$

Exercise 39. As in Definition 3.3, let $F(s)$ denote the generating Dirichlet series for $f(n)$, absolutely convergent for $\sigma > \sigma_f$. Then show that (5.108) and (5.109) may be expressed as

$$\zeta(s)F(s) = \sum_{n=1}^{\infty} \frac{Sf(n)}{n^s}, \quad \sigma > \max\{\sigma_f, 1\} \tag{5.114}$$

and

$$\frac{1}{\zeta(s)} F(s) = \sum_{n=1}^{\infty} \frac{S^{-1}f(n)}{n^s}, \quad \sigma > \max\{\sigma_f, 1\}, \tag{5.115}$$

respectively.

Exercise 40. Prove that

$$SB = AS, \tag{5.116}$$

whence that

$$BS^{-1} = S^{-1}A. \tag{5.117}$$

Solution. By definition

$$SBf(n) = \sum_{d|n} \varepsilon \left(\frac{N}{d}\right) f(d)$$

and the summand of the right-side sum is non-zero only if $d|N$, so that d must divide both n and N, or $d|(n,N)$, i.e.

$$S\dot{B}f(n) = \sum_{d|(n,N)} f(d) = ASf(n).$$

Hence (5.116) follows.

From (5.116) we have

$$B = S^{-1}AS \tag{5.118}$$

whence (5.117) follows.

We are in a position to prove the following theorem, which is a sort of generalization of (3.7).

Theorem 5.18. (Romanoff) *For any $f \in R$ and $k < n$, we have*

$$\sum_{d|n} \mu \left(\frac{n}{d}\right) f((d,k)) = 0. \tag{5.119}$$

Proof. We have on using (5.118) on the way,

$$\sum_{d|n} \mu \left(\frac{n}{d}\right) f((d,N)) = S^{-1}f((n,N)) = S^{-1}Af(n) = BS^{-1}f(n). \tag{5.120}$$

Putting $N = k < n$, we conclude (5.119) by (5.120). □

Exercise 41. Prove the relation

$$\sum_{d|n,(d,k)=\delta} \mu \left(\frac{n}{d}\right) = 0 \tag{5.121}$$

for each $\delta|k$, and

$$\sum_{d|n} \mu \left(\frac{n}{k}\right) f((d,k)) = \sum_{\delta} f(\delta) \sum_{d|n,(d,k)=\delta} \mu \left(\frac{n}{d}\right). \tag{5.122}$$

Theorem 5.18 follows also from (5.121) and (5.122).

Definition 5.15. (Romanoff [Romanoff (1951)]) Let H be a (complex) Hilbert space with inner product denoted by (f, g). Any sequence $\{f_n\} \subset H$ is said to have a *D-property*, or more precisely, D_g-property, if $(f_m, f_n) = g((m, n))$ for a given arithmetical function $g \in R$.

Exercise 42. Let $\{f_n\} \subset H$ have the D_g-property, and define a new sequence $\{\phi_n\} \subset H$ by

$$\phi_n = \sum_{d|n} \mu\left(\frac{n}{d}\right) f_d, \tag{5.123}$$

where the right-hand side may be viewed as a Dirichlet convolution $\mu * f$ by abuse of language. Then prove that the $\{\phi_n\}$ is orthogonal.

Solution. For $k < n$, we have

$$(\phi_n, f_k) = \left(\sum_{d|n} \mu\left(\frac{n}{d}\right) f_d, f_k\right) = \sum_{d|n} \mu\left(\frac{n}{d}\right) (f_d, f_k) \tag{5.124}$$

$$= \sum_{d|n} \mu\left(\frac{n}{d}\right) g((d, k)) = 0$$

by Theorem 5.18. Hence if $m < n$, then ϕ_m being a linear combination of $f_k, (k \le m < n)$, we have $(\phi_m, \phi_n) = 0$. Similarly, for $m > n$, $(\phi_m, \phi_n) = \overline{(\phi_n, \phi_m)} = 0$. Hence the result follows.

We write

$$G(n) := (\mu * g)(n) = \sum_{d|n} \mu\left(\frac{n}{d}\right) g(d) \tag{5.125}$$

throughout in this section. The Möbius inversion formula claims that Eqn (5.125) is equivalent to

$$g(n) = \sum_{d|n} G(d). \tag{5.126}$$

Exercise 43. Use (5.124) and (5.123) to establish that

$$(\phi_n, \phi_n) = G(n). \tag{5.127}$$

Solution. By (5.124), we have

$$(\phi_n, \phi_n) = \sum_{d|n} \mu\left(\frac{n}{d}\right) (\phi_n, f_d) = (\phi_n, f_n).$$

Hence, substituting (5.123) again,

$$(\phi_n, \phi_n) = \sum_{d|n} \mu\left(\frac{n}{d}\right) (f_d, f_n) = \sum_{d|n} \mu\left(\frac{n}{d}\right) g((d, n)),$$

which is the right-hand side of (5.127).

The following theorem gives a condition for D_g-property.

Theorem 5.19. *A necessary and sufficient condition for a sequence to exist in H having the D_g-property is that $g(n)$ satisfies the condition*

$$G(n) = (\mu * g)(n) \geq 0, \quad n \in \mathbb{N}. \tag{5.128}$$

Proof. Necessity follows from (5.127) and $(\phi_n, \phi_n) \geq 0$.

Sufficiency is proved in the proof of Theorem 5.20. $\qquad\square$

Theorem 5.20. *Let $G(n)$ be an arbitrary arithmetical function with $G(n) > 0$ and define $g(n)$ by (5.126). If two sequences $\{f_n\}$ and $\{\varphi_n\}$ are related by (cf. (5.123))*

$$f_n = \sum_{d|n} \sqrt{G(d)}\varphi_d. \quad n \in \mathbb{N}, \tag{5.129}$$

then $\{\varphi_n\}$ is an ONS if and only if $\{f_n\}$ has the D-property.

Proof. Suppose $\{f_n\}$ has the D-property. Then, since (5.129) amounts to

$$\varphi_n = \frac{1}{\sqrt{G(d)}} \sum_{d|n} \mu\left(\frac{n}{d}\right) f_d = \frac{1}{\sqrt{G(d)}}\phi_n, \tag{5.130}$$

with ϕ_n in (5.123). Since by Exercise 42, $\{\phi_n\}$ is orthogonal and by (5.127), $\{\varphi_n\}$ is a normalization of $\{\phi_n\}$, it follows that $\{\varphi_n\}$ is an ONS.

Conversely, if $\{\varphi_n\}$ is an ONS, then $\{f_n\}$ has the D-property. This is immediate since

$$(f_n, f_m) = \sum_{d|n} \sqrt{G(d)} \sum_{\delta|m} \sqrt{G(\delta)}(\varphi_d, \varphi_\delta) = \sum_{d|(n,m)} G(d) = g((n,m)),$$

by (5.126). This also proves sufficiency part of Theorem 5.19. $\qquad\square$

Theorem 5.21. *Let $\omega(n)$ be a completely multiplicative function, which takes non-zero values and square summable:*

$$\tilde{\sigma} := \sum_{k=1}^{\infty} |\omega(k)|^2 < \infty \tag{5.131}$$

and let $\{\alpha_n\}$ be an ONS in H. Define f_n by

$$f_n = \overline{\omega(n)^{-1}} \sum_{k=1}^{\infty} \omega(k)\alpha_{nk}, \quad n \in \mathbb{N}. \tag{5.132}$$

Then f_n has the D_g-property with

$$g(n) = \frac{\tilde{\sigma}}{|\omega(n)|^2}. \tag{5.133}$$

Proof. Noting that

$$f_n = \overline{\omega(n)^{-1}} \sum_{k \equiv 0 \, (\bmod \, n)} \omega\left(\frac{k}{n}\right) \alpha_k$$

and using the continuity of the inner product, we find that

$$(f_n, f_m) = \overline{\omega(n)^{-1}} \omega(m)^{-1} \sum_{k \equiv 0 \, (\bmod \, n)} \sum_{\ell \equiv 0 \, (\bmod \, m)} \omega\left(\frac{k}{n}\right) \bar{\omega}\left(\frac{\ell}{m}\right) (\alpha_k, \alpha_\ell),$$

$$(5.134)$$

whose right-hand side is

$$\overline{\omega(n)^{-1}} \omega(m)^{-1} \sum_{k \equiv 0 \, (\bmod \, [n,m])} \omega\left(\frac{k}{n}\right) \bar{\omega}\left(\frac{k}{m}\right)$$

$$= |\omega(n)|^{-2} |\omega(m)|^{-2} \sum_{k \equiv 0 \, (\bmod \, [n,m])} |\omega(k)|^2,$$

with $[n, m]$ denoting the l.c.m. of n and m. Factoring $|\omega([n, m])|^2$ out and using the well-known relation $mn = [n, m](n, m)$, we conclude that

$$(f_n, f_m) = \frac{\tilde{\sigma}}{|\omega((n, m))|^2}, \qquad (5.135)$$

which is $g((n, m))$ in (5.133), completing the proof. $\qquad \square$

We are now in a position to state main results of this section.

Theorem 5.22. *Notation being the same as in Theorem 5.21, define* φ_n *as in* (5.130) *with* f_n *as in Theorem 5.21:*

$$\varphi_n = (\tilde{\sigma}\Omega(n))^{-1/2} \sum_{d|n} \mu\left(\frac{n}{d}\right) \overline{\omega(d)^{-1}} \sum_{k=1}^{\infty} \omega(k)\alpha_{dk}, \quad n \in \mathbb{N}. \qquad (5.136)$$

Then φ_n *is an ONS, where*

$$\Omega(n) = \sum_{d|n} \mu\left(\frac{n}{d}\right) |\omega(d)|^{-2} = |\omega(n)|^{-2} \prod_{p|n} \left(1 - |\omega(p)|^2\right), \qquad (5.137)$$

with p *ranging over all prime divisors of* n.

Many identities can be derived using Theorem 5.22. A typical example of the choice of g and Ω is $\omega(n) = n^{-s}$ with $\sigma = \operatorname{Re} s > \frac{1}{2}$ and then $\Omega(n) = n^{2\sigma} \prod_{p|n} \left(1 - p^{-2\sigma}\right)$ and $\tilde{\sigma} = \zeta(2\sigma)$.

The last equality in (5.137) is the Euler product whose theory is given in §3.4.

Theorem 5.23. *Suppose* $\{\alpha_n\}$ *is an ONS in* H, *complete in a subspace* H' *of* H, *then* $\{\varphi_n\}$ *defined by* (5.136) *is a complete ONS in* H'.

Proof. Clearly $\{\varphi_n\} \subset H'$, so that the subspace H'' spanned by $\{\varphi_n\}$ is a subset of H'. To show that $H' = H''$ it is enough to show $\alpha_n \in H''$, which is characterized by the Parseval identity (5.25)

$$(1 =)(\alpha_n, \alpha_n) = \sum_{k=1}^{\infty} |(\alpha_n, \varphi_k)|^2. \tag{5.138}$$

Note that

$$(\alpha_1, \varphi_k) = \frac{\mu(k)}{\tilde{\sigma}(k)\Omega(k)} \tag{5.139}$$

whence that

$$\sum_{k=1}^{\infty} |(\alpha_1, \varphi_k)|^2 = \frac{1}{\tilde{\sigma}(k)} \sum_{k=1}^{\infty} \frac{\mu(k)^2}{\Omega(k)} = \sum_{k=1}^{\infty} \prod_p \left(1 + \frac{1}{\Omega(p)}\right) \tag{5.140}$$

by Theorem 3.6. Hence by Exercise 44, this leads to (5.138) with $n = 1$.

Hence α_1 can be approximated arbitrarily near by a linear combination of φ_k's and so by that of f_n's by (5.130). I.e. given $\varepsilon > 0$ there are c_i's such that

$$\left\| \alpha_1 - \sum_{k=1}^{N} c_k f_k \right\| < \varepsilon. \tag{5.141}$$

Finally we introduce the isometry on H'

$$L_n \left(\sum_{k=1}^{\infty} a_k \alpha_k \right) = \sum_{k=1}^{\infty} a_k \alpha_{kn}. \tag{5.142}$$

It is clear that

$$L_n \left(\alpha_1 - \sum_{k=1}^{N} c_k f_k \right) = \alpha_n - \sum_{k=1}^{Nn} c_k' f_k = \alpha_n - \sum_{k=1}^{Nn} c_k'' f_k \tag{5.143}$$

by (5.129). Hence (5.144) may be transformed into

$$\left\| \alpha_n - \sum_{k=1}^{Nn} c_k'' f_k \right\| < \varepsilon, \tag{5.144}$$

so that $\alpha_n \in H''$. $\qquad \square$

Exercise 44. Prove the sequel to (5.131):

$$\tilde{\sigma} = \prod_p \frac{1}{1 - |\omega(p)|^2}. \tag{5.145}$$

5.3   Kuznetsov trace formula

This section is based on the extended survey [Agarwal, Kanemitsu and Li (2016)] whose main body consists of the coming paper [MaA]. As in that paper, we mainly follow the notation of Iwaniec [Iwaniec (1995)]. We state some of the results in the general case of a Fuchsian group Γ of the first kind.

Definition 5.16. We realize the **Lobachevsky plane** (hyperbolic plane) as the upper half-plane with **Poincaré metric**

$$\mathrm{d}s^2 = \frac{1}{y^2}(\mathrm{d}x^2 + \mathrm{d}y^2). \tag{5.146}$$

A $\mathrm{PSL}_2(\mathbb{R})$-invariant measure on $\mathcal{H}$ is connected with the Poincaré metric and is given by

$$\mathrm{d}\mu(z) = \frac{1}{y^2}\mathrm{d}x\mathrm{d}y. \tag{5.147}$$

An invariant measure of a measurable subset $F \subset \mathcal{H}$ is defined by

$$|F| = \int_F \mathrm{d}\mu.$$

A discrete group of motions of the upper half-plane is a discrete subgroup of $\mathrm{PSL}_2(\mathbb{R})$ (discrete in the induced topology).

Definition 5.17. A **Fuchsian group Γ of the first kind** is a discrete group of motions of the upper-half plane for which there exists a fundamental domain F with finite invariant measure $\mathrm{d}\mu$.

Let V be a vector space of dimension n with hermitian inner product $(f_1, f_2)_V$ and let χ be a unitary representation of Γ in V. A vector-valued function $f : \mathcal{H} \to V$ is said to be strictly **automorphic** with respect to Γ and χ (or Γ-**automorphic**) if

$$f(\gamma z) = \chi(\gamma)f(z) \tag{5.148}$$

holds for all $\gamma \in \Gamma$.

If χ is trivial ($\chi \equiv 1$), f lives on a **fundamental domain** $F = \Gamma \backslash \mathcal{H}$. For a suitably chosen fundamental domain F, we let $H = H(\Gamma : \chi)$ be the Hilbert space $L^2(F; V, \mathrm{d}\mu)$ of all vector-valued functions with values in V, square integrable w.r.t. the measure $\mathrm{d}\mu$.

$$H(\Gamma : \chi) = L^2(F; V, \mathrm{d}\mu) = \{f : F \to V \big| \|f\| < \infty\},$$

where the norm $\|f\| = (f, f)$ is induced from the inner product

$$(f_1, f_2) = \int_F (f_1, f_2)_V \, d\mu. \tag{5.149}$$

If $V = \mathbb{C}$, then (5.149) amounts to **the Petersson scalar product**

$$(f, g) = \int_{\Gamma \backslash \mathcal{H}} f(z) \bar{g}(z) \frac{dxdy}{y^2}.$$

On the upper half-plane $\mathcal{H}$, the **non-Euclidean Laplace operator (Laplacian)** is given by

$$L = -y^2 \left(\frac{\partial^2}{\partial x^2} + \frac{\partial^2}{\partial y^2} \right) = -y^2 \Delta^e, \tag{5.150}$$

where Δ^e signifies the standard (Euclidean) Laplace operator on the complex plane (cf. (2.22)).

To define an automorphic Laplacian $\mathcal{A}(\Gamma; \chi)$ for the group Γ and χ, we first consider the action of L on the space D of all smooth automorphic functions $f : \mathcal{H} \to V$ such that both the restriction $f|_F$ and Lf lie in $L^2(F; V, d\mu)$. Then since D is dense in $H(\Gamma : \chi)$, L is a symmetric lower semi-bounded operator on $H(\Gamma : \chi)$. Hence Theorem 5.10 implies that L has a Friedrics extension, which can be proved to be unique. We call this self-adjoint non-negative unbounded operator the **automorphic Laplacian** denoted $\mathcal{A} = \mathcal{A}(\Gamma; \chi)$.

Let $\mathfrak{a}$ be a cusp of Γ and let $\Gamma_{\mathfrak{a}}$ be the parabolic subgroup which stabilizes $\mathfrak{a}$ and $\sigma_{\mathfrak{a}} \in \mathrm{PSL}(2, \mathbb{R})$ is such that

$$\sigma_{\mathfrak{a}} \infty = \mathfrak{a}, \sigma_{\mathfrak{a}}{}^{-1} \Gamma_{\mathfrak{a}} \sigma_{\mathfrak{a}} = B = \left\{ \begin{pmatrix} 1 & b \\ 0 & 1 \end{pmatrix} : b \in \mathbb{Z} \right\}. \tag{5.151}$$

The Kloosterman sum for Γ is defined by ([Iwaniec (1995), p. 51])

$$S_{\mathfrak{ab}}(m, n; c) = \sum{}^* e \left(\frac{dm + an}{c} \right), \tag{5.152}$$

where the summation is over

$$\begin{pmatrix} a & * \\ c & d \end{pmatrix} \in B \backslash \sigma_{\mathfrak{a}}^{-1} \Gamma \sigma_{\mathfrak{a}} / B.$$

In the case of the full modular group, (5.152) reduces to the classical **Kloosterman sum** ([Iwaniec (1995), p. 52])

$$S(m, n; c) = S_{\infty\infty}(m, n; c) = \sum_{a \bmod c}{}^* e \left(\frac{am + \bar{a}n}{c} \right), \tag{5.153}$$

where the summation is over relatively prime residues mod c. For the sake of simplicity, we confine ourselves to (5.153) and refer to it as the Kloosterman sum.

In this book we use the notation Δ for the hyperbolic Laplacian $-L$ and the automorphic Laplacian $\mathcal{A}(\Gamma; \chi)$:

$$\Delta = y^2 \left(\frac{\partial^2}{\partial^2 x} + \frac{\partial^2}{\partial^2 y} \right). \tag{5.154}$$

author	Laplacian	Laplace eqn	eigenvalues
Iwaniec [Iwaniec (1995)]	Δ	$(\Delta + \lambda)f = 0$	$\lambda = s(1-s)$
Liu [Liu]	$-\Delta$	$(\Delta - \lambda)f = 0$	$\lambda = s(1-s) = \frac{1}{4} + \nu^2$
Motohashi [Motohashi (1997)]	$-\Delta$	$(\Delta - \lambda)f = 0$	$\lambda = s(1-s) = \frac{1}{4} + \varkappa^2$

Table 5.2. Laplacian and eigenvalues

In [Iwaniec (1995), (1.26), p. 22] the **Whittaker function** is defined by

$$W_s(z) = 2\sqrt{y} K_{s-1/2}(2\pi y)e(x) \tag{5.155}$$

which is assumed to satisfy the symmetry condition

$$W_s(\bar{z}) = W_s(z). \tag{5.156}$$

Remark 5.3. One of the reasons why Iwaniec calls (5.155) a Whittaker function seems to be the following.

$$G_{1,2}^{2,0}\left(z \left|\begin{matrix} a \\ b, c \end{matrix}\right.\right) = \frac{1}{2\pi i} \int_{(c)} \frac{\Gamma(b+s)\Gamma(c+s)}{\Gamma(a+s)} z^{-s}\, ds = z^{\frac{1}{2}(b+c-1)} e^{-\frac{1}{2}z} W_{\varkappa,\mu}(z), \tag{5.157}$$

where

$$\varkappa = \frac{1}{2}(b+c) - a, \quad \mu = \frac{1}{2}(b-c) \tag{5.158}$$

which reduces to

$$G_{1,2}^{2,0}\left(z \left|\begin{matrix} \frac{1}{2} \\ b, -b \end{matrix}\right.\right) = \pi^{-\frac{1}{2}} e^{-\frac{1}{2}z} K_b\left(\frac{1}{2}z\right). \tag{5.159}$$

Hence

$$W_{-\frac{1}{2},b}(z) = \left(\frac{z}{\pi}\right)^{\frac{1}{2}} K_b\left(\frac{1}{2}z\right). \tag{5.160}$$

Then we may look at [Liu, §3] to deduce the Fourier expansion.

$$u_j(\sigma_{aj}z) = \rho_{aj}(0)y^{1-s_j} + \sum_{n\neq 0} \rho_{aj}(n)W_{s_j}(nz) \tag{5.161}$$

$$= \rho_{aj}(0)y^{1-s_j} + 2\sqrt{y}\sum_{n\neq 0} \rho_{aj}(n)K_{s_j-1/2}(2\pi ny)e(nx).$$

$$E_c(\sigma_{aj}z, s) = \delta_{ac}y^s + \varphi_{acj}(s)y^{1-s} + \sum_{n\neq 0} \varphi_{ac}(n,s)W_s(nz) \tag{5.162}$$

$$= \delta_{ac}y^s + \varphi_{acj}(s)y^{1-s} + 2\sqrt{y}\sum_{n\neq 0} \varphi_{ac}(n,s)K_{s-1/2}(2\pi|n|y)e(nx).$$

Here $\lambda_{aj}(n)$ and $\varphi_{ac}(n, 1/2 + it)$ are the Fourier coefficients of (a complete orthonormal system of) Maass forms and the eigen packet of Eisenstein series in $\mathcal{L}(\Gamma\backslash\mathcal{H})$ ([Iwaniec (1995), p. 117]). Hence, in particular,

$$\varphi_{\infty\infty}(n,s) = \frac{\pi^s}{\Gamma(s)\zeta(2s)}|n|^{-1/2}\sum_{ab=|n|}\left(\frac{a}{b}\right)^{s-1/2} = \frac{\pi^s}{\Gamma(s)\zeta(2s)}|n|^{s-1}\sigma_{1-2s}(|n|) \tag{5.163}$$

where σ indicates the sum-of-divisors function, [Iwaniec (1995), (3.25), p. 67].

On [Iwaniec (1995), p. 118], the normalization is introduced, which we will use in this paper:

$$\nu_{aj} = \left(\frac{4\pi|n|}{\cosh \pi\nu_j}\right)^{1/2}\rho_{aj}(n), \eta_{ac}(n,t) = \left(\frac{4\pi|n|}{\cosh \pi t}\right)^{1/2}\varphi_{ac}(n, 1/2 + it) \tag{5.164}$$

for $n \neq 0$.

We call any C^2-function $h(x)$ on $[0, \infty)$ a **test function** if it satisfies the condition

$$f(0) = 0, \quad f^{(j)}(x) << (x+1)^{-2-\varepsilon}, \quad j = 0, 1, 2. \tag{5.165}$$

On [Iwaniec (1995), p. 143, p. 147], this condition is referred to as (9.13), which is to read (9.14).

The following theorem is stated as the Bruggeman-Kuznetsov formula on [Iwaniec (1995), Theorem 9.3, p. 140].

Theorem 5.24. (Kuznetsov [Kuznetsov (1980)]) *Let* $\mathfrak{a}$, $\mathfrak{b}$ *be cusps of the Fuchsian group of the first kind* Γ *and* $mn \neq 0$. *Then for any test function* *h satisfying the condition* (5.165), *we have*

$$\sum_{j} h(t_j)\bar{\nu}_{aj}(m)\nu_{bj}(n) + \sum_{c} \frac{1}{4\pi} \int_{-\infty}^{\infty} h(t)\bar{\eta}_{ac}(m,t)\eta_{bc}(n,t)\, dt \qquad (5.166)$$

$$= \delta_{ab}\delta_{mn}h_0 + \sum_{c} c^{-1} S_{ab}(m,n;c)h^{\pm}\left(\frac{4\pi\sqrt{|mn|}}{c}\right).$$

The Kuznetsov trace formula is a sort of the Poisson summation formula ([Iwaniec (1995), p. 138]) which is equivalent to the functional equation and so Theorem 5.24 may be proved most naturally as a consequence of the functional equation in Theorem 5.25 (due to Fay [Fay (1997)]) for the **Kloosterman sums zeta-function with Bessel function weight** $Z_s(m,n)$ [Iwaniec (1995), pp. 138-139].

$Z_s(m,n)$ is defined by [Iwaniec (1995), (5.16), p. 81]

$$2\sqrt{|mn|}Z_s(m,n) = \sum_{c} c^{-1} S_{ab}(m,n;c)\tilde{J}_{2s-1}\left(\frac{4\pi}{c}\sqrt{mn}\right), \qquad (5.167)$$

where

$$\tilde{J}_{2s-1}(z) = \begin{cases} J_{2s-1}(z), & mn > 0, \\ I_{2s-1}(z), & mn < 0. \end{cases} \qquad (5.168)$$

On the other hand, the Kloosterman sums zeta-function $L_s(m,n)$ is defined by

$$L_s(m,n) = \sum_{c>0} c^{-2s} S_{ab}(m,n;c) \qquad (5.169)$$

and studied by Selberg [Selberg (1965)] and later by Goldfeld and Sarnak [Goldfeld and Sarnak (1983)]. Their relationship is given by

$$Z_s(m,n) = \pi(4\pi^2|mn|)^{s-1} \sum_{k=0}^{\infty} \frac{(4\pi^2 mn)^k}{k!\Gamma(k+2s)} L_{s+k}(m,n). \qquad (5.170)$$

Theorem 5.25. (Fay [Fay (1997), Corollary 3.6]) *The series $Z_s(m,n)$ has an analytic continuation over the whole s-plane and satisfies the functional equation*

$$Z_s(m,n) - Z_{1-s}(m,n) \qquad (5.171)$$

$$= \frac{1}{2\pi|n|}\delta_{ac}\delta_{mn}\sin\pi\left(s-\frac{1}{2}\right) - \frac{1}{2s-1}\sum_{c}\varphi_{ac}(m,1-s)\varphi_{ac}(n,s).$$

$Z_s(m, n)$ has simple poles at $s = s_j$ and $s = 1 - s_j$ with residue

$$-\frac{1}{2s_j - 1} \sum_{s_k = s_j} \bar{\rho}_{ak}(m) \rho_{bk}(n) \tag{5.172}$$

provided $s_j \neq \frac{1}{2}$.

Proof of Theorem 5.24 from Theorem 5.25. Let

$$f(s) = \frac{4\pi \left(s - \frac{1}{2}\right)}{\cos \pi \left(s - \frac{1}{2}\right)} h \left(i \left(s - \frac{1}{2}\right)\right). \tag{5.173}$$

Multiplying (5.171) by $f(s)$ and integrating along $\operatorname{Re} s = 1 - \varepsilon$, we obtain

$$\frac{1}{2\pi i} \int_{(1-\varepsilon)} Z_s(m, n) h(s) \, ds + \frac{1}{2\pi i} \int_{(\varepsilon)} Z_s(m, n) h(s) \, ds \tag{5.174}$$

$$= \frac{1}{2\pi |n|} \delta_{ac} \delta_{mn} \frac{1}{2\pi i} \int_{(1-\varepsilon)} \sin \pi \left(s - \frac{1}{2}\right) h(s) \, ds$$

$$- \frac{1}{2\pi i} \int_{(1-\varepsilon)} \sum_c \varphi_{ac}(m, 1 - s) \varphi_{ac}(n, s) \frac{h(s)}{2s - 1} \, ds.$$

On the left side of (5.174), the second integral transforms into the first plus the sum of residues

$$-4\pi \sum_j \frac{\bar{\rho}_{ak}(m) \rho_{bk}(n)}{\cosh(\pi t_j)} h(t_j). \tag{5.175}$$

Substituting the definition (5.167), these integrals become

$$\frac{2}{2\pi i} \int_{(1-\varepsilon)} Z_s(m, n) h(s) \, ds = \frac{1}{\sqrt{|mn|}} \sum_c c^{-1} S_{ab}(m, n; c) \tilde{h} \left(\frac{4\pi \sqrt{|mn|}}{c}\right), \tag{5.176}$$

where

$$\tilde{h}(x) = \frac{1}{2\pi i} \int_{(1-\varepsilon)} \tilde{J}_{2s-1}(x) f(s) \, ds. \tag{5.177}$$

Since the mid-term amounts to $\delta_{ab} \delta_{ab} h_0$, where

$$h_0 = \frac{1}{\pi} \int_{-\infty}^{\infty} t \tanh(\pi t) h(t) \, dt, \tag{5.178}$$

the proof follows under normalization (5.164).

5.3.1   *Kuznetsov formula reversed*

As described on [Iwaniec (1995), pp. 141-147] and on [Baker (2003), pp. 16-18], the reverse Kuznetsov trace formula is to be regarded as an expansion in J-Bessel functions due to Sears and Titchmarsh [Sears and Titchmarsh (1954), (4.4)] and in many literature this reversed form is referred to as the Kuznetsov trace formula, [Baker (2003)], [Miatello (1990)], [Venkov (1990)].

Let $f(x)$ be a continuous function of bounded variation on $\mathbb{R}_+$ such that

$$\int_0^\infty |f(x)| x^{-1/2}\, dx < \infty \tag{5.179}$$

in particular $f(x)$ may be an infinitely many times differentiable function with compact support. We follow [Iwaniec (1995)] which gives the clearest exposition thereof.

Let f^0 be the projection of f on the space spanned by odd indexed Bessel functions $\{J_{2n+1} | n \geq 0\}$ and is given by the **Neumann series**

$$f^0(x) = \sum_{n=0}^\infty 2(2n+1) J_{2n+1}(x) N_f(2n+1) \tag{5.180}$$

and

$$N_f(\lambda) = \int_0^\infty J_\lambda(y) f(y) \frac{dy}{y} \tag{5.181}$$

is the **Neumann integral**. In [Venkov (1990), p. 36] this formula is stated with a typo of $2ir$ which should be $2n+1$.

On the other hand, let

$$B_\nu(x) = \frac{1}{2\sin\frac{\pi}{2}\nu} \left(J_{-\nu}(x) - J_\nu(x)\right) \tag{5.182}$$

and define the **Titchmarsh integral** $T_f(t)$ by

$$T_f(t) = \int_0^\infty f(x) B_{2it}(x) \frac{dx}{x} = \int_0^\infty f(x) \frac{J_{-2it}(x) - J_{2it}(x)}{2\sinh\pi t} \frac{dx}{x}. \tag{5.183}$$

Then define the continuous superposition of projections of f on B_{2it} by

$$f^\infty(x) = \int_0^\infty T_f(t) B_{2it}(x) \tanh(\pi t)\, dt = \int_0^\infty T_f(t) \frac{J_{-2it}(x) - J_{2it}(x)}{2\cosh(\pi t)}\, dt. \tag{5.184}$$

Theorem 5.26. (Sears-Titchmarsh inversion) *We have the* **Sears-Titchmarsh inversion**

$$f = f^0 + f^\infty. \tag{5.185}$$

Also define the constant

$$f^\infty = \frac{1}{\pi} \int_0^\infty T_f(t) \tanh(\pi t) \, dt. \qquad (5.186)$$

Theorem 5.27. (Iwaniec [Iwaniec (1995), Theorem 9.5]) *Let* $\mathfrak{a}$, $\mathfrak{b}$ *be cusps of the Fuchsian group of the first kind* Γ *and let* $m, n > 0$ *be integers. Then for any test function* f *satisfying the condition* (5.165), *we have*

$$\delta_{\mathfrak{a}\mathfrak{b}}\delta_{mn}f^\infty + \sum_c c^{-1} S_{\mathfrak{a}\mathfrak{b}}(m,n;c) f^\infty \left(\frac{4\pi\sqrt{|mn|}}{c} \right) \qquad (5.187)$$

$$= \sum_j T_f(t_j)\bar{\nu}_{\mathfrak{a}}(m)\nu_{\mathfrak{b}j}(n) + \sum_c \frac{1}{4\pi} \int_{-\infty}^\infty T_f(t)\bar{\eta}_{\mathfrak{a}j\mathfrak{c}}(m,t)\eta_{\mathfrak{b}\mathfrak{c}}(n,t) \, dt.$$

It seems that the corresponding formulas in [Baker (2003)] are incorrect in comparison with other refs.

Let $\mathcal{S}_k(\Gamma)$ denote the space of cusp forms of weight k which is spanned by Poincaré series in contrast to (5.219):

$$U_{\mathfrak{a}m}(z) = \sum_{\gamma \in \Gamma_\mathfrak{a} \backslash \Gamma} j_{\sigma_\mathfrak{a}^{-1}\gamma}(z)^{-k} e(m\sigma_\mathfrak{a}^{-1}\gamma z), \qquad (5.188)$$

where j is the denominator appearing in

$$j_\gamma(z)^{-k} f(\gamma z) = f(z), \quad \gamma \in \Gamma, \qquad (5.189)$$

so that

$$\operatorname{Im}\gamma(z) = y^{-2}|j_\gamma(z)|^2. \qquad (5.190)$$

Let f_{jk} be an ONB of $\mathcal{S}_k(\Gamma)$. Let

$$U_{\mathfrak{a}m}(z) = \frac{(k-2)!}{(4\pi m)^{k-1}} \sum_j \overline{\hat{f}_{\mathfrak{a}jk}}(m) f_{jk}(z) \qquad (5.191)$$

be the expansion of Poincaré series with respect to this basis. The exact complement to Theorem 5.27 is

Theorem 5.28. (Iwaniec [Iwaniec (1995), Theorem 9.6]) *Let* $\mathfrak{a}$, $\mathfrak{b}$ *be cusps of the Fuchsian group of the first kind* Γ *and let* $m, n > 0$ *be integers. Then for any test function* f *satisfying the condition* (5.165), *we have*

$$-\delta_{\mathfrak{a}\mathfrak{b}}\delta_{mn}f^\infty + \sum_c c^{-1} S_{\mathfrak{a}\mathfrak{b}}(m,n;c) f^0 \left(\frac{4\pi\sqrt{|mn|}}{c} \right) \qquad (5.192)$$

$$= \sum_{k=1}^\infty i^{2k} N_f(2k-1)\bar{\psi}_{\mathfrak{a}j2k}(m)\psi_{\mathfrak{b}j2k}(n),$$

where $\psi_{\mathfrak{a}jk(m)}$ are the normalized Fourier coefficients

$$\psi_{\mathfrak{a}jk}(m) = \left(\frac{\pi^{-k}\Gamma(k)}{(4m)^{k-1}}\right)^{1/2} \hat{f}_{\mathfrak{a}jk}(m). \qquad (5.193)$$

Adding Theorems 5.27 and 5.28 in view of the Sears-Titchmarsh inversion give the reversed Kuznetsov sum formula in contrast to Theorem 5.24.

Theorem 5.29. (Iwaniec [Iwaniec (1995), Theorem 9.5]) *Let $\mathfrak{a}$, $\mathfrak{b}$ be cusps of the Fuchsian group Γ of the first kind and let $m, n > 0$ be integers. Then for any test function f satisfying the condition (5.165), we have*

$$\sum_c c^{-1}\mathcal{S}_{\mathfrak{a}\mathfrak{b}}(m,n;c)f\left(\frac{4\pi\sqrt{|mn|}}{c}\right) \qquad (5.194)$$

$$= \sum_j T_f(t_j)\bar{\nu}_{\mathfrak{a}j}(m)\nu_{\mathfrak{b}j}(n) + \sum_{\mathfrak{c}}\frac{1}{4\pi}\int_{-\infty}^{\infty} T_f(t)\bar{\eta}_{\mathfrak{a}\mathfrak{c}}(m,t)\eta_{\mathfrak{b}\mathfrak{c}}(n,t)\,\mathrm{d}t$$

$$+ \sum_{k=1}^{\infty} i^{2k}N_f(2k-1)\bar{\psi}_{\mathfrak{a}j2k}(m)\psi_{\mathfrak{b}j2k}(n).$$

Remark 5.4. We remark that Theorem 5.29 is to coincide with [Motohashi (1997), Theorem 2.3, p. 64]. But, since in the latter, the Neumann series part ([Motohashi (1997), (2.2.6), p. 51]) is replaced by [Motohashi (1997), (2.2.9), p. 51], it has a seemingly different outlook. On [Motohashi (1997), p. 92], it is claimed that the Neumann series expansion due to Titchmarsh and others ([Sears and Titchmarsh (1954)]) is dispensed with in his argument.

Using the Hankel transform, Theorem 5.29 may be clearly understood as a procedure corresponding to the mapping $x \leftrightarrow \frac{1}{x}$ under which the intervals $(0,1)$ and $(1,\infty)$ map each other.

The **Hankel transform of order** ν with $\mathrm{Re}\,\nu > -\frac{1}{2}$ is defined by

$$H_{f,\nu}(u) = \int_0^\infty f(x)J_\nu(ux)\,\mathrm{d}x \qquad (5.195)$$

for $y > 0$ which has admits the inversion formula

$$f(x) = \int_0^\infty H_{f,\nu}(u)J_\nu(ux)ux\,\mathrm{d}u. \qquad (5.196)$$

We use the $\nu = 0$ case $H_{f,0}$ and refer to it as the Hankel transform H_f.

Then the Neumann series (5.180) may be expressed as

$$f^0(x) = \int_0^1 H_f(u) J_0(ux) ux \, du. \tag{5.197}$$

Proof is given in [Iwaniec (1995), p. 231] and depends on the formula

$$\frac{2\nu}{xy} \frac{d}{dx} J_\nu(ux) J_\nu(uy) = u J_{\nu-1}(ux) J_{\nu-1}(uy) - u J_{\nu+1}(ux) J_{\nu+1}(uy). \tag{5.198}$$

Integrating over $\nu = \ell = 1, 3, 5, \cdots$ one has

$$\sum_{0 < \ell \text{odd}} 2\ell J_\ell(x) J_\ell(y) = xy \int_0^1 u J_0(ux) J_0(uy) \, du. \tag{5.199}$$

Multiplying (5.199) by $f(y)y^{-1}$ and integrating in y, we obtain

$$f^0(x) = \int_0^1 ux H_f(u) J_0(ux) \, du. \tag{5.200}$$

By the Sears-Titchmarsh inversion (5.26) and the Hankel inversion (5.196), we conclude that

$$f^\infty(x) = \int_1^\infty ux H_f(u) J_0(ux) \, du. \tag{5.201}$$

Hence we conclude that the Sears-Titchmarsh inversion is a counterpart of the division of the real line into two parts, which then is responsible for the functional equation. Thus the Sears-Titchmarsh is a counterpart of the functional equation and this explains the reverse Kuznetsov formula is also deduced from the zeta-symmetry as the Kuznetsov trace formula has been most naturally deduced from Fay's functional equation in Theorem 5.25.

We note the following correspondence between groups. The right half-plane is represented by the positive real axis, which is a multiplicative group, in view of **analytic continuation**. And the Spiegelung $\tau \leftrightarrow -\frac{1}{\tau}$, which is one of the generators of the modular group, corresponds to the inversion $x \leftrightarrow \frac{1}{x}$ under which the two intervals $(0,1)$ and $(1,\infty)$ maps into each other.

domain	positive real axis	upper half-plane	right-half plane
mapping	$x \leftrightarrow \frac{1}{x}$	$\tau \leftrightarrow -\frac{1}{\tau}$	$s \leftrightarrow 1-s$
group	$\mathbb{R}^\times$	Γ	symmetry: $s \leftrightarrow 1-s$

Table 5.3. Mapping and group structure

In the Riemann zeta-case the variables are connected by

$$s = -i\tau, \quad \operatorname{Im}\tau > 0, \quad \operatorname{Re}s > 1 \qquad (5.202)$$

because of the presence of a simple pole at $s = 1$, the right-half plane is narrowed down to $\operatorname{Re}s > 1$, the domain of absolute convergence, leaving the **critical strip** $0 < \sigma < 1$ so mysterious.

Proof of (5.200). We use two well-known formulas

$$\frac{2\nu}{z}J_\nu(z) = J_{\nu-1}(z) + J_{\nu+1}(z) \qquad (5.203)$$

and

$$J'_\nu(z) = \frac{1}{2}(J_{\nu-1}(z) - J_{\nu+1}(z)). \qquad (5.204)$$

Since

$$\frac{2\nu}{xy}\frac{\mathrm{d}}{\mathrm{d}x} = J'_\nu(ux)\frac{2\nu}{y}J_\nu(uy) + J'_\nu(ux)\frac{2\nu}{y}J_\nu(ux),$$

(5.200) follows on substituting above formulas.

5.3.2   *Proof of Theorem 5.29*

We shall prove Theorem 5.29 by modifying Motohashi's argument [Motohashi (1997), pp. 44-67] and under the assumption that

$$f^*(s) = \int_0^\infty f(x)\left(\frac{x}{2}\right)^{-2s}\mathrm{d}x \qquad (5.205)$$

or

$$f(x) = \frac{1}{2\pi i}\int_{(\alpha)} f^*(x)\left(\frac{x}{2}\right)^{2s-1}\mathrm{d}s. \qquad (5.206)$$

We appeal to

$$G_{0,2}^{1,0}\left(z\left|\begin{matrix} - \\ a,b \end{matrix}\right.\right) = \frac{1}{2\pi i}\int_{(c)} \frac{\Gamma(a+s)}{\Gamma(1-b-s)}z^{-s}\,\mathrm{d}s = z^{\frac{1}{2}(a+b)}J_{a-b}(2\sqrt{z})\,. \qquad (5.207)$$

Motohashi [Motohashi (1997), (2.2.17), p. 54] sticks to the special case

$$\frac{x}{2}G_{0,2}^{1,0}\left(\frac{x^2}{4}\left|\begin{matrix} - \\ \nu - 1/2, -\nu - 1/2 \end{matrix}\right.\right) = \frac{x}{2}\left(\frac{x^2}{4}\right)^{\frac{1}{2}(-1)}J_{2\nu}\left(2\sqrt{\frac{x^2}{4}}\right) = J_{2\nu}(x). \qquad (5.208)$$

But another essential case appears [Motohashi (1997), (2.4.13, p. 65)]:

$$G_{0,2}^{1,0}\left(\frac{x^2}{4}\ \bigg|\ \begin{matrix} - \\ k-1, -k \end{matrix}\right) = \left(\frac{x^2}{4}\right)^{\frac{1}{2}(-1)} J_{2k-1}\left(2\sqrt{\frac{x^2}{4}}\right) = \frac{2}{x} J_{2k-1}(x).$$

$$(5.209)$$

Another formula which is not explicitly mentioned in Motohashi [Motohashi (1997)] is

$$G_{0,2}^{2,0}\left(z\ \bigg|\ \begin{matrix} - \\ a, b \end{matrix}\right) = \frac{1}{2\pi i} \int_{(c)} \Gamma(a+s)\Gamma(b+s)z^{-s}\,ds = 2\,z^{\frac{1}{2}(a+b)} K_{a-b}\left(2\sqrt{z}\right).$$

$$(5.210)$$

This is applied in the special case

$$G_{0,2}^{2,0}\left(z\ \bigg|\ \begin{matrix} - \\ -\frac{1}{2}+ir, -\frac{1}{2}-ir \end{matrix}\right) = 2\,z^{-\frac{1}{2}} K_{2ir}\left(2\sqrt{z}\right). \qquad (5.211)$$

For the proof of Theorem 5.29 we need two more well-known formulas

$$K_\nu(z) = \frac{\pi}{2\sin(\pi\nu)}\left(I_{-\nu}(z) - I_\nu(z)\right) \qquad (5.212)$$

and

$$I_\nu(z) = e^{-\frac{\pi i\nu}{2}} J_\nu(iz), \qquad -\frac{\pi}{2} < \arg z < \pi. \qquad (5.213)$$

In Motohashi [Motohashi (1997)], the integral

$$I := \frac{1}{2\pi i} \int_{(\alpha)} \sin(\pi s)\Gamma\left(s - \frac{1}{2} + is_j\right)\Gamma\left(s - \frac{1}{2} - is_j\right)\left(\frac{x^2}{4}\right)^{-s}\,ds$$

$$(5.214)$$

is computed in an indirect way. A common procedure is to apply Euler's formula and rewrite it as

$$I = \frac{1}{2i}\left(\frac{1}{2\pi i} \int_{(\alpha)} \Gamma\left(s - \frac{1}{2} + is_j\right)\Gamma\left(s - \frac{1}{2} - is_j\right)\left(\frac{x^2}{4}e^{-\pi i}\right)^{-s}\,ds\right.$$

$$(5.215)$$

$$\left. - \frac{1}{2\pi i} \int_{(\alpha)} \Gamma\left(s - \frac{1}{2} + is_j\right)\Gamma\left(s - \frac{1}{2} - is_j\right)\left(\frac{x^2}{4}e^{\pi i}\right)^{-s}\,ds\right).$$

Putting $z = s - \frac{1}{2}$ and factoring out $\left(\frac{x^2}{4}e^{-\pi i}\right)^{-1/2}$ and $\left(\frac{x^2}{4}e^{-\pi i}\right)^{-1/2}$, respectively, and applying (5.210), we further transform it into

$$I = \frac{1}{2i}\frac{2}{x}i\left(2K_{2is_j}\left(e^{-\frac{1}{2}\pi i}x\right) + 2K_{2is_j}\left(e^{\frac{1}{2}\pi i}x\right)\right)$$

$$= \frac{1}{x}\frac{\pi}{\sin 2\pi i s_j}\left(I_{-2is_j}\left(e^{-\frac{1}{2}\pi i}x\right) - I_{2is_j}\left(e^{-\frac{1}{2}\pi i}x\right)\right. \tag{5.216}$$

$$\left. + I_{-2is_j}\left(e^{\frac{1}{2}\pi i}x\right) - I_{2is_j}\left(e^{\frac{1}{2}\pi i}x\right)\right)$$

by (5.212). Hence by (5.213)

$$I = \frac{1}{2\sin(\pi i 2s_j)x}\left(J_{-2is_j}(x)\left(\left(e^{-\frac{1}{2}\pi i}\right)^{-2is_j} + \left(e^{\frac{1}{2}\pi i}\right)^{-2is_j}\right)\right. \tag{5.217}$$

$$\left. - J_{2is_j}(x)\left(\left(e^{-\frac{1}{2}\pi i}\right)^{2is_j} + \left(e^{\frac{1}{2}\pi i}\right)^{2is_j}\right)\right).$$

Hence we arrive at

$$\frac{1}{2\pi i}\int_{(\alpha)}\Gamma\left(s - \frac{1}{2} + is_j\right)\Gamma\left(s - \frac{1}{2} - is_j\right)\left(\frac{x^2}{4}\right)^{-s}\,ds. \tag{5.218}$$

$$= \frac{1}{2\sinh(\pi s_j)}\left(J_{-2is_j}(x) - J_{2is_j}(x)\right)\frac{1}{x}.$$

On [Venkov (1981), p. 36] the **real analytic Poincaré series**

$$P_m(z, s) = \sum_{\sigma \in \Gamma_\infty \backslash \Gamma} \bar{\chi}(\sigma)y(\sigma z)^s e((m - \xi)\sigma z), z \in \mathcal{H} \tag{5.219}$$

is defined, where χ is a one-dimensional unitary representation of Γ. Here $\xi \in \mathbb{R}$ is defined as follows. For a cusp σ_a let σ_a is a scaling matrix which satisfies the following condition similar to (5.151):

$$\sigma_a \infty = a, \sigma_a^{-1}\Gamma_a\sigma_a = \langle\gamma\rangle, \gamma \equiv \begin{pmatrix} 1 & 1 \\ 0 & 1 \end{pmatrix} \bmod \pm 1. \tag{5.220}$$

Then

$$\chi(\sigma_a^{-1}\Gamma_a\sigma_a) = e(\xi). \tag{5.221}$$

In view of the definition of a general Fuchsian group of the first kind, it seems that

$$P_{\mathfrak{a}m}(z,\chi) = \sum_{\gamma \in \Gamma_\alpha \backslash \Gamma} \bar{\chi}(\gamma) y(\sigma_{\mathfrak{a}^{-1}}\gamma z)^s e(m\sigma_{\mathfrak{a}}^{-1}\gamma z) \qquad (5.222)$$

can be a target for study. But for simplicity's sake, we restrict to the case where the cusp is only at infinity, and $\xi = 0$. The series converges absolutely and uniformly on compact sets in the domain $\sigma > 1$ and $P_m(\cdot, s) \in H(\Gamma : \chi)$.

The key idea for trace formulas is due to Selberg [Selberg (1965)] and depends on two ways of expressing the inner product of two Poincaré series

$$(P_m(\cdot, s_1), P_m(\cdot, \bar{s}_2)) = \int_{\mathcal{F}} P_m(z, s_1) \overline{P_n(z, \bar{s}_2)}\, dz.$$

The inner product of Poincaré series is computed in Motohashi [Motohashi (1997), p. 44] with $\chi = 1$ and integers $m, n > 0$.

Lemma 5.6. (Unfolding), [Motohashi (1997), Lemma 2.1] *For* $\operatorname{Re} s_2 + \alpha > \operatorname{Re} s_1 > \alpha + \frac{1}{4}$ *we have*

$$(P_m(\cdot, s_1), P_m(\cdot, \bar{s}_2)) = \delta_{mn}\Gamma(s_1 + s_2 - 1)(4\pi m)^{1-s_1-s_2} \qquad (5.223)$$

$$+ 2^{2(1-s_2)}\pi^{s_1-s_2+1}\Gamma(s_1+s_2-1)\sum_{c=1}^{\infty} c^{-2s_1}S(m,n;c)W\left(\frac{4\pi}{c}\sqrt{mn}; s_1, s_2\right),$$

where

$$W(2z; s_1.s_2) = G_{1,3}^{2,0}\left(z \left| \begin{matrix} s_2 \\ 0, s_2 - s_1, 1 - s_1 \end{matrix}\right.\right). \qquad (5.224)$$

Lemma 5.7. (Parseval formula), [Motohashi (1997), Lemma 2.2] *Let* $\{u_j\}$ *be as in (5.161). Then for* $\operatorname{Re} s_j > \frac{3}{4}$, $j = 1, 2$ *we have*

$$(P_m(\cdot, s_1), P_m(\cdot, \bar{s}_2)) = \frac{\pi}{\Gamma(s_1)\Gamma(s_2)}(4\pi\sqrt{mn})^{1-s_1-s_2}\left(\frac{n}{m}\right)^{\frac{1}{2}(s_1-s_2)}$$

$$\times \left\{ \sum_{j=1}^{\infty} \overline{\rho_j(m)}\rho_j(n)\Theta(s_1, s_2; s_j) \right. \qquad (5.225)$$

$$\left. + \frac{1}{\pi}\int_{-\infty}^{\infty} \cosh(\pi r)\bar{\eta}_{\mathfrak{a}c}(m,r)\eta_{\mathfrak{b}c}(n,r)\Theta(s_1, s_2; r)\, dr \right\},$$

where

$$\Theta(s_1, s_2; r)$$

$$= \Gamma\left(s_1 - \frac{1}{2} + ir\right)\Gamma\left(s_1 - \frac{1}{2} - ir\right)\Gamma\left(s_2 - \frac{1}{2} + ir\right)\Gamma\left(s_2 - \frac{1}{2} - ir\right).$$

$$(5.226)$$

Remark 5.5. For Parseval identity cf. (5.25). $W(2z; s_1, s_2)$ appears on [Motohashi (1997), (2.1.3), p. 44] ([Motohashi (1997), (2.6.15), p. 77]) and (5.230) is its special case. $\mathfrak{p}(x, t)$ on [Motohashi (1997), (2.3.5), p. 75] is another form for the same G-function. [Motohashi (1997), (2.6.15), p. 77] is another form for the W-function. As he states on pp. 91–92, his Lemma 2.1 replaces Kuznetsov's heavy use of Bessel functions.

Another G-function that appears is [Motohashi (1997), (2.6.11), p. 75] with half-unit argument though, which is known as the Barnes integral.

$$
G_{1,4}^{4,0} \left(\frac{1}{2} \, \middle| \, \begin{matrix} (0,1) \\ \left(\frac{1}{2}(\mu+\nu), \frac{1}{2}\right), \left(\frac{1}{2}(\mu-\nu), \frac{1}{2}\right), \left(\frac{1}{2}(-\mu+\nu), \frac{1}{2}\right), \left(-\frac{1}{2}(\mu+\nu), \frac{1}{2}\right) \end{matrix} \right)
$$
$$
= K_\mu \left(\frac{1}{2} \right) K_\nu \left(\frac{1}{2} \right),
$$

$$(5.227)$$

which is then applied in deriving [Motohashi (1997), Lemma 3.5, p. 107] for the functional equation for the Rankin L-function associated with a pair of cusp forms.

Theorem 5.30. (Unfolding = Parseval formula), [Motohashi (1997), Lemma 2.2] *Let $\{u_j\}$ be as in (5.161). Then for s_j's satisfying $\operatorname{Re} s_2 + \alpha > \operatorname{Re} s_1 > \alpha + \frac{1}{4}$ and $\operatorname{Re} s_j > \frac{3}{4}$, $j = 1, 2$ we have*

$$
\delta_{mn} \Gamma(s_1 + s_2 - 1)(4\pi m)^{1-s_1-s_2} \tag{5.228}
$$
$$
+ 2^{2(1-s_2)} \pi^{s_1-s_2+1} \Gamma(s_1 + s_2 - 1) \sum_{c=1}^{\infty} c^{-2s_1} S(m, n; c) W\left(\frac{4\pi}{c} \sqrt{mn}; s_1, s_2 \right)
$$
$$
= \frac{\pi}{\Gamma(s_1)\Gamma(s_2)} (4\pi\sqrt{mn})^{1-s_1-s_2} \left(\frac{n}{m} \right)^{\frac{1}{2}(s_1-s_2)}
$$
$$
\times \left\{ \sum_{j=1}^{\infty} \overline{\rho_j(m)} \rho_j(n) \Theta(s_1, s_2; s_j) \right.
$$
$$
\left. + \frac{1}{\pi} \int_{-\infty}^{\infty} \cosh(\pi r) \bar{\eta}_{a\mathfrak{c}}(m, r) \eta_{b\mathfrak{c}}(n, r) \Theta(s_1, s_2; r) \, dr \right\}.
$$

As an important corollary, we prove

Corollary 5.1. *For integers $m, n > 0$ let $L_s(m, n)$ be the Kloosterman* *sums zeta-function defined by* (5.169). *Then for* $\operatorname{Re} s > \frac{1}{4}$ *we have*

$$\tilde{L}_{mn}(s) := (2\pi\sqrt{mn})^{2s-1} L_s(m,n) \tag{5.229}$$

$$= \frac{1}{2} \frac{1}{\cosh(\pi r)} \sum_{j=1}^{\infty} \overline{\rho_j(m)}\rho_j(n) \sin(\pi s) \Gamma\left(s - \frac{1}{2} + is_j\right) \Gamma\left(s - \frac{1}{2} - is_j\right)$$

$$+ \frac{1}{2\pi} \int_{-\infty}^{\infty} \overline{\eta}_{ac}(m,r)\eta_{bc}(n,r) \sin(\pi s) \Gamma\left(s - \frac{1}{2} + ir\right) \Gamma\left(s - \frac{1}{2} - ir\right) dr$$

$$+ \sum_{c=1}^{\infty} c^{-1} S(m,n;c) J_{2k-1}\left(\frac{4\pi}{c}\sqrt{mn}\right) \frac{\Gamma(k-1+s)}{\Gamma(k+1-s)} - \delta_{mn} \frac{1}{2\pi} \frac{\Gamma(s)}{\Gamma(1-s)}.$$

Proof. In Theorem 5.30 we choose $s_1 = 1$, $s_2 = s$ with $\operatorname{Re} s > 1 - \alpha$ for $0 < \alpha < \frac{1}{4}$ and invoke ([Motohashi (1997), (2.4.3), p. 63])

$$\frac{\Gamma(s)}{\Gamma(1-s)} W(z;1,s) = \left(\frac{z}{2}\right)^{2(s-1)} - \frac{2}{z} \sum_{k=1}^{\infty} (2k-1) \frac{\Gamma(k-1+s)}{\Gamma(k+1-s)} J_{2k-1}(z), \tag{5.230}$$

which provides an evaluation of $G_{1,3}^{2,0}\left(z \left|\begin{matrix} s \\ 0, 1-s, 0 \end{matrix}\right.\right)$.

(5.226) amounts to

$$\Theta(1,s;r) = \Gamma\left(\frac{1}{2} + ir\right) \Gamma\left(\frac{1}{2} - ir\right) \Gamma\left(s - \frac{1}{2} + ir\right) \Gamma\left(s - \frac{1}{2} - ir\right) \tag{5.231}$$

$$= \frac{\pi}{\cos(\pi ir)} \Gamma\left(s - \frac{1}{2} + ir\right) \Gamma\left(s - \frac{1}{2} - ir\right).$$

(5.228) reduces to

$$\delta_{mn}\Gamma(s)(4\pi m)^{-s} \tag{5.232}$$

$$+ 2^{2-2s}\pi^{2-s} n^{1-s}\Gamma(s) \sum_{c=1}^{\infty} c^{-2} S(m,n;c) W\left(\frac{4\pi}{c}\sqrt{mn};1,s\right)$$

$$= \frac{\pi}{\Gamma(1)\Gamma(s)} (4\pi\sqrt{mn})^{-s} \left(\frac{n}{m}\right)^{\frac{1}{2}(1-s)}$$

$$\times \left\{ \sum_{j=1}^{\infty} \overline{\rho_j(m)}\rho_j(n)\Theta(1,s;s_j) \right.$$

$$\left. + \frac{1}{\pi} \int_{-\infty}^{\infty} \cosh(\pi r)\overline{\eta}_{ac}(m,r)\eta_{bc}(n,r)\Theta(1,s;r) \, dr \right\}. \tag{5.233}$$

Dividing both sides by $\Gamma(1-s)$ we are led to the form to which we may apply (5.230).

$$\delta_{mn}\frac{\Gamma(s)}{\Gamma(1-s)}(4\pi m)^{-s} + \pi^s m^{s-1}L_s(m,n) \tag{5.234}$$

$$+ 2^{1-2s}\pi^{1-s}\frac{1}{\sqrt{m}}n^{1/2-s}\sum_{c=1}^{\infty}c^{-1}S(m,n;c)J_{2k-1}\left(\frac{4\pi}{c}\sqrt{mn}\right)\frac{\Gamma(k-1+s)}{\Gamma(k+1-s)}$$

$$= \sin(\pi s)(4\pi\sqrt{mn})^{-s}\left(\frac{n}{m}\right)^{\frac{1}{2}(1-s)}$$

$$\times\left\{\pi\sum_{j=1}^{\infty}\frac{\overline{\rho_j(m)}\rho_j(n)}{\cosh(\pi s_j)}\Gamma\left(s-\frac{1}{2}+is_j\right)\Gamma\left(s-\frac{1}{2}-is_j\right)\right.$$

$$\left.+ \int_{-\infty}^{\infty}\overline{\eta}_{\mathfrak{ac}}(m,r)\eta_{\mathfrak{bc}}(n,r)\Gamma\left(s-\frac{1}{2}+ir\right)\Gamma\left(s-\frac{1}{2}-ir\right)\,\mathrm{d}r\right\}.$$

Dividing both sides by $2^{1-2s}\pi^{1-s}\sqrt{m}^{-1/2}\sqrt{n}^{1/2-s}$, we rewrite (5.234) as

$$\delta_{mn}\frac{1}{2\pi}\frac{\Gamma(s)}{\Gamma(1-s)}\left(\frac{n}{m}\right)^{s-1/2} \tag{5.235}$$

$$+ (2\pi\sqrt{mn})^{2s-1}L_s(m,n) + \sum_{c=1}^{\infty}c^{-1}S(m,n;c)J_{2k-1}\left(\frac{4\pi}{c}\sqrt{mn}\right)\frac{\Gamma(k-1+s)}{\Gamma(k+1-s)}$$

$$= \frac{\sin(\pi s)}{2\pi}$$

$$\times\left\{\pi\sum_{j=1}^{\infty}\frac{\overline{\rho_j(m)}\rho_j(n)}{\cosh(\pi s_j)}\Gamma\left(s-\frac{1}{2}+is_j\right)\Gamma\left(s-\frac{1}{2}-is_j\right)\right.$$

$$\left.+ \int_{-\infty}^{\infty}\overline{\eta}_{\mathfrak{ac}}(m,r)\eta_{\mathfrak{bc}}(n,r)\Gamma\left(s-\frac{1}{2}+ir\right)\Gamma\left(s-\frac{1}{2}-ir\right)\,\mathrm{d}r\right\}.$$

Since

$$\delta_{mn}\left(\frac{n}{m}\right)^{s-1/2} = \delta_{mn},$$

(5.235) leads to (5.229) completing the proof. □

To prove Theorem 5.29, we multiply (5.229) by $\varphi^*(s)$ and integrate in the form (5.205) or substitute (5.229) in

$$\frac{1}{2\pi i}\int_{(\alpha)}\tilde{L}_{mn}(s)f^*(s)\,\mathrm{d}s = \sum_{c=1}^{\infty}c^{-1}S(m,n;c)f\left(\frac{4\pi}{c}\sqrt{mn}\right). \tag{5.236}$$

Then we interchange the order of integration and appeal to the known formulas for G-functions. We apply (5.210), (5.212), and (5.213) successively for the integrals of the form

$$\int_{(\alpha)} \Gamma\left(s - \frac{1}{2} + ir\right) \Gamma\left(s - \frac{1}{2} - ir\right) z^{-s}\,\mathrm{d}r$$

and we appeal to (5.209) for integrals of the form

$$\int_{(\alpha)} \frac{\Gamma(k - 1 + s)}{\Gamma(k + 1 - s)} z^{-s}\,\mathrm{d}r.$$

More rigorous treatment appears in [Motohashi (1997), pp. 64-67]. What we obtain is [Motohashi (1997), Theorem 2.3, p. 64] in which there appear the Fourier coefficients $q_{m,n}(k)$ of $U_{am}(z)$ in (5.222) appears. We may replace $q_{m,n}(k)$ by the coefficients in (5.191) by [Motohashi (1997), Lemma 2.3, p. 51] to arrive at Theorem 5.29.

Riemann

Chapter 6

Control systems and number theory

Abstract: This chapter is a continuation of Chapter 6, **I** and we keep the material up to J-lossless factorization. If we choose the Riemann zeta-function $\zeta(s+1)$ shifted by 1, where $s = \sigma + j\omega$, then we can see a very close correspondence between their region of stability and analyticity (both are $\sigma > 0$), the critical lines ($\sigma = 0$ and $\sigma = -\frac{1}{2}$, respectively), only the functional equation aspect being not clear in the former since the transfer functions are mostly just rational functions. Here, we might introduce a more sophisticated transfer function, as in the case of FOPID (Fractional Order Proportional-Integral-Differential) control, which turns out to be the Riemann-Liouville fractional integral transform (6.76), **I** of the input function as discussed §6.9, **I**. Namely, we might expect that a finer theory of transfer functions be developed which would be certain special functions whose avatars appear as rational functions (e.g. Padè approximation thereof).

On the other hand, the H^∞-norm or more generally, H^{2k}-norm problem appear as the **worst-case estimate** of effectiveness of control, thus the name **robust control**. In addition to the correspondence alluded to above, if we consider the power norm (6.87), we can see a very close correspondence between $2k$-mean values of the zeta-function and the H^{2k} control problem. Especially, we speculate that control theory is that of finite power signals while number theory is that of infinite power signals or more schematically, control theory is the world of Lindlöf hypothesis while number theory is in the world of the RH.

As our specific and main tool, we treat the chain scattering representation (of a plant) and H^∞-control problem. The homographic transformation of the former works as the action of the symplectic group on the Siegel upper half-space (in the case of constant matrices) and we hope to have a new look at the control as a group action. We may view in particular the unity feedback system as accommodated in the chain scattering representation, giving a better insight into the structure of the system. In the unity feedback system included is the PID-compensator whose generalization to fractional order calculus is the FOPID controller mentioned above.

Some new materials are added. One is illustrative examples of smart grid circuits. By viewing the connections of impedances as concatenation of operators, we can elucidate complicated circuits.

The second is the robust control system and generalized Nevanlinna-Pick interpolation, in which we essentially use the chain scattering representation (of 1 variable) to generalize Kimura's results. In this way we resurrect the theory of complex functions in one variable in control theory setting.

Apart from this we add control theory in the unit disc, §6.6 as opposed to "number theory in the unit disc", Chapter 4 in which the most striking theorem is that taking arcwise limit of the integral of the theta-function one may deduce both the differentiability of the Riemann's function and the quadratic reciprocity law at a stretch. In the case of control theory it seems still in the state of a germ and some threshold entities involving the Beurling inner function, Blascke products etc. which may be treated by number-theoretic methods. The boundary behavior is also of interest with A. Wintner as one the earliest precursor.

Our main concern being the exhibition of similarities between control systems and number theory and incorporate specific controllers in chain scattering representation, we state only fragments of control theory itself and some remain speculative (but hopefully more accessible to non-specialists). The interested reader may consult more specified books including [Helton (1998)], [Kimura (1997)] etc.

6.1   Introduction and preliminaries

It turns out that there is great similarity in control theory and number theory in their treatment of the signals in time domain (t) and frequency domain (ω) or an expanded frequency domain $(s = \sigma + j\omega)$ which is conducted by the Laplace transform in the case of control theory while in the theory of zeta-functions, this role is played by the Mellin transform, both of which convert the signals in time domain to those in the right half-plane (expanded frequency domain). Following the tradition of electrical engineering, we use the symbol j to indicate the imaginary unit i hereafter, so that the complex variable is always in the form

$$s = \sigma + j\omega. \tag{6.1}$$

For integral transforms, cf. §6.9, **I**.

6.2   State space representation and the visualization principle

Let $x = x(t) \in \mathbb{R}^n$, $u = u(t) \in \mathbb{R}^r$ and $y = y(t) \in \mathbb{R}^m$ be the **state** function, **input** function and **output** function, respectively. We write $\dot{x}$ for $\frac{\mathrm{d}}{\mathrm{d}t}x$. The system of (differential equations) DEs

$$\begin{cases} \dot{x} = Ax + Bu, \\ y = Cx + Du \end{cases} \tag{6.2}$$

is called a **state equation** for a **linear system**, where $A \in M_{n,n}(\mathbb{R})$, B, C, D are given constant matrices.

The state x is not visible while the input and output are so, and the state may be thought of as an interface between the past and the present information since it contains all the information contained in the system from the past. The x being invisible, (6.2) would read

$$y = Du, \tag{6.3}$$

which appears in many places in literature in disguised form. All the subsequent systems e.g. (6.30) are variations of (6.3). And whenever we would like to obtain the state equation, we are to restore the state x to make a recourse to (6.2), which we would call the **visualization principle**. In the case of feedback system, it is often the case that (6.3) is given in the form of (6.37). It is quite remarkable that this controller S works for the matrix variable in the symplectic geometry (cf. §6.3.1).

Definition 6.1. A linear system with the input $u = o$

$$\dot{x} = \frac{d}{dt}x = Ax, \tag{6.4}$$

called an **autonomous system**, is said to be **asymptotically stable** if for all initial values, $x(t)$ approaches a limit as $t \to \infty$.

It is well-known that the solution of (6.4) is given by

$$x = e^{At}x(0), \tag{6.5}$$

where $e^{At} = \sum_{n=0}^{\infty} \frac{1}{n!}A^n t^n$ is the matrix exponential function. Hence the system is asymptotically stable if and only if

$$\|e^{At}\| \to 0 \quad \text{as} \quad t \to \infty. \tag{6.6}$$

A linear system is said to be **stable** if (6.6) holds, which is the case if all the eigenvalues of A have negative real parts. Cf. §6.5 in this regard. It also amounts to saying that the step response of the system approaches a limit as time elapses, where **step response** means a response

$$y(t) = \int_0^t e^{A(t-\tau)}u(\tau)\,d\tau, \tag{6.7}$$

with the **unit step function** $u = u(t)$ as the input function, which is 0 for $t < 0$ and 1 for $t \geq 0$.

Using the Lagrange constant variation method, the first equation in (6.2) can be solved as follows. In (6.5) we put $\boldsymbol{x}(t) = e^{At}\boldsymbol{c}(t)$. Then we have

$$A\boldsymbol{x} + B\boldsymbol{u} = e^{At}(A\boldsymbol{c}(t) + \dot{\boldsymbol{c}}),$$

or

$$\dot{\boldsymbol{c}}(t) = e^{-At}B\boldsymbol{u}(t),$$

whence

$$\boldsymbol{c}(t) = B \int e^{-A\tau}\boldsymbol{u}(\tau)\,\mathrm{d}\tau + C,$$

with an arbitrary constant C. Substituting this and specifying the constant by the initial value, we deduce that

$$\boldsymbol{x} = \boldsymbol{x}(t) = e^{At}\boldsymbol{x}(0) + Be^{At}\int_0^t e^{-A\tau}\boldsymbol{u}(t)\,\mathrm{d}\tau. \tag{6.8}$$

Up here, the things are happening in the time domain. We now move to a frequency domain. For this purpose, we refer to the Laplace transform discussed in §6.9, **I**. It has the effect of shifting from the time domain to the frequency domain and vice versa. For more details, see e.g. [Kimura (1997)]. Taking the Laplace transform of (6.2) with $\mathbf{x}(0) = \mathbf{o}$, we obtain

$$\begin{cases} sX(s) = AX(s) + BU(s) \\ Y(s) = CX(s) + DU(s), \end{cases} \tag{6.9}$$

which we solve as

$$Y(s) = G(s)U(s), \tag{6.10}$$

where

$$G(s) = C(sI - A)^{-1}B + D, \tag{6.11}$$

where I indicates the identity matrix, which is sometimes denoted I_n to show its size.

In general, supposing that the initial values of all the signals in a system are 0, we call the ratio of output/input of the signal, the **transfer function**, and denote it by $G(s)$, $\Phi(s)$, etc. We may suppose so because if the system is in equilibrium, then we may take the values of parameters at that moment as standard and may suppose the initial values to be 0.

(6.11) is called the **state space representation** (form, realization, description, characterization) of the transfer function $G(s)$ of the system (6.2), and is written as

$$G(s) = \left(\begin{array}{c|c} A & B \\ \hline C & D \end{array}\right). \tag{6.12}$$

According to the visualization principle above, we have the **embedding principle**: Given a state space representation of a transfer function $G(s)$, it is to be embedded in the **state equation** (6.2).

Example 6.1. If

$$G(s) = \left(\begin{array}{c|c} A & B \\ \hline C & D \end{array}\right) = \left(\begin{array}{cc|c} 0 & 1 & 0 \\ -2 & -3 & 1 \\ \hline -10 & -2 & 2 \end{array}\right), \tag{6.13}$$

then it follows from (6.11) that

$$G(s) = -2\frac{s+5}{(s+1)(s+2)} + 2 = \frac{2(s+3)(s-1)}{(s+1)(s+2)}. \tag{6.14}$$

The principle above will establish the most important **cascade connection** (concatenation rule) [Kimura (1997), (2.13), p. 15]: Given two state space representations

$$G_k(s) = \left(\begin{array}{c|c} A_k & B_k \\ \hline C_k & D_k \end{array}\right), \quad k = 1, 2, \tag{6.15}$$

their cascade connection $G(s) = G_1(s)G_2(s)$ is given by

$$G(s) = G_1(s)G_2(s) = \left(\begin{array}{cc|c} A_1 & B_1C_2 & B_1D_2 \\ O & A_2 & B_2 \\ \hline C_1 & D_1C_2 & D_1D_2 \end{array}\right). \tag{6.16}$$

Proof of (6.16)   We have the input/output relation (6.11)

$$Y(s) = G_1(s)U(s), \quad U(s) = G_2(s)V(s) \tag{6.17}$$

which means that

$$\begin{cases} \dot{\boldsymbol{x}} = A_1\boldsymbol{x} + B_1\boldsymbol{u}, \\ \boldsymbol{y} = C_1\boldsymbol{x} + D_1\boldsymbol{u} \end{cases} \tag{6.18}$$

and

$$\begin{cases} \dot{\boldsymbol{\xi}} = A_2\boldsymbol{\xi} + B_2\boldsymbol{v}, \\ \boldsymbol{u} = C_2\boldsymbol{\xi} + D_2\boldsymbol{v} \end{cases} \tag{6.19}$$

Eliminating $\boldsymbol{u}$, we conclude that

$$\begin{cases} \dot{\boldsymbol{x}} = A_1\boldsymbol{x} + B_1C_2\boldsymbol{\xi} + B_1D_2\boldsymbol{v}, \\ \boldsymbol{y} = C_1\boldsymbol{x} + D_1C_2\boldsymbol{\xi} + D_1D_2\boldsymbol{v} \end{cases} \tag{6.20}$$

Hence

$$\begin{cases} \begin{pmatrix} \dot{\boldsymbol{x}} \\ \dot{\boldsymbol{\xi}} \end{pmatrix} = \begin{pmatrix} A_1 & B_1C_2 \\ O & A_2 \end{pmatrix} \begin{pmatrix} \boldsymbol{x} \\ \boldsymbol{\xi} \end{pmatrix} + \begin{pmatrix} B_1D_2 \\ B_2 \end{pmatrix} \boldsymbol{v}, \\ \boldsymbol{y} = \begin{pmatrix} C_1 & D_1C_2 \end{pmatrix} \begin{pmatrix} \boldsymbol{x} \\ \boldsymbol{\xi} \end{pmatrix} + D_1D_2\boldsymbol{v} \end{cases} \tag{6.21}$$

whence we conclude (6.16).

Example 6.2. Given two state space representations (6.125), their parallel connection $G(s) = G_1(s) + G_2(s)$ is given by

$$G(s) = G_1(s) + G_2(s) = \left(\begin{array}{cc|c} A_1 & O & B_1 \\ O & A_2 & B_2 \\ \hline C_1 & C_2 & D_1 + D_2 \end{array} \right). \tag{6.22}$$

Indeed, we have (6.18) and for (6.19), we have

$$\begin{cases} \dot{\boldsymbol{\xi}} = A_2\boldsymbol{\xi} + B_2\boldsymbol{u}, \\ \boldsymbol{y} + \boldsymbol{z} = C_2\boldsymbol{\xi} + D_2\boldsymbol{u} \end{cases} \tag{6.23}$$

Hence for (6.21), we have

$$\begin{cases} (\boldsymbol{x} + \boldsymbol{\xi})^{\cdot} = A_1\boldsymbol{x} + A_2\boldsymbol{\xi} + (B_1 + B_2)\boldsymbol{u}, \\ \boldsymbol{y} = C_1\boldsymbol{x} + C_2\boldsymbol{\xi} + (D_1 + D_2)\boldsymbol{v} \end{cases} \tag{6.24}$$

whence (6.22) follows.

As an example, combining (6.16) and (6.22) we deduce

$$I - G_1(s)G_2(s) = \left(\begin{array}{ccc|c} I & O & & O \\ O & -A_1 & -B_1C_2 & -B_1D_2 \\ & O & -A_2 & -B_2 \\ \hline O & -C_1 & -D_1C_2 & V^{-1} \end{array} \right). \tag{6.25}$$

Example 6.3. For (6.2), we consider the inversion $U(s) = G^{-1}(s)Y(s)$. Solving the second equality in (6.2) for $\mathbf{u}$ we obtain

$$\mathbf{u} = -D^{-1}C\mathbf{x} + D^{-1}\mathbf{y}.$$

Substituting this in the first equality in (6.2), we obtain

$$\dot{\mathbf{x}} = (A - BD^{-1}C)\mathbf{x} + BD^{-1}\mathbf{y},$$

whence

$$G^{-1}(s) = \left(\begin{array}{c|c} A - BD^{-1}C & -BD^{-1} \\ \hline -D^{-1}C & D^{-1} \end{array}\right). \tag{6.26}$$

Example 6.4. If the transfer function

$$\Theta(s) = \begin{pmatrix} \Theta_{11} & \Theta_{12} \\ \Theta_{21} & \Theta_{22} \end{pmatrix} \tag{6.27}$$

has a state space representation

$$\Theta(s) = \left(\begin{array}{c|c} A & B \\ \hline C & D \end{array}\right) = \left(\begin{array}{c|cc} A & B_1 & B_2 \\ \hline C_1 & D_{11} & D_{12} \\ C_2 & D_{21} & D_{22} \end{array}\right), \tag{6.28}$$

then we are to embed it in the linear system

$$\begin{cases} \dot{\mathbf{x}} = A\mathbf{x} + \begin{pmatrix} B_1 & B_2 \end{pmatrix} \begin{pmatrix} \mathbf{b}_1 \\ \mathbf{b}_2 \end{pmatrix}, \\ \begin{pmatrix} \mathbf{a}_1 \\ \mathbf{a}_2 \end{pmatrix} = \mathbf{y} = \begin{pmatrix} C_1 \\ C_2 \end{pmatrix} \mathbf{x} + \begin{pmatrix} D_{11} & D_{12} \\ D_{21} & D_{22} \end{pmatrix} \begin{pmatrix} \mathbf{b}_1 \\ \mathbf{b}_2 \end{pmatrix}. \end{cases} \tag{6.29}$$

6.3 Chain scattering representation

Following [Kimura (1997), p. 7, p. 67], we first give the definition of a chain scattering representation of a system.

Suppose $\mathbf{a}_1 \in \mathbb{R}^m$, $\mathbf{a}_2 \in \mathbb{R}^q$, $\mathbf{b}_1 \in \mathbb{R}^r$ and $\mathbf{b}_2 \in \mathbb{R}^p$ are related by

$$\begin{pmatrix} \mathbf{a}_1 \\ \mathbf{a}_2 \end{pmatrix} = P \begin{pmatrix} \mathbf{b}_1 \\ \mathbf{b}_2 \end{pmatrix}, \tag{6.30}$$

where

$$P = \begin{pmatrix} P_{11} & P_{12} \\ P_{21} & P_{22} \end{pmatrix}. \tag{6.31}$$

According to the embedding principle, this is to be thought of as $\mathbf{y} = S\mathbf{u}$ corresponding to the second equality in (6.2).

(6.30) means that

$$a_1 = P_{11}b_1 + P_{12}b_2, \quad a_2 = P_{21}b_1 + P_{22}b_2. \tag{6.32}$$

Assume that P_{21} is a *(square) regular* matrix (whence $q = r$). Then from the second equality of (6.32), we obtain

$$b_1 = P_{21}^{-1}(a_2 - P_{22}b_2) = -P_{21}^{-1}P_{22}b_2 + P_{21}^{-1}a_2. \tag{6.33}$$

Substituting (6.33) in the first equality of (6.32), we deduce that

$$a_1 = (P_{12} - P_{11}P_{21}^{-1}P_{22})b_2 + P_{11}P_{21}^{-1}a_2. \tag{6.34}$$

Hence putting

$$\Theta = CHAIN(P) = \begin{pmatrix} P_{12} - P_{11}P_{21}^{-1}P_{22} & P_{11}P_{21}^{-1} \\ -P_{21}^{-1}P_{22} & P_{21}^{-1} \end{pmatrix} \tag{6.35}$$

$$= \begin{pmatrix} \Theta_{11} & \Theta_{12} \\ \Theta_{21} & \Theta_{22} \end{pmatrix},$$

which is usually referred to as a **chain scattering representation** of P, we obtain an equivalent form of (6.30)

$$\begin{pmatrix} a_1 \\ b_1 \end{pmatrix} = CHAIN(P) \begin{pmatrix} b_2 \\ a_2 \end{pmatrix} = \begin{pmatrix} \Theta_{11} & \Theta_{12} \\ \Theta_{21} & \Theta_{22} \end{pmatrix} \begin{pmatrix} b_2 \\ a_2 \end{pmatrix}. \tag{6.36}$$

Suppose that a_2 is fed back to b_2 by

$$b_2 = Sa_2, \tag{6.37}$$

where S is a **controller**. Multiplying the second equality in (6.32) by S and incorporating (6.37), we find that

$$b_2 = Sa_2 = SP_{21}b_1 + SP_{22}b_2,$$

whence $b_2 = (I - SP_{22})^{-1}SP_{21}b_1$.

Let the **closed-loop transfer function** Φ be defined by

$$a_1 = \Phi b_1. \tag{6.38}$$

Φ is given by

$$\Phi = P_{11} + P_{12}(I - SP_{22})^{-1}SP_{21}. \tag{6.39}$$

(6.39) is sometimes referred to as a **linear fractional transformation** and denoted by

$$LF(P; S).$$

Substituting (6.37), (6.36) becomes

$$\begin{pmatrix} a_1 \\ b_1 \end{pmatrix} = \begin{pmatrix} \Theta_{11}S + \Theta_{12} \\ \Theta_{21}S + \Theta_{22} \end{pmatrix} a_2,$$

whence we deduce that

$$\Phi = (\Theta_{11}S + \Theta_{12})(\Theta_{21}S + \Theta_{22})^{-1} = \Theta S, \tag{6.40}$$

the linear fractional transformation (which is also referred to as a **homographic transformation** and denoted by $HM(\Phi; S)$), where in the last equality we mean the action of Θ on the variable S. We must impose the non-constant condition $|\Theta| \neq 0$. Then $\Theta \in GL_{m+r}(\mathbb{R})$.

If S is obtained from S' under the action of Θ', $S = \Theta'S'$, then its composition Θ'' with (6.40) yields $\Theta''S' = \Phi\Phi' = \Theta\Theta'S'$, i.e.

$$\Theta'' = \Theta\Theta', \quad HM(\Theta; HM(\Theta'; S)) = HM(\Theta\Theta'; S), \tag{6.41}$$

which is referred to as the **cascade connection** or the cascade structure of Θ and Θ'.

Thus the chain-scattering representation of a system allows us to treat the feedback connection as a cascade connection. An FOPID controller in §6.6, **I**, being a unity feedback connection, is also accommodated in this framework.

Suppose a closed-loop system is given with $z = a_1 \in \mathbb{R}^m$, $y = a_2 \in \mathbb{R}^q$, $w = b_1 \in \mathbb{R}^r$ and $u = b_2 \in \mathbb{R}^p$ and Φ given by (6.31).

§6.5 introduces the Hardy space H^∞ which consists of functions analytic in $\mathcal{RHP}$—right half-plane $\sigma > 0$ and are in L^∞ on the boundary $\sigma = 0$, the norm being defined by (6.83) below.

H^∞-control problem

Find a controller S such that the closed-loop system is internally stable and the transfer function Φ satisfies

$$\|\Phi\|_\infty < \gamma \tag{6.42}$$

for a positive constant γ.

Finally we briefly refer to the dual chain-scattering representation of the plant P in (6.31). We assume P_{12} is a square invertible matrix (whence $m = p$). Then the argument goes in parallel to that leading to (6.36). Defining the **dual chain scattering matrix** by

$$DCHAIN(P) = \begin{pmatrix} P_{12}^{-1} & P_{11}P_{12}^{-1} \\ -P_{12}^{-1}P_{22}P_{21} - P_{22}P_{12}^{-1}P_{11} \end{pmatrix}, \tag{6.43}$$

we obtain

$$CHAIN(P) \cdot DCHAIN(P) = E. \tag{6.44}$$

6.3.1   *Siegel upper space*

Let $*$ denote the conjugate transpose of a square matrix: $S^* = {}^t\bar{S}$ and let the imaginary part of S defined by $\operatorname{Im} S = \frac{1}{2j}(S - S^*)$. Let $\mathcal{H}_n$ be the **Siegel upper half-space** consisting of all the matrices S (recall Eq. (6.37)) whose imaginary parts are positive definite ($\operatorname{Im} S > 0$—imaginary parts of all eigen values are positive) and satisfies $S = {}^tS$:

$$\mathcal{H}_n = \{S \in \mathrm{M}_n(\mathbb{C}) \,|\, \operatorname{Im} S > 0,\ S = {}^tS\} \tag{6.45}$$

and let $\mathrm{Sp}(n, \mathbb{R})$ denote the **symplectic group** of order n:

$$\mathrm{Sp}(n, \mathbb{R}) \tag{6.46}$$

$$= \left\{ \Theta = \begin{pmatrix} \Theta_{11} & \Theta_{12} \\ \Theta_{21} & \Theta_{22} \end{pmatrix} \middle| \begin{pmatrix} \Theta_{11} & \Theta_{12} \\ \Theta_{21} & \Theta_{22} \end{pmatrix}^{-1} = \begin{pmatrix} \Theta_{22} & -{}^t\Theta_{12} \\ -{}^t\Theta_{21} & \Theta_{11} \end{pmatrix} \right\}.$$

The **action** of $\mathrm{Sp}(n, \mathbb{R})$ on $\mathcal{H}_n$ is defined by (6.40) which we restate as

$$\Theta S = (\Theta_{11}S + \Theta_{12})(\Theta_{21}S + \Theta_{22})^{-1} (= \Phi). \tag{6.47}$$

Theorem 6.1. *For a controller S living in the Siegel upper space, its rotation $Z = -jS$ lies in the right half-space $\mathcal{RHS}$. i.e. being stable because it has positive real parts. For the controller Z, the feedback connection*

$$-j\mathbf{b}_2 = Z(-j\mathbf{a}_2) \tag{6.48}$$

is accommodated in the cascade connection (6.41) of the chain scattering representation Θ , which is then viewed as the action (6.41) of $\Theta \in \mathrm{Sp}(n, \mathbb{R})$ on $S \in \mathcal{H}_n$:

$$(\Theta\Theta')S = \Theta(\Theta'S); \quad \text{or} \quad HM(\Theta; HM(\Theta'; S)) = HM(\Theta\Theta'; S), \tag{6.49}$$

where Θ is subject to the condition

$${}^t\bar{\Theta}U\Theta = U, \tag{6.50}$$

with $U = \begin{pmatrix} O & I_n \\ -I_n & O \end{pmatrix}$.

Remark 6.1. With action, we may introduce the orbit decomposition of $\mathcal{H}_n$ and whence the fundamental domain. We note that in the special case of $n = 1$, we have $\mathcal{H}_1 = \mathcal{H}$ and $\mathrm{Sp}(1, \mathbb{R}) = \mathrm{SL}_n(\mathbb{R})$ and the theory of modular forms of one variable is well-known. Siegel modular forms are generalizations of the one variable case into several variables. This corresponds to the relation between SISO (single input, single output) and MIMO (multiple input, multiple output) systems. As in the case of the *sushmna principle* in

[Kanemitsu (2007)], there is a need to rotate the upper half-space into the right half-space $\mathcal{RHS}$, which is a counter part of the right-half plane $\mathcal{RHP}$. In the case of Siegel modular forms, the matrices are constant, while in control theory, they are analytic functions (mostly rational functions analytic in $\mathcal{RHP}$). Cf. the passage in [Kimura (1997), p. 103, ll. 2-3]. The most essential and restrictive assumption in §6.3 is that P_{21} is a *regular* matrix and it is stated [Kimura (1997), p. 103] that Xin and Kimura [Xin and Kimura (1994)] succeeded in treating the case of non-singular matrices. A general theory would be useful for control theory. See §6.7 for physically realizable cases. There are many research problems lying in this direction.

The following table shows the correspondence between the control system and the Riemann zeta-function ($\zeta(s)$, $s = \sigma + j\omega$). If a shift of 1 is made, then the region of convergence coincide. In the case of the Riemann zeta-function $\zeta(s)$, the critical line used to be $\sigma = 1$ and non-vanishing on $\sigma = 1$ is equivalent to the prime number theorem (PNT). For the refined form of the PNT, the critical line is $\sigma = \frac{1}{2}$. Cf. Proposition 6.1 below.

system	functions	action	region of convergence	critical line
S	rational	symplectic	$\sigma > 0$	$\sigma = 0$
s or τ	meromorphic	modular	$\sigma > 1$	$\sigma = 1$ or $\frac{1}{2}$

Table 6.1. Correspondence between control systems and zeta-functions

6.3.2 *(Unity) feedback system*

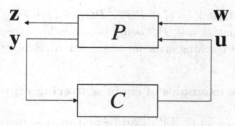

Fig. 6.1: Unity feedback system

The synthesis problem of a controller of the **unity feedback system**, depicted in Figure 6.1, refers to the **sensitivity reduction problem**, which asks for the estimation of the **sensitivity function** $S = S(s)$ multiplied by an appropriate frequency weighting function $W = W(s)$:

$$S = (I + CP)^{-1}, \tag{6.51}$$

is a transfer function from r to e, where $C = K$ is a **compensator** and P is a plant. The problem consists in reducing the magnitude of S over a specified frequency range Ω, which amounts to finding a compensator C stabilizing the closed-loop system such that

$$\|WS\|_\infty < \gamma \tag{6.52}$$

for a positive constant γ.

To accommodate this in the H^∞ control problem (6.30), we choose the matrix elements P_{ij} of P in such a way that the closed-loop transfer function Φ in (6.39) coincides with WS. First we are to choose $P_{22} = -P$. Then we choose $P_{12} = -W, P_{21}P$ so that $P_{12}P_{21} = -WP$. Then Φ becomes $P_{11} + W(I + CP)^{-1}CP = P_{11} - W + W(I + CP)^{-1}$. Hence choosing $P_{11} = W$, we have $\Phi = WS$. Hence we may choose e.g.

$$P = \begin{pmatrix} P_{11} & P_{12} \\ P_{21} & P_{22} \end{pmatrix} = \begin{pmatrix} W & P \\ -W & -P \end{pmatrix}. \tag{6.53}$$

Proof of (6.51).

Denoting the Laplace transforms by the corresponding capital letters and taking the disturbance d into account, we obtain since $U = CE = C(R - Y)$

$$Y = PU + PD = PC(R - Y) + D,$$

whence it follows that $Y = (I + PC)^{-1}PCR + (I + PC)^{-1}D$. In the case where $d = 0$, so that $D = 0$, PC being the open loop transfer function, we see that SR is the tracking error for the input R. Hence (6.51) holds true.

6.4   Illustrative examples of chain scattering representation

This corresponds to §3 of [Lifh (2015)] and unifies many examples given there about smart grid circuits.

The three main ingredients in (electrical) circuits are coil (L), condenser (C) and resistance (R). The inverse electro-motive force generated by these component is given respectively by

$$e_L = -L\frac{di}{dt}, \quad e_C = -\int_0^t i\,dt, \quad e_R = -Ri,$$ (6.54)

where $i = i(t)$ and $e = e(t)$ with subscript indicates the current and the voltage of the prescribed component resp. Hereafter we write u for the voltage e.

The governing law of the circuits is the Kirchhoff laws which have two versions. The first law is the one used for node analysis to the effect that the sum of currents flowing into a node is 0 while the second law is the one used for loop analysis which is to the effect that the sum of all electro-motive forces in a closed circuit is 0.

We view (6.54) as **impedance operators** $Z = Z(t)$ (eventually as distributions [Rosenfel'd and Yahinson (1966)]) with current I flowing through it, i.e.

$$Z_L I = -L\frac{dI}{dt}, \quad Z_C I = -\int_0^t I\,dt, \quad Z_R = -RI,$$ (6.55)

where L, C, and R indicate the coil, condenser and resistance, respectively.

We consider the cascade connection $\mathcal{Z}$ of two impedances $Z_i = Z_i(t)$, $i = 1, 2$ with the potential u and u_0 and with the current I flowing from u to u_0, thus the voltage difference is $u_0 - u$.

Fig. 6.2: Cascade connection of two impedances

Then we have

$$u_0 - u = (Z_1 + Z_2)I,$$ (6.56)

where the addition of impedances is in the sense of additive operators, i.e. $Z_1 I + Z_2 I$.

Suppose Z_2 is a parallel connection of the coil with inductance L and a resistor with resistance r and that the current flowing the coil is i and that Z_1 is a resister with resistance R_1. Then

$$Z_2 I = -L\frac{di}{dt} = -r(I - i), \quad u_0 - u = -R_1 I - r(I - i). \tag{6.57}$$

Substituting the first equality in (6.57) into (6.56), we obtain

$$u_0 - u = -(R_1 + r)I + ri$$

or

$$I = \frac{r}{R_1 + r} i - \frac{1}{R_1 + r}(u_0 - u). \tag{6.58}$$

Substituting this in (6.57), we obtain

$$L\frac{di}{dt} = r\left(\frac{r}{R_1 + r} - 1\right) i - \frac{r}{R_1 + r}(u_0 - u),$$

whence

$$\frac{di}{dt} = -\frac{R_1 r}{(R_1 + r)L} i - \frac{r}{(R_1 + r)L}(u_0 - u). \tag{6.59}$$

More generally, we consider the combination of two such cascade connections at node u_0. Two impedances $Z_i = Z_i(t)$, $i = 3, 4$ with the potential $u = 0$ and u_0 and with the current I_2 flowing from u to u_0.

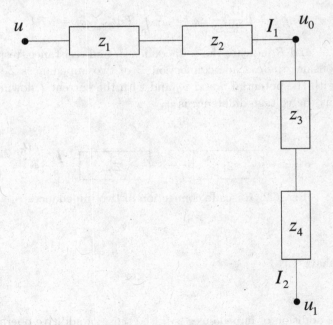

Fig. 6.3: Cascade connection of two impedances

Then we have

$$u_0 = (Z_3 + Z_4)I_2. \tag{6.60}$$

We choose $Z_3 = Z_C = u_C$ (condenser) and Z_4 a resister with resistance R_4. Then (6.60) becomes

$$u_0 = u_C + R_4(I_1 - I_2), \tag{6.61}$$

where I_1 is the current flowing the impedances Z_1, Z_2 in (6.56). Hence substituting (6.61) in (6.59), we conclude

Theorem 6.2. *If two cascade connections of two impedances Z_i, $i = 1, \cdots, 4$ are connected at the node u_0 with voltage difference $u_0 - u$ and i flowing through it, then (6.57) in the form*

$$u_0 - u = -R_1 I - L\frac{di}{dt} \tag{6.62}$$

describes the whole paradigm as either

$$\frac{di}{dt} = -\frac{1}{(R_1 + r)L}(rR_1 i - R_4 I_1) - \frac{R_4}{(R_1 + r)L}I_2 - \frac{1}{(R_1 + r)L}(u_C - u) \tag{6.63}$$

or

$$\frac{di}{dt} = -\frac{R_1 + R_4}{L}I_1 + \frac{R_4}{L}I_2 - \frac{1}{L}(u_C - u), \tag{6.64}$$

where u_C is described in Corollary 6.1.

Corollary 6.1. *The cascade connection (6.59) of two impedances Z_1 (resistance) and Z_2 (the parallel connection of a coil and a resistance) is a special case of (6.63) with $I_1 = I_2$ and (therefore) $u_C = u_0$. A special case with $I_1 = i_1$, $I_2 = i_2$ is given in [Lifh (2015)].*

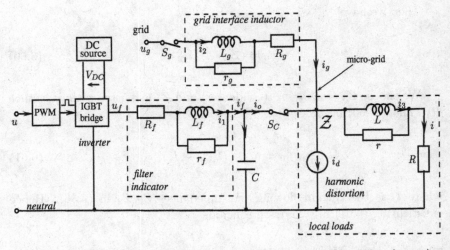

Fig. 6.4: The three-phase inverter system (represented as a single one)

Example 6.5. We consider the three cascade connections $\mathcal{Z}_i$, $i = 1, 2, 3$ connected at a node $\mathcal{Z}$ with the flowing-in currents i_f, i_g and flowing-out currents i, i_d (d is for harmonic distortion). And the configurations of each $\mathcal{Z}_i$ are similar. $\mathcal{Z}_1$ indicates the filter inductor with Z_2 a parallel connection of the coil with inductance L_f and a resistor with resistance r_f and that the current flowing the coil is i_1 and that Z_1 is a resister with resistance R_f. Other two are similar as given in Table 6.2 below.

In this case, (6.58) reads

$$i_f = \frac{r_f}{R_f + r_f}i_1 - \frac{1}{R_f + r_f}(u_0 - u), \quad i_g = \frac{r_g}{R_g + r_g}i_2 - \frac{1}{R_g + r_g}(u_0 - u_g),$$

$$(6.65)$$

$$i = \frac{r}{R + r}i_3 + \frac{1}{R + r}u_0.$$

Hence from Theorem 6.2 it follows that

$$\frac{di_1}{dt} = -\frac{R_f r_f}{(R_f + r_f)L_f} i_1 - \frac{r_f}{(R_f + r_f)L_f}(u_0 - u), \qquad (6.66)$$

$$\frac{di_2}{dt} = -\frac{R_g r_g}{(R_g + r_g)L_g} i_2 - \frac{r_g}{(R_g + r_g)L_g}(u_0 - u_g),$$

$$\frac{di_3}{dt} = -\frac{Rr}{(R + r)L} i_3 + \frac{r}{(R + r)L} u_0.$$

We want to add one more state variable u_0 which is the electro-motive force u_c generated by the condenser C. Since the current flowing the condenser is $i_f - i_0$, we have

$$\frac{d}{dt} u_0 = \frac{1}{C}(i_f - i_0). \qquad (6.67)$$

At the node $\mathcal{Z}$, we have by the Kirchhoff law,

$$i_0 + i_g = i + i_d,$$

whence

$$i_0 = i + i_d - i_g. \qquad (6.68)$$

Hence (6.67) amounts to

$$\frac{d}{dt} u_0 = \frac{1}{C}(i_f + i_g - i - i_d). \qquad (6.69)$$

Substituting (6.65) in this, we deduce that

$$\frac{d}{dt} u_0 = \frac{1}{C}\left(\frac{r_f}{R_f + r_f} i_1 + \frac{r_g}{R_g + r_g} i_2 - \frac{r}{R + r} i_3 \right. \qquad (6.70)$$

$$- \left(\frac{r_f}{R_f + r_f} + \frac{r_g}{R_g + r_g} + \frac{r}{R + r} \right) u_0$$

$$\left. - i_d + \frac{1}{R_g + r_g} u_g + \frac{1}{R_f + r_f} u \right).$$

We put

$$\boldsymbol{x} = \begin{pmatrix} i_1 \\ i_2 \\ i_3 \\ u_0 \end{pmatrix}, \quad \boldsymbol{u} = \begin{pmatrix} \boldsymbol{w} \\ u \end{pmatrix}, \quad \boldsymbol{w} = \begin{pmatrix} i_d \\ u_g \\ i_{\text{ref}} \\ u \end{pmatrix}, \quad \boldsymbol{y} = \begin{pmatrix} e \\ i_0 \end{pmatrix}, \qquad (6.71)$$

where $e = u_{\text{ref}} - u_0$. Then, putting

$$A = \begin{pmatrix} -\frac{R_f r_f}{(R_f+r_f)L_f} & 0 & 0 & -\frac{r_f}{(R_f+r_f)L_f} \\ 0 & -\frac{R_g r_g}{(R_g+r_g)L_g} & 0 & -\frac{r_g}{(R_g+r_g)L_g} \\ 0 & 0 & -\frac{Rr}{(R+r)L} & -\frac{r}{(R+r)L} \\ \frac{r_f}{R_f+r_f} & \frac{r_g}{R_g+r_g} & -\frac{r}{R+r} & -\frac{1}{C}\left(\frac{r_f}{R_f+r_f} + \frac{r_g}{R_g+r_g} + \frac{r}{R+r}\right) \end{pmatrix},$$

(6.72)

$$B = (B_1, B_2), \quad B_1 = \begin{pmatrix} 0 & 0 & 0 \\ 0 & \frac{r_g}{(R_g+r_g)L_g} & 0 \\ 0 & 0 & 0 \\ \frac{1}{C} & \frac{1}{(R_g+r_g)C} & 0 \end{pmatrix}, \quad B_2 = \begin{pmatrix} \frac{r_f}{(R_f+r_f)L_f} \\ 0 \\ 0 \\ \frac{1}{(R_f+r_f)C} \end{pmatrix},$$

$$C = \begin{pmatrix} C_1 \\ C_2 \end{pmatrix}, \quad C_1 = (0,0,0,-1), \quad C_2 = \left(0, -\frac{r_g}{R_g+r_g}, \frac{r}{R+r}, \frac{1}{R+r} + \frac{1}{R_g+r_g}\right),$$

$$D = \begin{pmatrix} D_{11} & D_{12} \\ D_{21} & D_{22} \end{pmatrix} = \begin{pmatrix} 0 & 0 & 1 & 0 \\ 1 & -\frac{1}{R_g+r_g} & 0 & 0 \end{pmatrix},$$

(6.66) and (6.70) lead to

$$\frac{\mathrm{d}}{\mathrm{d}t}x = Ax + Bu = Ax + (B_1, B_2)\begin{pmatrix} w \\ u \end{pmatrix}, \tag{6.73}$$

$$y = Cx + Du = \begin{pmatrix} C_1 \\ C_2 \end{pmatrix} x + \begin{pmatrix} D_{11} & D_{12} \\ D_{21} & D_{22} \end{pmatrix} \begin{pmatrix} w \\ u \end{pmatrix}.$$

connection	coil	c. resist.	c. current	resistance	current
$\mathcal{Z}_1$	L_f	r_f	i_1	R_f	i_f
$\mathcal{Z}_2$	L_g	r_g	i_2	R_g	i_g
$\mathcal{Z}_3$	L	r	i_3	R	i_0

Table 6.2. Components of $\mathcal{Z}_i$'s

6.5 Finite vs. infinite power signals

More general treatment of norm is given in Chapter 5, §5.1.

Definition 6.2. The norm of $x = \begin{pmatrix} x_1 \\ \vdots \\ x_n \end{pmatrix} \in \mathbb{C}^n$ is defined to be the standard

Euclidean norm

$$\|\boldsymbol{x}\| = \|\boldsymbol{x}\|_2 = \sqrt{\sum_{j=1}^{n} |x_j|^2} \tag{6.74}$$

or

$$\|\boldsymbol{x}\|_1 = \|\boldsymbol{x}\|_1 = |x_1| + \cdots + |x_n|, \tag{6.75}$$

or by the maximum norm

$$\|\boldsymbol{x}\| = \|\boldsymbol{x}\|_\infty = \max\{|x_1|, \cdots, |x_n|\}, \tag{6.76}$$

or anything that satisfies the axioms of the norm. They introduce the same topology on $\mathbb{C}^n$. We refer to these as **Euclidean norm** denoted without suffices.

The definition of the norm of a matrix of degree d, say should be given in a similar way by viewing its elements as an n^2-dimensional vector, i.e. embedding it in $\mathbb{C}^{n^2}$. If $A = (a_{ij})$, $1 \le i, j \le n$, then

$$\|A\| = \|A\|_2 = \sqrt{\sum_{i,j=1}^{n} |a_{ij}|^2} \tag{6.77}$$

or otherwise.

Exercise 45. The maximum norm is a limit of the p-norm as $p \to \infty$: For $\boldsymbol{x} = (z_1, \cdots, x_n)$,

$$\lim_{p \to \infty} \|\boldsymbol{x}\|_p = \lim_{p \to \infty} \left(\sum_{k=1}^{n} |x_k|^p \right)^{\frac{1}{p}} = \|\boldsymbol{x}\|_\infty. \tag{6.78}$$

Solution. Suppose $|x_1| = \max_{1 \le k \le n}\{|x_k|^p\}$. Then for any $p > 0$ $|x_1| = (|x_1|^p)^{\frac{1}{p}} \le (\sum_{k=1}^{n} |x_k|^p)^{\frac{1}{p}}$.

On the other hand, since $|x_1| \ge |x_k|$, $1 \le k \le n$, we obtain

$$\left(\sum_{k=1}^{n} |x_k|^p \right)^{\frac{1}{p}} = |x_1| \left(1 + \sum_{k=2}^{n} \left| \frac{x_k}{x_1} \right|^p \right)^{\frac{1}{p}} \le |x_1|(1 + n - 1)^{\frac{1}{p}}. \tag{6.79}$$

For $p > 1$, the Bernoulli inequality gives $(1 + n - 1)^{\frac{1}{p}} \le 1 + \frac{n-1}{p} \to 1$ as $p \to \infty$. Hence the right-hand side of (6.79) tends to $|a_1|$.

The proof of (6.78) can be readily generalized to give

$$\lim_{p \to \infty} \|f\|_p = \|f\|_\infty = \sup_{t \ge 0} |f(t)|. \tag{6.80}$$

The p-norm in (6.80) is defined by (5.11). Note that the functions are not ordinary functions but classes of functions which are regarded as the same if they differ only at measure 0 set. L^p is a Banach space and in particular L^2 is a Hilbert space, cf. Theorem 5.1.

Definition 6.3. Let $X = (X, \mu)$ be a Lebesgue measurable set. A measurable function $f(x)$ on X is said to be **essentially bounded** if there exists a constant $c > 0$ such that

$$|f(x)| \leq c \tag{6.81}$$

a.e. The infimum of such c's is called the **essential norm** of f denoted

$$\operatorname{ess\,sup}_{x \in X} |f| = \inf\{c | |f(x)| \leq c\}. \tag{6.82}$$

Let $a = \operatorname{ess\,sup}_{x \in X} |f|$ and for $0 < b \leq a$ let

$$E_b = \mu(X(|f| > b)).$$

Then $E_a = 0$ and $E_b = \mu(X(|f| > b)) > 0$ for $0 < b < a$.

Theorem 6.3. *The set L^∞ of all (equivalence classes of) functions that are essentially bounded on X forms a Banach space w.r.t. the infinity norm*

$$\|f\|_\infty = \operatorname{ess\,sup}_{x \in X} |f|. \tag{6.83}$$

Further, if $\mu(X) < \infty$, then

$$L^\infty \subset L^p \tag{6.84}$$

for any $p \geq 1$ and

$$\lim_{p \to \infty} \|f\|_p = \operatorname{ess\,sup}_{x \in X} |f| = \|f\|_\infty. \tag{6.85}$$

Proof. If $a = \operatorname{ess\,sup}_{x \in X} |f| < \infty$, then

$$\|f\|_p \leq \left(\int_X a^p \, ds \right)^{1/p} = a\mu(X)^{1/p} \to a$$

as $p \to \infty$, whence $\limsup_{p \to \infty} \|f\|_p \leq a$.

On the other hand, if $0 < b < a$, then by the above remark,

$$\|f\|_p \geq \left(\int_{E_b} b^p \, ds \right)^{1/p} = b\mu(X)^{1/p} \to b$$

as $p \to \infty$, whence $\liminf_{p \to \infty} \|f\|_p \geq b$. Since b can be arbitrarily near a, we conclude (6.85). $\qquad\square$

For the p norm to be finite, it is necessary that $\|f(t)\| \to 0$ as $t \to \infty$, which excludes signals of infinite duration such as unit step signals or periodic ones from L_p. To circumvent the inconvenience, the notion of averaged norm, or the mean $2k$th power value

$$M_{2k}(f) = M_{2k}(f,T) = \frac{1}{T} \int_0^T \|f(t)\|^{2k} \, dt \qquad (6.86)$$

or similar, has been introduced and intensively studied both in analytic number theory and control theory. In the latte the **power norm** is often used with $k = 1$:

$$\text{power}_{2k}(f) = \lim_{T \to \infty} M_{2k}(f,T)^{\frac{1}{2k}} = \lim_{T \to \infty} \left(\frac{1}{T} \int_0^T \|f(t)\|^{2k} \, dt \right)^{\frac{1}{2k}}. \qquad (6.87)$$

Remark 6.2. In mathematics and in particular in analytic number theory, studying the mean square or higher power moments in the form of a sum or an integral is quite common. Especially, this idea is applied to finding out the true order of magnitude of the error term on average. Such an average result will give a hint on the order of the error term itself.

Let $\zeta(s)$ denote the Riemann zeta-function defined by (3.31) having a simple pole at $s = 1$. It is essential that it does not vanish on the line $\sigma = 1$ for the prime number theorem (PNT; cf. (3.35)) to hold. The plausible best bound for the error term for the PNT is equivalent to the celebrated **Riemann hypothesis (RH)** to the effect that the Riemann zeta-function does not vanish on the critical line $\sigma = \frac{1}{2}$. Since the values on the critical line are expected to be small, the averaged norm $M_2(\zeta)$ or $M_4(\zeta)$, i.e. the mean value

$$M_{2k}(\zeta, \sigma) = \frac{1}{T} \int_0^T \left| \zeta(\sigma + it) \right|^{2k} \, dt \qquad (6.88)$$

for $\sigma \geq \frac{1}{2}$, $k \geq 1$ is of great interest and there have appeared a great deal of research work on the subject. The first result for $M_4\left(\zeta, \frac{1}{2}\right)$ is due to Ingham who used the approximate functional equation for the Riemann zeta-function to obtain (6.91).

Proposition 6.1. *Let $d_k(n)$ denote the Piltz divisor function (k-fold divisor function) generated by*

$$\zeta^k(s) = \sum_{n=1}^{\infty} \frac{d_k(n)}{n^{2s}}, \qquad \sigma > 1 \qquad (6.89)$$

(cf. (3.148)). *Then as* $T \to \infty$ *we have*

$$M_{2k}(\zeta, \sigma) = \frac{1}{T} \int_0^T \left| \zeta(\sigma + it) \right|^{2k} dt = \sum_{n=1}^{\infty} \frac{d_k^2(n)}{n^{2\sigma}} \tag{6.90}$$

for $\sigma > 1 - \frac{1}{k}$ *and every integer* $k > 2$ *while*

$$M_4\left(\zeta, \frac{1}{2}\right) = \frac{1}{T} \int_0^T \left| \zeta\left(\frac{1}{2} + it\right) \right|^4 dt = \frac{1}{4\pi^2} \log^4 T (1 + o(1)). \tag{6.91}$$

See e.g. [Bellman (1980)]. The main interest in such estimates as (6.91) lies in the fact that estimates for all $k \in \mathbb{N}$

$$M_{2k}\left(\zeta, \frac{1}{2}\right) = O(T^a) \tag{6.92}$$

would imply the weak Lindelöf hypothesis (LH) in the form

$$\zeta\left(\frac{1}{2} + it\right) = O(T^{\frac{a}{2k} + \varepsilon}) \tag{6.93}$$

for every $\varepsilon > 0$. It is apparent that the RH implies the LH.

Cf. [Ivić (1985)] and [Titchmarsh (1986)] for more details on power moments.

The class of functions f with power norm $\text{power}_2(f) = 0$, called the class of **finite energy signals**, has been extensively studied and which clearly contains the subclass L^2, which we may call **finite energy functions**. As is mentioned in Papoulis [Papoulis (1962)], it would be worth considering the class of functions with non-zero average power, i.e. $0 < \text{power}_{2k}(f) < \infty$, the **finite power signals** class. Such functions do not have Fourier transforms in general. An important feature is that this class contains all periodic functions. Indeed, let $f(t)$ be a periodic function given by

$$f(t) = \sum_{n=-\infty}^{\infty} a_n e^{in\omega_0 t} \tag{6.94}$$

with $\sum_{n=-\infty}^{\infty} |a_n|^2 < \infty$, so that $f \in \ell^2$. Then

$$\text{power}_{2k}(f) = \frac{2\pi}{\omega_0} \int_{-\frac{\pi}{\omega_0}}^{\frac{\pi}{\omega_0}} |f(t)|^2 dt = \sum_{n=-\infty}^{\infty} |a_n|^2. \tag{6.95}$$

The class of finite power signals also contains non-periodic function which are usually specified by some averages. The most common one is the **autocorrelation** defined by

$$R_f(t) = \lim_{T \to \infty} \frac{1}{2T} \int_{-T}^{T} \bar{f}(\tau) f(t + \tau) \, d\tau. \tag{6.96}$$

Proposition 6.1 implies that the Riemann zeta-function has positive power norm value for $\sigma > \frac{1}{2}$ so that it is a finite power signal for $\sigma > \frac{1}{2}$ and that $\zeta\left(\sigma + \frac{1}{2}\right)$ is not a finite but an **infinite power signal**. This suggests the formidable difficulty of the LH, and *a fortiori* the RH.

signal	finite-energy functions	finite power signal	infinite power signal
power	$\mathrm{power}_2(f) = 0$	$0 < \mathrm{power}_{2k}(f) < \infty$	$\mathrm{power}_{2k}(f) = \infty$
example	periodic functions	$\zeta(1 + it)$	$\zeta\left(\frac{1}{2} + it\right)$

Table 6.3. Classification of signals.

The **Hardy space** H^p (cf. e.g. [Kimura (1997), p. 39]) is the main ingredient in modern control theory. It consists of all $f(s)$ which are analytic in $\mathcal{RHP}$—right half-plane $\sigma > 0$ such that $f(j\omega) \in L^p$; in particular, H^∞ with sup norm (6.85). Thus H^∞-control problem is about those (rational) functions which are analytic in $\mathcal{RHP}$, *a fortiori* stable, with regard to the sup norm. Thus the above-mentioned mean-value problem for the Riemann zeta-function corresponds to the H^{2k}-control problem with transfer functions which are mostly rational functions. Thus we might say that the $2k$-th mean values of the finite Dirichlet series (main ingredients in the approximate functional equation) corresponds to the H^{2k}-control problem. As is known, if all the mean values are $O(T^\varepsilon)$, then the (strong) Lindelöf hypothesis would follow. Thus, all these are centered around the LH. Asking for (almost) all individual values, the H^∞-control problem corresponds to the weak LH and eventually to the RH.

Remark 6.3. The H^∞-control is often referred to as *robust* suggesting that the error estimate is invulnerable because it estimates all *individual values* rather than average values as in the case of H^{2k}. However, as Theorem 6.3 asserts, they are equivalence classes of functions rather than ordinary functions and those values on measure 0 sets are discarded.

Furthermore, what prevails here is finitely many sample values taken at a certain time interval which are then approximated by the pre-assigned finitely many scales, thus making everything *quantized*. However, as the following real world sampling theorem asserts, we can restore the digital data to that extent which suits for daily use.

The following theorem ([Weaver (1983), Theorem 5.2, pp. 119-120]) explains why the sampling rate may be taken as the **10 samples per cycle**

of the highest frequency. Recall that a signal is called **almost band-limited** if the support of its Fourier transform is contained in a finite interval, say $[-\Omega.\Omega]$, Ω being called the band-length.

Theorem 6.4. (Real world sampling theorem) *If the function $f(t)$ is almost band-limited with band-length 2Ω and* **almost time-limited** *with time limit $2X$:*

$$\int_{-\infty}^{X} |f(t)| \, dt < \varepsilon, \quad \int_{X}^{\infty} |f(t)| \, dt < \varepsilon, \tag{6.97}$$

then $f(t)$ can be recovered to any desired accuracy from its sampled values at uniformly-spaced intervals $|\Delta t| < \frac{1}{10\Omega}$ apart:

$$f(t) = \sum_{n=-M}^{M} f(n\Delta t) \frac{\sin(2\pi 5\Omega(t - n\Delta t))}{2\pi 5\Omega(t - n\Delta t)} + R_\varepsilon(t), \tag{6.98}$$

where

$$|R_\varepsilon(t)| \ll \varepsilon, \quad M\Delta t > X, \quad \Delta t < \frac{1}{10\Omega}. \tag{6.99}$$

Remark 6.4. An example of a function which is both almost band-limited and almost time-limited is the Gaussian function or its several-variable version

$$f(t) = e^{-at^2}, \quad a > 0. \tag{6.100}$$

This is a density function of the normal (Gaussian) distribution as well as an almost unique example of a rapidly decreasing function. Phenomena whose distribution is in terms of this density function is often referred to as Gaussian.

Comparing the sample frequency 44100 with (6.99), it seems that $\Omega = 4410$ only and that much lower frequencies are cut.

In practice therefore, what is needed is *craftsman's intuition* for choosing right scales, and above all things, the proper threshold number (a transfer function in the case of control theory).

In scientific disciplines, there are some empirical *threshold numbers* which are used often with great efficiency, although the reason why such thresholds appear is not theoretically clear. Below we shall mention 4 such examples of threshold numbers.

The first is the sampling rate which may be taken as the **10 samples per cycle of the highest frequency**. By modifying the argument in

[Kanemitsu (2007)], we shall deduce the sampling theorem. Use is made of the Fourier expansion of the Fourier transform.

The second is related to the famous *central limit theorem* (abbreviated as CLT), which asserts that as the number n becomes infinity of samples X_i with the common mean μ and the common variance σ^2, then the *sample mean* $\bar{X} = \frac{X_1 + \cdots + X_n}{n}$ approaches the *normal distribution* (or Gaussian) $N(\mu, \frac{\sigma^2}{n})$. In the book of Sakata et al [Sears and Titchmarsh (1954), p. 70], iHt is stated if $n \geq 30$, then the sample can be thought of as a large sample and can be approximated by the normal distribution (by the CLT).

The third is related to an empirical value for the volume per molecular weight of a protein molecule. On [Morita (1977), p. 67] it is stated that the volume per molecular weight of a protein molecule contained in a unit lattice is empirically known to be $1.8 \sim 3.0 A^3/Da$. It is quite interesting to work with the solid structure of proteins through the notion of point groups, a subject in bioinformatics.

The fourth is related to the choice of two large primes p, q in the RSA cryptosystem. It is stated that the bigger prime should be taken a few decimal digits longer than the smaller one. One reason is stated in the Exercise based on a possible factorization by the Fermat primes.

The Reynolds number which is the basic threshold quantity in fluid mechanics determining the boundary at which the flow becomes turbulent.

6.6 Control theory in the unit disc

This section is a counterpart of control theory in contrast with number theory in the unit disc, which is in contrast with function theory in the complex plane and that in the unit ball [Rudin (2008)]. The following table is a sequel to Table 1 in Preface.

domain	$\mathcal{D}$	$\mathcal{H}$	$\mathcal{RHR}$
name	unit disc	upper half-plane	expanded frequency domain
variable	$|z| < 1$	$\operatorname{Im}\tau > 0$	$\sigma = \operatorname{Re}s > 0$
mapping	$z \leftrightarrow \frac{1}{z}$	$\tau \leftrightarrow -\frac{1}{\tau}$	$s \leftrightarrow 1 - s$
object	Lambert series	modular form	zeta-function

Table 1 (cont. from Preface). Domains and group structure

Here the variables are connected by

$$s = -i\tau, \quad w = e^{i\tau}, \quad \operatorname{Im}\tau > 0. \tag{6.101}$$

By the second transformation, the upper half-plane is mapped into the inside of the unit circle. This may be replaced by the transformation

$$w = e^{i\psi_0}\frac{z - z_0}{z - \bar{z}_0}, \quad z.z_0 \in \mathcal{H} \tag{6.102}$$

and ψ_0 is any real constant. [Kanemitsu (2015), §3.3.1, pp. 73-78] and [Kanemitsu (2015), §6.3.4, pp. 206-209].

As is suggested by the title of the book of Rudin [Rudin (2008)], "function theory in the unit ball" and that in the $\mathcal{RHR}$ can go parallel. This corresponds to the relation between the Lambert series and zeta-function developed in Chapter 4. This section describes some relevance of control theory in the unit disc developed in [Helton (1998), pp. 120-123] to number theory in the unit disc.

Definition 6.4. A complex function $f(s)$ (defined on a symmetric domain w.r.t. the real axis) satisfying the condition

$$f(s) = \overline{f(\bar{s})} \tag{6.103}$$

for all s is called **real** (on the real axis).

We denote by $\mathcal{A}_{\mathcal{RHP}}$ the (real) vector space of all functions $f(s)$ which are analytic in the $\mathcal{RHP}$ and which are real, bounded and continuous in the closed $\mathcal{RHP}$. By $\mathcal{A}$ we denote the disc counterpart of $\mathcal{A}_{\mathcal{RHP}}$, i.e. the vector space of functions which are analytic in the $\mathcal{D}$ and which are real, bounded and continuous in the closed $\bar{\mathcal{D}}$.

Let H^∞ be the algebra of analytic functions on the unit disc $D \subset \mathbb{C}$. A function in H^∞ is called **inner** if it has a radial limit a.e. on the unit circle. Let

$$f(a, z) = \frac{\bar{a}}{|a|}\frac{a - z}{1 - \bar{a}z} \tag{6.104}$$

and let $\{z_n\} \subset D$ satisfy the Blaschke condition

$$\sum_n (1 - |z_n|) < \infty. \tag{6.105}$$

Then the **Blaschke product**

$$B(z) = e^{ic} z^m \prod f(z_n, z) \qquad (6.106)$$

is an inner function.

6.7 *J*-lossless factorization and interpolation

In this section we partially follow Helton ([Helton (1980)], [Helton (1982)], [Helton (1998)]) who uses the unit ball in place of $\mathcal{RHP}$. They shift to each other under the complex exponential map. For conventional control theory, the unit ball is to be replaced by the critical line ($\sigma = 0$). In practice what appears is the algebra of functions ([Helton (1982), p. 2]).

$\mathcal{R} = \{$functions defined on the unit ball having the rational continuation $\qquad (6.107)$

   to the whole space$\}$

or still larger algebra ψ consisting of those functions which have (pseudo-) meromorphic continuations ([Helton (1982), footnote 6, p. 27]). The occurrence of the gamma function [Helton (1982), Fig. 2.5, p. 17] justifies our incorporation of more advanced special functions and ultimately zeta-functions in control theory (see §6.7, **I**).

Along with the algebra $\mathcal{R}$, one considers

$$\mathcal{B}H^\infty = \{F | \text{analytic on the unit ball having the supremum norm} < 1\}. \qquad (6.108)$$

Then the only mapping $\Theta \in \mathcal{R}U(m, r)$ acting on $\mathcal{B}H^\infty$ must satisfy the *J*-lossless property: Let Θ denote an $(m + r) \times (m + r)$ matrix. Then

$$\Theta^* J_{mr} \Theta \leq J_{mr}, \qquad (6.109)$$

where $J_{mr} = \begin{pmatrix} I_m & O \\ O & -I_r \end{pmatrix}$ is a **signature matrix**, with I_m indicating the identity matrix of degree m. (6.109) with equality is interpreted to be the power preservation of the system in the chain scattering representation (6.35) ([Kimura (1997), p. 82]). To explain these we state.

Definition 6.5. If in the system $\Sigma = \Sigma(s)$ in (6.30) is stable and the input power is always equal to the output power, i.e.

$$\|\boldsymbol{b}_1(j\omega)\|^2 + \|\boldsymbol{b}_2(j\omega)\|^2 = \|\boldsymbol{a}_1(j\omega)\|^2 + \|\boldsymbol{a}_2(j\omega)\|^2, \qquad (6.110)$$

then the system is said to be **lossless**.

In terms of (6.36), (6.110) reads

$$\|a_1(j\omega)\|^2 - \|b_1(j\omega)\|^2 = \|b_2(j\omega)\|^2 - \|a_2(j\omega)\|^2,$$

which in turn amounts to (6.109) with equality sign:

$$^t\Theta(-j\omega)J_{mr}\Theta(j\omega) = J_{pr}. \tag{6.111}$$

In view of the signature matrix J, a power-preserving system Σ is said to be J-lossless, or more precisely, the matrix Θ satisfying both (6.111) and

$$\Theta^*(s)J_{mr}\Theta(s) \leq J_{pr}, \tag{6.112}$$

in the $\mathcal{RHP}$, is called (J_{mr}, J_{pr})-lossless or (J, J')-lossless.

Remark 6.5. As is mentioned in Abstract, it is not clear what the counterpart is of the functional equation in control theory since the transfer function is too simple to have a symmetry property. A plausible candidate for the symmetry looks like J-lossless conjugation.

Definition 6.6. A J-loss matrix $\Theta = \Theta(s)$ is called a **stabilizing conjugator** of a transfer function $G = G(s)$ if
(i) ΘG is stable
and
(ii) the degree of Θ is equal to the number of unstable poles of G, multiplicity being counted.

Condition (i) means that all unstable poles of G are cancelled by the zeros of Θ and (ii) means that the degree of Θ is minimal to achieve (i), i.e. the zeros of G are not cancelled by the poles of Θ. The latter situation does not match the case of the Riemann zeta-function ζ, where the zeros at negative even integers of ζ are *cancelled by* the poles of the gamma function $\Gamma\left(\frac{s}{2}\right)$. The shifting effect of Condition (i) of an unstable pole into stable one seems to have a similar effect as the reflection of with respect to the critical line; however, this is done by multiplication of a stabilizing conjugator, which has some arbitrariness. This probably means that we are to seek the *niryana* of the transfer function and the gamma factor if any whose approximation works as the stabilizing conjugator. As can be seen in [Helton (1982)], there is associated the notion of hyperbolic geometry to control systems, as is the same with automorphic forms, there seems to lie a rich field of research in the direction.

6.8 Robust controller and interpolation

In [Ban, Ogawa, Ono, and Ishida (2009)], a servo control system is designed for a DC motor using the LSDP (Loop Shaping Design Procedure) method of McFarlane and Glover [MacFarlane (1996)]. The transfer function is given by

$$G(s) = \frac{b}{s(s+a)}, \tag{6.113}$$

where

$$b = \frac{K_m}{J_{eq}R_m}, \quad a = K_m b = \frac{K_m^2}{J_{eq}R_m}, \tag{6.114}$$

both of which are positive constants, where J_{eq} is the moment of inertia, R_m the armature resistance, and K_m the torque of the motor.

This is exactly the case considered by Kimura [Kimura (1984)] as an extension of the robust stabilizability of a plant whose unstable poles lie in the right half-plane $\mathcal{RHP} = \{s \in \mathbb{C} | \operatorname{Re} > 0\}$ to the one with integrator.

Definition 6.7. A minimal phase factor $f_m(s)$ of $f(s)$ is one having no zero in $\operatorname{Re} s \geq 0$ and such that $|f_m(j\omega)| = |f(j\omega)|$.

Let $p_0(s)$ be the nominal model of the plant dynamics with a simple pole at $s = 0$ and let r be the uncertainty band function defined by

$$r(s) = \frac{r'_m(s)}{s}, \tag{6.115}$$

where r'_m is the minimal phase function (where we follow the notation of Kimura [Kimura (1984)] and the prime does not mean the derivative).

A transfer function $p(s)$ is said to belong to the class $C(p_0, r)$ if

(i) $p(s)$ has the same number of unstable poles as $p_0(s)$, where $p_0(s)$ is the nominal model of the plant dynamics with a simple pole at $s = 0$.

(ii) $|p(j\omega) - p_0(j\omega)| \leq |r(j\omega)|$, $|r(j\omega)| > 0$ for $\forall \omega \in \mathbb{R}$.

Theorem 6.5. ([Kimura (1984), Theorem 2]) *Suppose*

(A1) All the unstable poles (those lying in $\mathcal{RHP}$) of $p_0(s)$ are simple and $p_0(s)$ has a simple pole at $s = 0$.

(A2) The relative degree of the denominator of $r(s)$ is at most 1.

Then the class of plants are robustly stabilizable if and only if the Pick matrix P in (6.120) is positive definite and

$$|\beta_0| = |r'_m(0)/\tilde{p}_0(0)| < 1, \tag{6.116}$$

where $\tilde{p}_0$ is defined by (6.157) and is the stable nominal model of the plant arising from $p_0(s)$.

6.9 A generalized Nevanlinna-Pick interpolation problem

We follow the notation in Kimura [Kimura (1997)]. Let H^∞ denote the space of all functions analytic in the right half-plane $\mathcal{RHP}$ with H^∞ norm $\|f\|_\infty \mathrm{esssup} |f(j\omega)| < \infty$. BH^∞ indicates the subspace of H^∞ whose elements have the H^∞ norm < 1, i.e. $|f(j\omega)| < 1$ for all ω ([Kimura (1997), p. 42]).

The generalized Nevanlinna-Pick problem (GNP) states that given $(\alpha_i, \beta_i) \in \mathbb{C}$, $1 \le j \le \ell$ and additionally $\beta_0, \beta_{\ell+1} \in \mathbb{C}$ such that

$$\mathrm{Re}\,\alpha_i > 0, \quad |\beta_i| < 1, \quad 1 \le i \le \ell, \tag{6.117}$$

one is supposed to find a function $u \in BH^\infty$ satisfying the interpolation conditions

$$u(\alpha_i) = \beta_i, \quad 1 \le i \le \ell, \quad u(0) = \beta_0,\ u(\infty) = \beta_{\ell+1}. \tag{6.118}$$

Remark 6.6. The additional interpolation conditions in (6.118) need some explanation. The value at infinity pertains to the properness of the controller $c(s)$ expressed in (A1) in terms of $u(s)$ by (6.162) and $\beta_{\ell+1} = 0$ will suffice.

On the other hand, Lemma 6.5 below implies that all the unstable poles of $p_0(s)$ are to be canceled out by the zeros of $q(s)$, including the simple pole at $s = 0$, which imposes the other interpolation condition on $u(s)$, i.e.

$$|u(0)| = |\beta_0|, \tag{6.119}$$

which is the essential addition to the original NP problem.

Since the origin and the point at infinity are symmetric points (relative to the fractional transformations, cf. [Ahlfors (1996)]), the values at those points may be conveniently used to determine the linear fractional transformations (cf. Lemma 6.2 below).

The Hermitian matrix P defined by

$$P = \begin{pmatrix} \frac{1-\beta_1\bar{\beta}_1}{\alpha_1+\bar{\alpha}_1} & \cdots & \frac{1-\beta_1\bar{\beta}_\ell}{\alpha_1+\bar{\alpha}_{ell}} \\ \cdots & \cdots & \cdots \\ \frac{1-\beta_\ell\bar{\beta}_1}{\alpha_\ell+\bar{\alpha}_1} & \cdots & \frac{1-\beta_\ell\bar{\beta}_\ell}{\alpha_\ell+\bar{\alpha}_{ell}} \end{pmatrix} \tag{6.120}$$

is called the Pick matrix. Let

$$B_i(s) = \frac{\alpha_i - s}{\bar{\alpha}_i + s}, \ 1 \le i \le \ell, \tag{6.121}$$

which is a BH$^\infty$-function satisfying $|B_i(j\omega)| = 1$ for all $\omega \in \mathbb{R}$. Let $\beta_{i,k+1}$ denote the Fenyves array defined by

$$\beta_{i,1} = \beta_i, \ 1 \le i \le \ell, \tag{6.122}$$

$$\beta_{i,k+1} = \frac{(\alpha_i + \bar{\alpha}_i)(\beta_{i,k} - \beta_{k,k})}{(\alpha_i - \alpha_k)(1 - \bar{\beta}_{k,k}\beta_{i,k})}, \ 1 \le k \le i-1 \le \ell-1.$$

Let

$$\rho_k = \beta_{k,k}, \ 1 \le k \le \ell. \tag{6.123}$$

It is well-known that P is positive definite if and only if $|\beta_{i,h}| < 1$. Define

$$\Theta_k(s) = \frac{1}{\sqrt{1 - |\rho_k|^2}} \begin{pmatrix} -B_k(s) & \rho_k \\ -\bar{\rho}_k B_k(s) & 1 \end{pmatrix}, \ 1 \le k \le \ell. \tag{6.124}$$

In its classical version, the Nevanlinna-Pick theorem reads

Theorem 6.6. *The solvability of the NP problem is equivalent to the positive definiteness of the Pick matrix.*

Since

$$G(s) = \left(\begin{array}{c|c} A & B \\ \hline I & O \end{array} \right) = I(sI - A)^{-1}B + O = (sI - A)^{-1}B, \tag{6.125}$$

we say that the J-lossless matrix $\Theta = \Theta(s)$ is a J-lossless stabilizing conjugator of the pair (A, B) ([Kimura (1997), p. 113]) if it stabilizes G in the sense of [Kimura (1997), Definition 5.1, p. 108], i.e. if $\Theta(s)$ is a rational function with minimum degree which cancels all the unstable poles of $G(s)$.

The following lemma is a combination of [Kimura (1997), Theorem 5.2] and [Kimura (1997), Lemma 5.6].

Lemma 6.1. *The pair (A, B) in (6.125) has a J-lossless stabilizing (anti-stabilizing) conjugator $\Theta(s)$ if and only if the algebraic Riccati equation*

$$XA + {}^tAX - XBJ^tBX = O \tag{6.126}$$

has solution $X \ge 0$ such that $A - XBJ^tBX$ is stable (anti-stable).

Further, if A is stable ($-A$ is anti-stable), then the solution X of (6.126) is invertible.

We recall the definition of the homographic transformation in §6.3. On a controller S, the chain scattering matrix (6.35) acts as in (6.40), i.e.

$$\Theta S = (\Theta_{11}S + \Theta_{12})(\Theta_{21}S + \Theta_{22})^{-1} =: HM(\Theta, S). \tag{6.127}$$

Then we have the concatenation rule (6.41).

In our present case, the controller is a scalar $s \in \mathbb{C}$ and (6.127) amounts to

$$HM(\Theta, S) = \Theta S = \frac{\Theta_{11}s + \Theta_{12}}{\Theta_{21}s + \Theta_{22}}, \tag{6.128}$$

where $\Theta_{k\ell}$ are also scalars.

Lemma 6.2. *Suppose*

$$\Theta = \begin{pmatrix} t_{11} & t_{12} \\ t_{21} & t_{22} \end{pmatrix} \tag{6.129}$$

is $\in \mathrm{GL}_2(\mathbb{C})$,

$$u_1 = \frac{t_{11}u_{\ell+1} + t_{12}}{t_{21}u_{\ell+1} + t_{22}} \tag{6.130}$$

and that the inverse transformation of (6.130) is given by

$$u_{\ell+1}(s) = f_1(s)\frac{u_1(s) + f_3(s)}{1 + f_2(s)u_1(s)}, \tag{6.131}$$

where

$$f_1(s) = \frac{t_{22}(s)}{t_{11}(s)}, \quad f_2(s) = -\frac{t_{21}(s)}{t_{11}(s)}, \quad f_3(s) = -\frac{t_{12}(s)}{t_{22}(s)}. \tag{6.132}$$

Let

$$\delta_0 (= u_{\ell+1}(0)) = f_1(0)\frac{u_1(0) + f_3(0)}{1 + f_2(0)u_1(0)} \tag{6.133}$$

$$\delta_\infty (= u_{\ell+1}(\infty)) = f_1(\infty)\frac{u_1(\infty) + f_3(\infty)}{1 + f_2(\infty)u_1(\infty)}$$

and suppose

$$|\delta_0| < 1, \quad |\delta_{\ell+1}| < 1. \tag{6.134}$$

Then the functions $u_{\ell+1} \in \mathrm{BH}^\infty$ *satisfying (6.136) are given by*

$$u_{\ell+1}(s) = \frac{cu_{\ell+1}(\infty)s + du_{\ell+1}(0)}{cs + d}. \tag{6.135}$$

Proof. By (6.134) we may choose the values of $u_{\ell+1}$ at $0, \infty$ as in (6.133) to obtain

$$|u_{\ell+1}(0)| < 1, \quad |u_{\ell+1}(\infty)| < 1, \tag{6.136}$$

whence we only need to determine the form of the linear transformation $u_{\ell+1}$. This is a well-known procedure (e.g. [Ahlfors (1996)], $0, \infty$ are symmetrical points) and we readily conclude (6.135). $\qquad\square$

Kimura gives the example with $c = d = 1$:

$$u_{\ell+1}(s) = \frac{u_{\ell+1}(\infty)s + u_{\ell+1}(0)}{s+1} = \frac{\delta_\infty s + \delta_0}{s+1}.$$

We need the following basic lemma which we will take for granted. It depends on the well-known bounded real lemma ([Kimura (1997), Theorem 4.5]) which connects boundedness of a transfer function and the algebraic Riccati equation.

Lemma 6.3. ([Kimura (1997), Theorem 4.5]) *Suppose Θ is (J, J')-unitary. Then there exists a termination u for which $HM(\Theta, u) \in BH^\infty$ if and only if Θ is (J, J')-lossless. If this is the case, then $HM(\Theta, s) \in BH^\infty$ if and only if $u \in BH^\infty$.*

The following theorem is a slightly modified version of [Kimura (1997), Theorem 5.8, p.115] in conjunction with [Kimura (1984)].

Theorem 6.7. *The GNP is solvable if and only if the pair of matrices*

$$A = \begin{pmatrix} \alpha_1 & 0 & \cdots & 0 \\ 0 & \alpha_2 & \cdots & 0 \\ & & \cdots & \\ 0 & 0 & \cdots & \alpha_\ell \end{pmatrix}, \ B = \begin{pmatrix} 1 - \beta_1 \\ 1 - \beta_2 \\ \cdots \\ 1 - \beta_\ell \end{pmatrix} \tag{6.137}$$

has a J-lossless stabilizing conjugator Θ for the signature matrix

$$J = \begin{pmatrix} I & O \\ O & -I \end{pmatrix}, \tag{6.138}$$

and

$$|\beta_0| < 1, \quad |\beta_\ell| < 1. \tag{6.139}$$

If this is the case, then all the parametric solutions u_1 are obtained by the procedure

$$u_k = HM(\Theta_k, u_{k+1}), \quad 1 \le k \le \ell \tag{6.140}$$

starting from any $u_{\ell+1} \in \mathrm{BH}^\infty$ *satisfying Condition* (6.151), *where* Θ_k *is defined by* (6.124). *I.e. if* Θ *is defined by*

$$\Theta = \prod_{k=1}^{\ell} \Theta_k, \tag{6.141}$$

then $HM(\Theta, u_{\ell+1})$ *solves the NP problem with the additional interpolation conditions enunciated in* (6.118)

$$u_1(0) = \beta_0, \quad u_1(\infty) = \beta_{\ell+1}. \tag{6.142}$$

Proof. We mix the two proofs given by Kimura [Kimura (1984)], [Kimura (1997)]. Let Θ be defined by (6.129). Then by (6.140), (6.141) and the concatenation rule we have ((6.130))

$$u_1 = HM(\Theta, u_{\ell+1}) = \frac{t_{11} u_{\ell+1} + t_{12}}{t_{21} u_{\ell+1} + t_{22}}. \tag{6.143}$$

Sufficiency follows from the following observation. The product of $\Theta(s)$ and $G(s)$ in (6.125) is stable by definition. Hence multiplying this by $s - \alpha_i$ and letting $s \to \alpha_i$ eliminates all but one term $(1, -\beta_i)\Theta(\alpha_i)$, which is 0, implying that

$$t_{11}(\alpha_i) = \beta_i t_{21}(\alpha_i), \quad t_{12}(\alpha_i) = \beta_i t_{22}(\alpha_i). \tag{6.144}$$

By (6.130), this gives the interpolation condition (6.118) save for the last two (6.142). This is also satisfied since $u_{\ell+1}$ satisfies Condition (6.151), which is equivalent to (6.142). Then by Lemma 6.3 we may conclude that $u_1 \in \mathrm{BH}^\infty$ and whence the sufficiency follows.

Now we prove necessity. It suffices to show that the J-lossless $\Theta_k(s)$ cancels the unstable pole s_k. For this, a modification of [Kimura (1997), (5.27), p. 117] works, which reads in our situation

$$\frac{1}{s - \alpha_i}(1 - \beta_i)\Theta_k(s) = \left(\frac{\nu_i}{s + \bar{\alpha}_k} \, 0 \right) + \frac{\mu_i}{s - \alpha_i}(1 - \beta_i'), \tag{6.145}$$

where

$$\nu_i = \frac{1 - \beta_i \bar{\beta}_k}{\sqrt{1 - |\beta_k|^2}} \frac{\alpha_k + \bar{\alpha}_k}{\alpha_k + \bar{\alpha}_k}, \quad \mu_i = \frac{1 - \beta_i \bar{\beta}_k}{\sqrt{1 - |\beta_k|^2}} \frac{\alpha_i - \alpha_k}{\alpha_k + \bar{\alpha}_k}. \tag{6.146}$$

Hence in particular,

$$\mu_k = 0, \tag{6.147}$$

i.e., Θ_k cancels the unstable pole at α_k. Hence Θ in (6.141) cancels all the unstable poles of $G = (A, B)$.

Moreover, from the very definition (6.124), it follows that

$$\Theta_k^*(j\omega)J\Theta_k(j\omega) = J, \tag{6.148}$$

whence

$$\Theta^*(j\omega)J\Theta(j\omega) = J, \tag{6.149}$$

i.e. Θ is J-lossless.

(6.149) implies as in [Kimura (1984)]

$$|f_1(j\omega)| = 1, \ |f_2(j\omega)| = |f_3(j\omega)| < 1, \ f_2(j\omega) = \bar{f}_3(j\omega). \tag{6.150}$$

By (6.139), (6.142) and (6.150), as in [Kimura (1984)] we conclude (6.134).

Hence, by Lemma 6.2, we may choose a suitable $u_{\ell+1}(s) \in \mathrm{BH}^\infty$ such that

$$u_{\ell+1}(0) = \delta_0, \quad u_{\ell+1}(\infty) = \delta_\infty \tag{6.151}$$

as in (6.133), then $u_1(s)$ satisfies (6.142) since (6.130) and (6.131) are equivalent.

$$\square$$

Remark 6.7. Since A in Theorem 6.7 is anti-stable, Lemma 6.1 implies that the solution X of the Riccati equation (6.126) has the inverse $P = X^{-1}$ satisfying the Lyapnov type equation

$$AP + PA^* = BJB^*. \tag{6.152}$$

It can be seen readily that the solution of (6.152) is identical to the Pick matrix, so that Theorem 6.7 is another expression of Theorem 6.6 plus the additional interpolation condition (6.142).

6.10   Proof of Theorem 6.5

First we quote the standard result from [Doyle and Stein (1981)]. Σ indicates the unity feedback system Σ in §6.3.2.

The synthesis problem of a controller C for the unity-feedback system, Σ say in §6.3.2 refers to the sensitivity reduction problem, which asks for the estimation of the sensitivity function $S = S(s)$ multiplied by an appropriate frequency weighting function $W = W(s)$, cf. (6.52). We recall that $S = (I + PC)^{-1}$ in (6.51) is a transfer function from r to e.

[Zames and Francis (1992)] introduced the function

$$q(s) = c(s)(1 + p_0(s)c(s))^{-1} \tag{6.153}$$

with nominal function $p_0(s)$ introduced in Definition 6.7, which can be solved for the stabilizer $c(s)$:

$$c(s) = q(s)(1 - p_0(s)q(s))^{-1}. \tag{6.154}$$

Lemma 6.4. *The controller $c(s)$ is a robust stabilizer for the class $C(p_0, r)$ if and only if Σ is stable for $p = p_0$ and*

$$|r(j\omega)q(j\omega)| < 1 \tag{6.155}$$

for all $\omega \in \mathbb{R}$.

Lemma 6.5. *The unity feedback system Σ is stable for $p = p_0$ if and only if*

(i) $q(s)$ is stable.

(ii) $1 - p_0(s)q(s)$ has the zeros at the unstable poles of $p_0(s)$ up to multiplicity. This implies that all the unstable poles of $p_0(s)$ are canceled out by the zeros of $q(s)$.

We are now in a position to prove Theorem 6.5. Suppose $\alpha_1, \cdots, \alpha_\ell$ are all the unstable poles of $p_0(s)$. Let $B(s)$ be the product of all $B_i(s)$'s in (6.121)—the Blaschke product (for which see e.g. [Marshall (1992)])

$$B(s) = \prod_{i=1}^{\ell} B_i(s), \tag{6.156}$$

which is a BH^∞-function satisfying $|B(j\omega)| = 1$ and cancels all the unstable poles of $p_0(s)$. We modify $p_0(s)$ by eliminating all the poles by multiplying the BH^∞-function ([Kimura (1984), (35)]):

$$\tilde{p}_0(s) := sB(s)p_0(s) \tag{6.157}$$

is a stable function. Coupled with this, we introduce the function $\tilde{q}(s)$ so that we keep the relation

$$\tilde{p}_0(s)\tilde{q}(s) = p_0(s)q(s), \tag{6.158}$$

i.e.

$$\tilde{q}(s) := \frac{q(s)}{sB(s)} = \frac{p_0(s)q(s)}{\tilde{p}_0(s)}. \tag{6.159}$$

If $c(s)$ is a stabilizer of Σ, then by Lemma 6.5, (ii), $\tilde{q}(s)$ is also stable and furthermore, we have

$$\tilde{p}_0(\alpha_i)\tilde{q}(\alpha_i) = 1, \ 1 \leq i \leq \ell. \tag{6.160}$$

By the definition (6.153), we also have

$$\tilde{p}_0(0)\tilde{q}(0) = 1. \tag{6.161}$$

Defining

$$u(s) := r_m(s)\tilde{q}(s) = \frac{r(s)q(s)}{B(s)}, \tag{6.162}$$

we see that $|u(s)| = |r(s)q(s)|$ and so from (6.155) and the maximum modulus principle that $u(s)$ is a BH$^\infty$ function. Also from (6.159), (6.160) and (6.161) we have

$$\beta_i := u(\alpha_i) = r'_m(\alpha_i)\tilde{q}(\alpha_i) = \frac{r'_m(\alpha_i)}{\tilde{p}_0(\alpha_i)}, \ 1 \le i \le \ell \tag{6.163}$$

with the additional condition

$$\beta_0 := u(0) = \frac{r'_m(0)}{\tilde{p}_0(0)}. \tag{6.164}$$

Thus we are led to the GNP problem whose solution is given by Theorem 6.7.

6.11 Another statement

Theorem 6.8. *Suppose the notation and conditions in Theorem 6.5 all hold true. I.e.*

(A1) All the unstable poles of $p_0(s)$ are simple and lie in $\mathcal{RHP}$ and $p_0(s)$ has a simple pole at $s = 0$.

(A2) The relative degree of the denominator of $r(s)$ is at most 1.

Then the class of plants are robustly stabilizable if and only if the pair of matrices

$$A = \begin{pmatrix} \alpha_1 & 0 & \cdots & 0 \\ 0 & \alpha_2 & \cdots & 0 \\ & & \cdots & \\ 0 & 0 & \cdots & \alpha_\ell \end{pmatrix}, \ B = \begin{pmatrix} 1 - \beta_1 \\ 1 - \beta_2 \\ \cdots \\ 1 - \beta_\ell \end{pmatrix} \tag{6.165}$$

has a J-lossless stabilizing conjugator Θ for the signature matrix (6.138) and

$$|\beta_0| = |r'_m(0)/\tilde{p}_0(0)| < 1. \tag{6.166}$$

Example 6.6. Numerical data is given in [Ban, Ogawa, Ono, and Ishida (2009)].

$$a = 10.8, b = 228,$$ (6.167)
$$a = 8.64, b = 274,$$

but we shall not dwell on this side, leaving the design of the robust controller ([MacFarlane (1996)]) for another occasion.

Riemann

Bibliography

Agarwal P., Kanemitsu S. and Li H.(2016). *On the Kuznetsov sum formula*, Proc. ICSFA. to appear.

Ahlfors L.(1996). *Comples analysis*, (McGraw-Hill, New York).

Apostol T. M.(1950), *Generalized Dedekind sums and the transformation formula of certain Lambert series*, Duke Math. J. **17** pp. 147-157.

Apostol T. M.(1951), *On the Lerch zeta function*, Pacific J. Math. **1** pp.161-167.

Apostol T. M.(1952). *Addendum to 'On the Lerch zeta function'*, Pacific J. Math. **2** pp. 10.

Apostol T. M. (1951), *Remark on the Hurwitz zeta function*, Proc. Amer. Math. Soc. **2** pp. 690–693.

Apostol T. M.(1957), *Mathematical analysis*, Addison-Wesley Publishing Company, Inc., Reading, Mass.

Apostol T. M.(1964). *A short proof of Shô Iseki's functional equation*, Proc. Amer. Math. Soc. **15**. pp. 618-622.

Apostol T. M.(1976). *Modular functions and Dirichlet series in number theory*, Springer Verl.

Arias-de-Reyna J. (2004), *Riemann's fragements on limit values of elliptic modular functions*, Ramanujan J. **8**. pp.57-123.

Artin E.(1923). *Über eine neue Art von L-Reihen (On a new kind of L series)*, Abh. Math. Sem, Univ. Hamburg **3** pp. 89-108 Collected Papers, Addison-Wesley 1965, pp.105-124.

Artin E.(1930), *Zur Theorie der L-Reihen mit allgemeinen Gruppencharacteren (On the theory of L-series with general characters)*, Abh. Math. Sem, Univ. Hamburg **8** (1930), 292-306 Collected Papers, Addison-Wesley 1965, 165-179.

Asano M.(2003), *Report on multiple zeta-functions and Dedekind sums*, Master thesis. (in Japanese).

Ashnel M. and Goldfeld D.(1997), *Zeta-functions, one-way functions and pseudorandom number generators*, Duke Math. J. **88** pp. 371-390.

Ayoub R.(1963) *An introduction to the analytic theory of numbers* AMS, Providence, R I.

Baker R. C.(2003). *Kloosterman Sums and Maass Forms*, Vol. **I**, Kendrick Press,

Heber City.

Balasubramanian R., Kanemitsu S. and Ramachandra K.(2003), *On ideal-function-like functions*, J. Comput. Appl. Math. **160** pp. 27-36.

Balasubramanian R. and Ramachandra K.(1998), *On the number of integers n such that nd(n) ≤ x*, Acta Arith. **49** pp. 313-322.

Balasubramanian R., Ding L. P., Kanemitsu S. and Tanigawa Y.(2008). *On the partial fraction expansion for the cotangent-like function*, Proc.Int.Conf.-Number Theory and Discrete Geometry, Ramanujan Mathematical Society Lecture Notes Series, No. **6** pp. 19-34.

Ban N., Ogawa H., Ono M., and Ishida Y.(2009). *A servo control system using the loop shaping procedure*, World Academy of Sci., Engrg and Techn. **60**. pp. 150-153.

Banerjee D., Chakraborty K., Kanemitsu S. and Magi B.(2016).*Abel-Tauber process and asymptotic formulas*, Kyushu J. Math. **71** no.2.

Basu A.(1999). *Alternate proof of Shô Iséki's functional equation*, Ramanujan J. **3**. pp. 297-301.

Bellman R.(1980), *Analytical number theory (an introduction)*, The Benjamin/Cummings Publ. Co. Inc. New York.

Berndt B. C.(1973), *A generalization of a theorem of Gauss on sums involving* [x], Amer.Math. Monthly **82**. pp. 44-51.

Berndt B. C.(1973), *Generalized Dedekind eta-function and generalized Dedekind sums*, Trans. Amer. Math. Soc. **178**. pp.495-508.

Berndt B. C.(1975), *Generalized Eisenstein series and modified Dedekind sums*, J. Reine Anger. Math. **272**. pp.182-193.

Berndt B. C.(1975), *Character analogues of the Poisson and Euler-Maclaurin summation formula with applications*. J. Number Theory **7**. pp. 413-445.

Berndlt B. C.(1977). *Modular transformations and generalization of several formulas of Ramanujan*, Rocky Mount. J. Math.**7**. pp. 147-189.

Berndt B. C.(1989). *Ramanujan's Notebooks, Part II*, Springer Verl., Heidelberg etc.

Berndlt B. C.(1991). *Ramanujan's Notebooks*, Part III, Springer Verl., Heidelberg etc.

Berndt B. C. and Kim S.(2015). *Logarithmic means and double series of Bessel functions*, Int. J. Number Theory, **11**(5). pp.1535-1556.

Berndt B. C. and Knopp M. I.(2008). *Hecke's theory of modular forms and Dirichlet series*, World Sci., Singapore etc.

Berndlt B. C. and Venkatachaliengar K.(2001). *On the transformation formula for the Dedekind eta-function, Symbolic Computation, Number Theory, Special Functions*, Physics and Combinatorics Fank G. Garvan, Mourad E. H. Ismail (eds), DEVM, Vol. **4** pp. 73-77.

Beurling A.(1949), *On two problems concerning linear transformations in Hilbert space*, Acta Math. **81**. pp. 239-255.

Birch B. J. and Stephens N. M.(1966), *The parity of te rank of the Mordell-Weil group*, Topology **5**. pp.295-299.

Blomer V.(2005). *Rankin-Selberg L-functions on the critical line*, Manuscripta Math., **117**. pp. 111-133.

Blum M. and Micali S.(1994). *How to construct cryptographically strong sequences of pseudorandom bits*, SIAM J. Comput. **13**. pp. 850-864.

Bodendeck R. and Halbritler U.(1972), *Über die Transformationsformel von* $\log \eta(\tau)$ *and gewisser Lambertscher Reihen*, Abh. Math. Sem. Univ. Hamburg **38**.pp.147-167.

Bombieri E. and Hejal D. A.(1995). *On the distribution of zeros of linear combination of Euler prducts*, Duke Math. J. **80**. pp. 821-862.

Borel A.(1999), *Automorphic forms on* $SL_2(\mathbb{R})$, Cambridge UP, Cambridge.

Borevič Z. and Šafarevič I.(1964), *The theory of numbers*, Izd. Nauka, Moscow 1964; German transl. Zahlentheorie, Birkhäuser, Basel and Stuttgart, 1966.

Briggs W. E.(1962). *Some Abelian results for Dirichlet series*, Mathematika **9**. pp. 49-53.

Bruggeman R.(1978). *Fourier coefficients of cusp forms*, Invent. Math., **45**. pp.1-18.

Bruggeman R.(1981), *Fourier coefficients of automorphic forms,* Lect. Notes Math., **865**, Springer Verl., New York/Berlin.

Bump D.(1996). *Automorphic forms and representations*, Cambridge UP, Cambridge.

Burr S. A.(1991), *The unreasonable effetiveness of number theory*, AMS, Prpvidence, R. I.

Buschman R. G.(1959). *Asymptotic expressions for* $\sum n^a f(n \log^r n)$, Pacific J. Math. **9**. pp. 9-12.

Butzer P. L. and Stark E. L.(1986). *"Riemann's example" of a continuous nondifferentiable function in the light of two letters (1865) of Christoffel to Prym*, Bull. Soc. Math. Belg. Sèr. A, **38**. pp. 45–73.

Carbone A. and Gromov M.(2001), *A mathematical slices of molecular biology, Supplement to volume 88 of Gazette des Mathématiciens*, French Math. Soc. (SMF), Paris.

Cauchy A. L.(1840), *Methode simple et nouvelle pour la determination complete des sommes alternee, formees avec les racines primitives des equattions binomes* J. Math. Pure Appl. (Liouvilee). **5**. pp. 154–183.

Chakraborty K., Kanemitsu S. and Kuzumaki K(2013), *Arithmetical class number formula for certain quadratic fields*, Hardy-Ramanujan J. **36**. pp.1-7.

Chakraborty K., Kanemitsu S. and Laurinčikas A., *On complex powers of L-functions and integers without small prime factors*, Acta Math. Hungar., to appear.

Chakraborty K., Kanemitsu S. and Maji B.(2016), *Modular-type relations associated to the Rankin-Selberg L-function*, to appear in *Ramanujan J.* (2016) DOI: 10.1007/s11139-015-9759-8.

Chakraborty K., Kanemitsu S. and Tsukada H.(2009), *Vistas of Special Functions II*, World Scientific, London-Singapore-New Jersey.

Chan H. H.(1988). *On the equivalence of Ramanujan's Partition identities and a connenction with the Rogers-Ramanujan continued fraction*, J. Math. Anal. Appl. **198**. pp. 111-120.

Chan H. H.(2006), Oral communication around Feb, 17, 2006.

Chandrasekharan K. and Minakshisundaram S.(1952), *Typical means*, Oxford UP, Oxford.

Chandrasekharan K. and Raghavan Narasimhan(1962). *Functional equations with multiple gamma factors and the average order of arithmetical functions*, Ann. of Math. (2) **76**. pp. 93–136.

Chen N. -X.(2010). *Möbius inversion inn physics*, World Sci., New Jersey etc.

Chowla S.(1937), *On some infinite series involving arithmetic functions*, Proc. Indian Acad. Sci. A **5**. pp. 511-513=Collected Papers of Sarvadaman Chowla, Vol. II, 489-491.

Chowla S. and Walfisz A.(1935), *Über eine Riemannsche Identität*, Acta Arith. **1**. pp. 87-112=Collected Papers of Sarvadaman Chowla, Vol.

Chowla P.(1968), *On the class-number of real quadratic fields*, J. Reine Angew. Math. **230**. pp. 51-60.

Chowla P. and Chowla S.(1968), *Formulae for the units and class-numbers of real quadratic fields*, J. Reine Angew. Math. **230**. pp.61-65.

Cohen H.(1988), *q-identities for Maass forms*, Invent. Math. **91**. pp.409-422.

Conrey J. B. and Gosh A.(1993). *On the Selberg class of Dirichlet series: small degrees*, Duke Math. J. **72**. pp. 673-693.

Damgard I. B.(1988). *On the randomness of Legendre and Jacobi sequencess*, Adv. Cryptology–CRYPTO '88 (Santa Barbara Calif. 1988) Lect. Notes Computer Sic. **403**, Springer Verl. Berlin pp.163-172.

Davenport H.(1937), *On some infinite series involving arithmetic functions*, Quart. J. Math. **8**. p. 8-13=Collected papers of H. Davenport IV, Academic Press London etc. 1977. pp.1781-1786.

Davenport H.(1937), *On some infinite series involving arithmetic functions II*, Quart. J. Math. **8**. pp. 313-320=Collected papers of H. Davenport, IV, Academic Press London etc. 1977, pp. 1787–1794.

Davenport H.(2000). *Multiplicative Number Theory*, Markham 1967, second ed. Springer 1982, third ed. (revised by H. L. Montgomery). Springer 2000.

David C., Fearnley J. and Kisilevsky H.(2004). *On the vanishing of twisted L-functions of elliptic curves*, Exper. Math. **13**(2). pp. 185-198.

Davis P. J. and Hersh R.(1986). *Descartes' dream—The world according to mathematics*, Harcourt Brace Jovanovich Publ., San Diego etc.

Dedekind R.(1930). *Erläuterungen zu zwei Fragmenten von Riemann*, Math. Werke Bd. **1**. pp. 159-173, Braunschweich. (in: Bernhard Riemanns gesammelte mathematische Werke und wissenschaftlichen Nachlass, 2, Aufl., 466-478, 1892).

de Haan L.(1970). *On regular variation and its applications to the weak convergence of sample extremes*, Math. Center Tracts **32**, Amsterdam.

de Koninck J. M. and Ivić A.(1980), *Topics in arithmetical functions*, North Holland, Amsterdam.

Delange H.(1959). *Sur des formules due a Atle Selberg*, Bull. Sci. Math. (2) **83**. pp.101-111.

Delange H.(1971). *Sur des formulae de Atle Selberg*, Acta Arith. **19**. pp.105-146.

de la Bretèche R. and Tennenbaum G.(2004), *Séries trigonométriques à coefficients arithmétiques*, J. Anal, Math. **92**. pp.1-79.

Deshouillers J. M. and Iwaniec H.(1982). *Kloostermann sums and Fourier coefficients of cusp forms,* Invent. Math., **70**. pp.212-288.

Dieter U.(1971). *Pseudo-random numbers: The exact distribution of pairs,* Math. Comput. **25**. pp.855-883.

Doyle J. C. and Stein G.(1981). *Multivariable feedback design: Concepts for classical/modern sysnthesis.* IEEE Trans. Automatic Control, **26**. pp. 4-16.

Duistermaat J. J.(1991). *Self-similarity of "Riemann's nondifferentiable function",* Nieuw Arch. Wisk. **9**. pp. 303-337.

Edwards H. M.(2003). *Riemann's zeta function,* Dover, New York.

Eisenstein G.(1975). *Aufgaben und Lehrsätze,* J. Reine Angew. Math. **27** (1975). pp. 281-283=Math. Werke, Vol. 1, 1975, Chelsea, (1975).pp. 108-110.

Elstoldt J., *A very simple proof of the eta transformation formula,* Manuscipt.

Erdélyi A., Magnus W., Oberhettinger F. and Tricomi F. G.(1953)(ed), Higher transcendental functions, Vol 1, McGraw-Hill. New York.

Falb P.(1990). *Methods of algebraic geometry in control theory, Part I: Scalar linear systems and affine algebraic geometry,* (Birhauser, Boston-Basel-Berlin).

Falb P.(1999). *Methods of algebraic geometry in control theory, Part II: Multivariable linear systems and projective algebraic geometry,* (Birhauser, Boston-Basel-Berlin).

Fawaz A. Y.(1951). *The explicit formula for $L_0(x)$,* Proc. London Math. Soc. (3) **1**. pp. 86-103.

Fay J. D.(1997), *Fourier coefficients of the resolvent for a Fuchsian group,* J. Reine Angew. Math. **293/294**,pp. 143-203.

Fergusson R.(1963), *An application of Stieltjes integrals to power series coefficients of the Riemann zeta-function,* Amer. Math. Monthly **70**,pp. 60-61.

Fischer W. (1951), *On Dedekind's function $\eta(\tau)$,* Pacific J. Math. **1**, pp. 83-95.

Fogels E.(1965), On the zeros of *L*-functions, Acta. Arith. **11** (1965), 67-96.

Fogels E.(1963). *Über die Ausnahmenullstelle der Heckeschen L-Funktionen,* Acta Arith. **8**,pp. 307-309.

Friedlander J.(1972), *On the number of ideals free from large prime divisors,* J. Reine Angew. Math. **255**,pp. 1-7.

Funakura T.(1990). *On Kronecker's limit formula for Dirichlet series with periodic coefficients,* Acta Arith. **55**, pp. 59-73.

Garbanati D.(1981). *Class field theory summarized,* Rocky Mount. J. Math. **11**,pp. 195-225.

Gerver J.(1970), *The differentiability of the Riemann function at certain rational multiples of π,* Amer. J. Math. **92**,pp. 33-55.

Gerver J.(1970), *More on the differentiability of the Riemann function,* Amer. J. Math. **93**,pp. 33-41.

Gerver J.(2003). *On cubic lacunary Fourier series,* Trans. Amer. Math. Soc. **355**. pp. 4297-4347.

Goldfeld O. and Hoffstein J.(1993), *On the number of Fourier coefficients that determine a modular form,* in A tribute to Emil Grosswald: Number theory and related analysis, Contemp. Math. **143** AMS, Providence,pp. 385-393.

Goldfeld D. and Sarnak P.(1983), *Sums of Kloosterman sums,* Invent. Math.

71,pp. 243-250.

Goldreich O., Krawczyk H. and Luby M.(1971). *On the existence of pseudo-random generators*, SIAM J. Comput. **22** (1993),pp. 1163-1175.

Goldstein L. J.(1971), *Analytic number theory*, (Prentice-Hall, New Jersey).

Goldstein L. J. and de la Torre P.(1974). *On the transformation formula of* $\log \eta(\tau)$, Duke Math. J. **41**,pp.291-297.

Goodman R. and Wallach N. R.(1980). *Whittaker vectors and conical vectors*, J. Funct. Anal. **39**, pp. 199-279.

Grosswald E.(1972), *Comments on some formulae of Ramanujan*, Acta Arith. **21**,pp. 25-34.

Guthman A.(1996). *The Riemann-Siegel integral formula for Dirichlet series associated to cusp forms*, in Analytic and Elementary Number Theory, Vienna, pp. 53–69.

Guthman A.(1997), *Die Riemann-Siegel-Integralformel für die Mellintransformation von Spitzenformen*. (in German) Arch. Math. (Basel) **69**, pp. 391–402.

Guthman A.(1997), *New integral representations for the square of the Riemann zeta-function*. Acta Arith. **82** pp. 309–330.

Guthman A.(1999). *Asymptotic expansions for Dirichlet series associated to cusp forms*. Publ. Inst. Math. (Beograd) (N.S.)\ **65** (79) pp. 69–96.

Hafner J. and Stopple J.(2000). *A heat kernel associated to Ramanujan's tau function*, Ramanujan J. **4** pp. 123-128.

Halmos P. R.(1950), *Introduction to Hilbert space*, (Chelsea, New York).

Hardy G. H.(1903), *Note on the limiting values of the elliptic modular-functions*, Quart. J. Math.(2) **34**,pp. 76-86; Collected Papers of G. H. Hardy IV, Oxford UP, Oxford 1966,pp. 351-361.

Hardy G. H.(1916), *Weierstrass non-differentiable functions*, Trans. Amer. Math. Soc. **17**,pp. 301–325.

Hardy G. H.(1949), *Divergent series*, (Oxford UP. Oxford).

Hardy G. H. and Littlewood J. E.(1914). *Some problems of Diophantine approximations, I. The fractional part of* $n^k\theta$, Acta Math.**37**,pp. 155-191=Collected Papers of G. H. Hardy Vol. I, Oxford University Press, Oxford, 28-63 (the original page 191 containing the contents of the paper is omitted in the Collected Papers).

Hardy G. H. and Littlewood J. E.(1914), *Some problems of Diophantine approximation, The trigonometrical series associated with the elliptic θ-function*, Acta Math.**37** (1914),pp. 193-239=Collected Papers of G. H. Hardy Vol. I, Oxford University Press, Oxford 1966, 67-112 (page 239 in the original paper which contains contents of the paper is omitted and as a result, the page numbers are stated in the Collected papers incorrectly as 193-239).

Hardy G. H. and Littlewood J. E.(1923), *Some problems of Diophantine approximation*, Trans. Cambridge Philos. Soc. **27** pp. 519-534=Collected Papers of G. H. Hardy Vol. I, Oxford UP, Oxford 1966, 212-226 (page 534 in the original paper which contains contents of the paper is omitted and as a result, the page numbers are stated in the Collected papers incorrectly as 519-533).

Hardy G. H. and Littlewood J. E.(1936), *Note on the theory of series (XX):*

On Lambert series, Proc. London Math. Soc (2) **41**,pp. 257-270; Collected Papers of G. H. Hardy IV, (Oxford UP, Oxford).

Hardy G. H. and Riesz M.(1915), *The general theory of Dirichlet's series*, (CUP. Cambridge; reprint, Hafner, New York 1972.)

Hardy G. H.(1932) and Wright E. M.(1932),*Introduction to the theory of numbers*, (Oxford UP, Oxford).

Hartman P. and Wintner A.(1938), *On certain Fourier series involving sums of divisors*, Trudy Tbillis. Mat. Obsc. im. G. A. Razmadze **3**,pp. 113-119.

Hashimoto M., Kanemitsu S.and Li H. L.(2009). *Examples of the Hurwitz transform*, J. Math. Soc. Japan **61**,pp. 651-660.

Hasse H. and Suetuna Z.(1931), *Ein allgemeines Teilerproblem der idealtheorie*, J. Fac. Sci. Imp. Univ. Tokyo **2**,pp.133-154; Collected papers of Zyoiti Suetuna, Vol. III, Nansosha, 1989, 231-252.

Hecke E.(1970). Math. Werke, Second Ed. (Göttingen: Vanderhoeck u. Ruprecht 1970, pp. 411 ff. and pp. 469 ff.)

Heilbronn H.(1967), *Zeta-functions and L-functions*, in Algebraic Number theory ed. (Cassels and Frölich Academic Press, New York 1967,pp. 204-230).

Hejhal D. A. Friedman J., Gutzwiller M. C. and OdlyzkoA. M.(1999). Emerging applications of number theory, Springer Verl., New York, Berlin etc.

Helton J. W.(1980), *The distance of a function to H^∞ in the Poincaré metric; elctrical power transfer*, J. Funct. Anal. **38**,pp. 273-314.

Helton J. W.(1982). *Non-Euclidean funcitional analysis and electronics*, Bull. Amer. Math. Soc. **7** pp.1-64.

Helton J. W. and Merino O.(1998), *Classical control using H^∞ methods, Theory, optimization and design*, (SIAM, Philadelphia).

Hinz J. G.(1980). *Eine Erweiterung des nullstellenfreie Bereiches der Heckeschen Zetafunktionen und Primdeale in Idealklassen*, Acta Arith. **38**,pp. 209-254.

Hiramatsu T. and Takada I.(1985), *Dedekind sums and automorphic forms*, RIMS Kokyuroku **572**,pp. 151-175.

Holschneider M. and Tchamichian P.(1991), *Pointwise analysis of Riemann's nondifferentiable function*, Invent. Math. **105**,pp. 157–275.

Hurwitz A.(1881), *Grundlagen eigener independenten Theorie der elliptischen Modulfunktionen und Theorie der Multiplikator-Gleichungen erster Stufe*, Math. Ann. **18**, pp. 528-592 (=Math. Werke Vol. 1, 1932, pp. 1-66).

Hurwitz A(1904). *Über die Theorie der elliptischen Modulfunktionen*, Math. Ann. **58**,pp. 343-360 (=Math. Werke Vol. 1. 1932).

Indlekofer K. H. (1991),*A survey of Turan's equivalent power series*, Number Theory, Analysis, and Combinatorics: Proceedings of the Paul Turan Memorial Conference held August 22-26, 2011 in Budapest Ed. by Pintz, János / Biró, András / Gy?ry, Kálmán / Harcos, Gergely / Simonovits, Miklós / Szabados, József, **7**,pp. 127-144.

Ingham A. E.(1964), *The distribution of prime numbers*. Cambridge Tracts Math. Math. Phys., No. **30** (Stechert-Hafner, Inc., New York.)

Iseki K.(1952), *A proof of a transformation formula in the theory of partitions*, J. Math. Soc. Japan **4**, pp. 14-26.

Iseki S.(1957). *The transformation formula for the Dedekind modular function and related functional equation,* Duke Math. J. **24**, pp. 653-662.

Iseki S.(1961). *A proof of a functional equation related to the theory of partitions,* Proc. Amer. Math. Soc. **12**, pp. 502-505.

Itatsu S.(1981), *Differentiability of Riemann's function,* Proc. Japan Acad. Ser. A Math. Sci. **57**,pp. 492–495.

Ivić A.(1985), *The Riemann zeta-function,* (Wiley, New York)

Iwaniec H.(1995), *Introduction to the Spectral Theory of Automorphic Forms,* Biblioteca de la Revista Matemática Iberoamericana, Madrid, 1995. Second edition: *Spectral Methods of Automorphic Forms,* Amer. Math. Soc. and Revista Matemática Iberoamericana, Providence, 2002.

Iwaniec H. and Kowalski E.(2004), *Analytic Number Theory,* Amer. Math. Soc. Colloquium Publ. **53**, Amer. Math. Soc., Providence.

Iwasawa K. (1972), *Lectures on p-adic L-functions,* Ann. Math. Studies 74, Princeton UP, Princeton.

Jacobi C. G. J. (1829), *Fundamenta Nova,* Königsberg, 1829, in Latin. Reprinted with corrections in: C. Jacobi. *Ges. Werke.* **8** volumes, Berlin. 1881-1891. Vol. 1. pp. 49-239. Reprinted New York (Chelsea, 1969).

Jacquet H. (1972), *Automorphic form on GL(2). part II,* Lect. Notes Math. **278**, (Springer, Berlin).

Jaffard S.(1996), *The spectrum of sigularities of Riemann's function,* Rev. Mat. Iberoamericana **12**, pp. 441–460.

Joris H. (1977), *On the evaluation of Gaussian sums for non-primitive Dirichlet characters,* Enseign. Math. (2) **23**,pp. 13-18.

Joyner D.(1986) *Distribution Theorems of L-functions,* Longman, Essex pp. 15-23.

Jutila M. and Motohashi Y.(2006), *Uniform bounds for Rankin-Selberg L-Functions,* in: Multiple Dirichlet Series, Automorphic Forms, and Analytic Number Theory, edited by Friedberg, Bump, Goldfeld, and Hoffstein, *Proc. Symp. Pure Math.,* volume 75, 2006, pp. 243-256.

Kahane J. P.(1964), *Lacunary Taylor and Fourier series,* Trans. Amer. Math. Soc. **79** pp. 199–213.

Kanemitsu S.(1981), *On the Riesz sums of some arithmetical functions,* in *p*-adic *L*-functions and algebraic number thery Surikaiseki Kenkyusho Kokyuroku, **411**, pp. 109-120.

Kanemitsu S.(2007), *Class field theory through examples,* A mini course at Shandong University, March and August 2007 (unpublished).

Kanemitsu S., Kumagai H., and Shirasaka S.(2010), *Ciphers from historical point of view,* Mem. Kagoshima Inst. Tech.pp. 51-56.

Kanemitsu S., Kumagai H., and Shirasaka S.(2011), *Ciphers IIfrom mathematical point of view,* Mem. Kagoshima Inst. Tech. pp. 39-43.

Kanemitsu S. and Kuzumaki T.(2009),*Transformation formulas for Lambert series,* Siaulai Math. Sem., **4**, pp. 105-123.

Kanemitsu S., Kuzumaki T. and Tanigawa Y., *On the Wintner-Ingham-Segal summability,* to appear.

Kanemitsu S., Laurinčikas A. and Ma J.(2010), *Square-free integers as ideal norms,* Acta Math. Sinica, English Ser. **26**,pp. 621-628.

Kanemitsu S., Tanigawa Y. and Yoshimoto M.(2003), *Ramanujan's formula and modular forms,* Proc. of the second China-Japan Sem., Kluwer, Dordrecht etc., pp. 159-212.

Kanemitsu S. and Tsukada H.(2007), *Vistas of special functions,* (World Scientific, Singapore-London-New York).

Kanemitsu S. and Tsukada H.(2015), *Contributons to the theory of zeta-functions: The modular relation supremacy,* World Scientific, Singapore etc.

Kanemitsu S. and Yoshimoto M. (2000), *Euler products, Farey series and the Riemann hypothesis,* Publ. Math. (Debrecen) **59**.pp. 431-449.

Katok S.(1972), *Fuchsian groups,* Lect. Notes Math. **278**, (Springer, Berlin).

Kim H. and Sarnak P. (2003), Appendix 2: *Refined estimates towards the Ramanujan and Selberg conjectures,* J. Amer. Math. Soc. **16**, pp. 175-181.

Kimura H.(1984), *Robust stabilizability for a class of transfer functions,* IEEE Trans. Automatic Control **27**, pp. 788-793.

Kimura H.(1997), *Chain scattering approach to H^∞-control,* Birkhäuser, Boston-Basel-Berlin.

Kitaoka Y., *Introduction to algebraic number theory,* to appear.

Knopp K.(1913), *Über Lambertsche Reihen,* J. Reine Angew. Math. **142**, pp. 283-315.

Knopp K. (1921), *Theory and applications of infinite series,* Blackie and Sons, London 1951 (first German edition published in 1921).

Knopp M. I. (1970), *Modular functions in analytic number theory,* Markahm, Chicago.

Knopmacher J.(1973), *A prime-divisor counting function,* Proc. Amer. Math. Soc. **40**,pp. 373-377.

Knuth D.(1997), *The art of computer programming,* 3rd ed. (Addison-Wesley, Reading).

Koecher M.(1953), *Ein newer Beweis der Kroneckerscher Grenzformel,* Arch. Math. **4**, pp. 310-321.

Kohnen W. (2005), *A very simple proof for the q-product expansion of the Δ-function,* Ramanujan J. **10** , pp. 71-73.

Koshlyakov N. S. (1954), *Investigation of some questions of analytic theory of the rational and quadratic fields, I (Russian),* Izv. Akad. Nauk SSSR, Ser. Mat. **18**, pp. 113–144, Errata: ibid. **19** (1955), 271 (in Russian).

Kotre J.(1995). *White gloves—How we create ourselves from our memory,* (The Free Press, New York etc.)

Kubert D. S. and Lang S.(1981), *Modular units,* Springer Verl. (Berlin-Heidelberg etc.)

Kubota T. (1973), *Elementary theory of Eisenstein series,* (Kodansha, Tokyo).

Kuroš Kuroš A. G. (1974), *The theory of groups,* Vol. II, (AMS-Chelsea, Providence, R.I.).

Kuznetsov N. V. (1980), *Petersson's conjecture for cusp forms of weight zero and Linnik's conjecture. Sums of Kloosterman sums,* Mat. Sb. (N.S.), **111 (153)** (1980),pp. 334-383, 479 (in Russian). English translation: Math. USSR Sbornik, **39** (1981),pp. 299-342.

Lagarias J. C. and Odlyzko A. M.(1997), *Effective version of the Chebotarev density theorem*, Algebraic Number Fields (Frölich, ed.), Academic Press, New York pp. 409-464.

Lambert J. H. (1771), *Anlage zur Archtektonik oder Theorie des Einfachen und Ersten in der philosophischen und mathematischen Erkenntnis*, **2** Bände, Riga, Vol. **2**, §875 (p.507).

Landau E. (1915), *Über die Anzahl der Gitterpunkte in gewissen Bereichen* (Zweite Mit.), *Nachr. Ges. Wiss. Göttingen, Math.-Phys. Kl.* pp. 209-243=Collected Works Vol. 6, Thales Verl., Essen 1985, 308-342.

Landau E. (1918). *Einführung in die elementare und analytische Theorie der algebraischen Zahlen und der Ideale*, Teubner, Leipzig 1918, (reprint, Chelsea, New York 1949).

Lang S.(1970), *Algebraic number theory*, (Addisen-Wesley, Reading Msaa.)

Lang S.(1973), *Elliptic functions*, Addison -Wesley.

Lau Y. K.(2002), *Summatory formula of the convolution of two arithmetical functions*, Monatsh. Math. **136** pp. 35-45.

Lau Y. K., Liu J. Y., and Ye Y. B.(2006). *Subconvexity bounds for Rankin-Selberg L-functions for congruence subgroups*, J. Number Theory, **121**, pp. 204-223.

Lau Y. K., Liu J. Y., and Ye Y. B.(2006). *A new bound $k^{2/3+\varepsilon}$ for Rankin-Selberg L-functions for Hecke congruence subgroups*, Intern. Math. Res. Papers, **7** , Article ID 35090,pp. 1-78.

Lehmer D. H. (1975), *Euler constants for arithmetic progressions*, Acta Arirh. **27**,pp. 125-142;(Collected Papers of Lehmer D. H., Charles Babbage.)

Lewittes J. (1972), Analytic continuation of Eisenstein series, Trans. Amer. Math. Soc. **171**,pp. 469-490.

Li F. H, Galhotra S. and Kanemitsu S.(2015), *Emerging importance of EVs in the Green Grid Era* Pure Appl. Math. J. Special issue: Mathematical aspects of engineering disciplines, Volume **4**, Issue 5-1, pp., 38-45 DOI: 10.11648/j.pamj.s.2015040501.18.

Li F. H., Wang N. L. and Kanemitsu S.(2012), *Number theory and its applications*, World Sci., London-Singapore-New Jersey, Nov. 28, 2012 (abbreviated as **I**).

Li F. H., *Evs. in Green Grid Era and climate change—Many a little makes a miracle*, in preparation.

Li H. L. Ma J. and Zhang W. P.(2010). *On some Diophantine Fourier series*, Acta Math. Sinica, English series, **26**, no. **6**, pp. 1125-1132.

Li H. L. and Yang Q. L.(2015). *On the Riesz sums in number theory*, Pure Appl. Math. J. Volume **4**, Issue 5-1, Special issue: Mathematical Aspects of Engineering Disciplines pp. 15-19; DOI: 10.11648/j.pamj.s.2015040501.13.

Liu J. Y., *Intorduction to Maass forms*, to appear.

Luther W. (1986), *The differentiability of Fourier gap series and Riemanns Example of a continuous, nondifferentiable function*, J. Approx. Theory **48** pp. 303321.

Ma J. and Agarwal P., *On the Kuznetsov sum formula*, in prepartion.

Maass H. (1949), *Über eine neue Art von nichtanalytischen automorphen Funk-*

tionen und die Bestimmung Dirichletscher Reihen durch Funktionalgle-ichungen, Math. Ann. **121**,pp. 141-183.

MacFarlane D. and Glover K.(1996), *A loop shaping design procedure using H control,* IEEE Trans. Automatic Control **37**, pp. 491-499.

McCarthy P. J. (1986), *Introduction to arithmetical functions,* (Springer Verl, New York etc.)

Marshall D. E. and A. Stray A.(1992), *Interpolating Blaschke products,* Pacific J. Math. **173**, pp. 759-769.

Martinet J. (1977), *Character theory and Artin L-functions,* Algebraic number fields ed. by A. Frölich, Academic Press, London pp. 1-87.

Mathews G. B. (1986), *Theory of numbers,* (reprinted Chelsea Publ.)

Mercier A. (1981), *Sums of the form* $\sum \frac{g(n)}{f(n)}$, Canad. Math. Bull. **24**,pp. 299-307.

Meyer C.(1967), *Über die Dedekindsche Transformationsformal für* $\log \eta(\tau)$, Abh. Math. Sem. Univ. Hamburg **30**,pp. 129-164.

Meyer Y.(1994), *Le traitment du signal et l'analyze mathématique,* Ann. Inst. Fourier Grenoble **50** pp. 593–632.

Miatello R. and Wallach N. R.(1990), *Kuznetsov formulas for real rank one groups,* J. Funct. Anal. **93**,pp. 171-206.

Miatello R. and Wallach N. R.(1989), *Construction of automorphic forms by means of Whittaker vector,* J. Funct. Anal. **86**,pp. 411-467.

Mikolás M.(1956). *Mellinsche Transformation und Orthogonalität bei* $\zeta(s,u)$; *Verallgemeinerung der Riemannschen Funkutionalgleichung von* $\zeta(s)$, Acta Sci. Math. (Szeged) **17**,pp. 143–164.

Mikolás M. (1957), *On certain sums generating the Dedekind sums,* Pacific J. Math. **7**,pp. 1167-1178.

Mikolás M.(1957), *Über gewisse Lambertsche Reihen, I: Verallgemeinerung der Modulfunktionen* $\eta(\tau)$ *und ihrer Dedekindschen Transformationformel,* Math. Z. **68**,pp. 100-110.

Milnor J. (1983), *On polylogarithms, Hurwitz zeta-functions and the Kubert identities,* Enseign. Math. (2) **29**,pp. 281-322.

Moore P. (1070), *Generalized Eisenstein series: incorporation of a non-trivial representation of* Γ, PhD Dissertation, Univ. of Washington Seattle 1070.

Morita Y.(1977), *On the Hurwitz-Lerch L-function,* J. Fac. Sci. Univ. Tokyo Sect. IA Math. **24**, pp. 29-43.

Motohashi Y.(1997), *Spectral theory of the Riemann zeta-function,* (Cambridge UP, Cambridge etc.)

Murty R. and Pacelli A(2004), *Quadratic reciprocity via theta functions,* Proc. Int. Conf.–Number Theory, **No. 1**, pp. 107–116.

Nakaya H. (1991), *The generalized division problem and the Riemann hypothesis,* Nagoya Math. J. **122**,pp. 149-159.

Nakaya H. (1992), *On the generalized division problem in arithmetic progressions,* Sci. Rep. Kanazawa Univ. **37**,pp. 23-47.

Narkiewicz W. (2004), *Elementary and analytic theory of algebraic numbers,* (1st ed. PWN, Warszawa 1974) and Springer Verl. 2004.

Neukirch J.(1988), *The Beilinson conjecture for algebraic number fields,* in

"*Beilinson's conjectures on special values of L-functions*," ed. by M. Rapoport et al, (Academic Press, Boston etc. pp. 193-247).

Neuenschwander E. (1978), *Riemanns example of a continuous, nondifferentiable function*, Math. Intelligencer **1**, pp. Issue 1, 4044.

Newman M, Ryabec C and Shure B. N.(1979), *The use of integral operators in number theory*, J. Number Theory **32**, pp. 123-130.

Nicolau A. (1994), *Finite products of interpolating Blaschke products*, J. London Math. Soc. (2) **50**, pp. 520-531.

Nowak W. G.(1991). *On the average number of finite abelian groups of a given order*, Ann. Sci. Math. Québec **15**, pp. 193-202.

Nowak W. G.(1991), *Sums of reciprocals of general divisor functions and the Selberg division problem*, Abh. Math. Sem. Univ. Hamburg **61**, pp. 163-173.

Nowak W. G.(1992), *Divisor problems in special sets of positive integers*, Acta Math. Univ. Comenianae Hamburg **41**,pp. 101-115.

Ogg A. P.(1969). Modular forms and functions, (Benjamin, New York.)

Palka B. P. (2009), *Operator theory and function theory for the unit ball of* $\mathbb{C}^d$, NFS Grant Award Abstract #0901642.

Papoulis A.(1962), *The Fourier integral and its applications*, (McGraw-Hill.)

Petersson H.(1982), Modulfunktionen und quardratiche Formen, Springer Verl. 1982.

Postnikov A. G.(1988), *Introduction to analytic number theory*, Amer. Math. Soc., (Providence, R.I., Transl. from the Russian).

Prachar K. (1988), *Primzahlverteilung*, (Springer Verl, Berlin).

Proskurin N. V.(1979), *Sum formulas for Kloosterman sums*, Zap. Naučn. Sem. LOMI **82**,pp. 103-135.

Rademacher H. (1931), *Zur Theorie der Modulfunktionen*, J. Reine Angew. Math. **167** pp. 312-336; Collected Papers of H. Rademacher I, 1974, pp. 652-677.

Rademacher H. (1973), *Topics in analytic number theory*, (Springer, Berlin.)

Rademacher H. (1974), *Collected Papers of Hans Rademacher I, II*, (MIT Press, Cambridge Mass.)

Ramachandra K., Sankaranarayanan A. and Srinivas K.(1976), *Addendum to K. Ramachandras paper: Some problems of analytic number theory* Acta Arith. **31**(4), pp. 313-324.

Rieger J. G.(1965), *Zum Teilerproblem von Atle Selberg*, Math. Nachr. **30**, pp.181-192.

Riemann B.(1854), Über die Darstellbarkeit einer Function durch eine trigonometrische Reihe, Habilitationsschrift, **13** Bd. Abh. Köonig. Ges. Wiss. Göttingen 1854; in [Riemann (1953)],pp. 227-265.

Riemann B.(1859), *Über die Anzahl der Primzahlen unter einer gegebenen Grösse*, Monatsber. Berlin. Akad. pp. 671–680.

Riemann B.(1953), *Fragmente über Grenzfälle der ellipitischen Modulfunctionen*, in [Riemann (1953)],pp. 455-465.

Riemann B.(1953), *Collected Works of Bernhard Riemann*, ed. by H. Weber, 2nd ed. (Dover, New York).

Romanoff N. P.(1951). *Hilbert spaces and number theory I, II*, Izv. Akad. Nauk

SSSR, Ser. Mat. **10** (1945),pp. 3–24 **15**, (1951),pp. 131–152.

Rosenfel'd A. S. and Yahinson V. I.(1966), *Transitional proceses and generalized functions,* (Nauka, Moscow).

Roy A. Zaharescu A. and Zaki M.(2016), *Some identities involving convolutions of Dirichlet characters and the Möbius function,* Proc. Math. Sci. **126**, pp. 21–33.

Rudin W.(2008), *Function theory in the unit ball of* $\mathbb{C}^n$, (Springer Verl., Berlin-Heidelberg).

Rudnick Z. and Sarnak P.(1994), *The behavior of eigenstates of arithmetic hyperbolic manifolds,* Comm. Math. Phys. **161**,pp.195-213.

Sakata T., Takada Y. and Hyakuake H.(1992), *Basic statistics,* Asakura-shoten, Tokyo.

Sarnak P.(1995), *Arithmetic quantum chaos, the Schur Lectures (Tel Aviv 1992),* Israel Math. Conf. vol.8, Bar Ilan,pp.183-236.

Sarnak P.(2001), *Estimates for Rankin-Selberg L-functions and quantum unique ergodicity,* J. Functional Analysis, **184**,pp. 419-453.

Schoenberg B.(1968). *Über das unendliche Produkt* $\prod_{k=1}^{\infty} (1 - x^k)$, Mitt. Math. Ges. Hamburg **9**, pp.4-11.

Schoenberg B.(1974), *Elliptic modular functions, Springer Verl.,* (Berlin-Heidelberg.)

Schoenfeld L.(1944). *A transformation formula in the theory of partitions.* Duke Math. J. **11**, pp. 873-887.

Scourfield E.(1991), *On some sums involving the largest prime divisions of* n, Acta Arith. **59**, pp. 339-363.

Sczech R.(1978), *Ein einfacher Beweis der Transformationsformel* für $\log \eta(\tau)$, Math. Ann. **237**,pp. 161-166.

Sears D. B. and Titchmarsh E. C.(1954), *Some eigenfunction formulae,* Proc. London Math. Soc. (2) **1**, pp. 165-175.

Segal S. L.(1975), *Ingham's summability and Riemann's hypothesis,* Proc. London Math. Soc. (3) **30**,pp. 5139-142.

Segal S. L.(1978), *Riemann's example of a continuous "non-differentiable" function continued,* The Math. Intelligencer, **1**, Issue **2**, pp. 81–82.

Selberg A.(1954), *Note on a paper by L. G. Sathe,* J. Indian Math. Soc. **18**,pp. 83-87; (Collected Papers, Vol. I. Springer Verl. 1989,pp. 418-422.)

Selberg A.(1956), *Harmonic analysis and discontinuous groups in weakly symmetric Riemannian spaces with applications to Dirichlet series,* J. Indian Math. Soc. **20**,pp. 47-87.

Selberg A.(1965), *On the estimation of Fourier coefficients of modular forms,* Proc. Symp. Pure Math. AMS (1965), v. **8**,pp. 1-15.

Selberg A.(1989), *Old and new conjectures and results about a class of Dirichlet series,* Proc. Amalfi Conf., Analytic Number Theory, Maiori 1989, E. Bombieri et al (Eds), Università di Salerno (1992),pp. 367-385; Coll. Papers of Atle Selberg, Vol. I, Springer, Berlin-Heidelberg-New York, 1989,pp. 47-63.

Seneta E.(1976), *Regularly varying functions,* Lect. Notes Math. **508**, (Springer Verl., Berlin-Heidelberg-New York).

Serre J. P.(1973). *A course in arithmetic*, Springer Verlag.

Shoenheimer R.(1942), *The dynamic state of body constituents*, Harvard Univ. Press. Massachusetts, 1942.

Siegel C. L.(1966), *Symplectic geometry*, Academic Press, New York 1964=*Amer. J. Math.* **65** (1943), 1-86; *Ges. Abh.* II, Springer, Berlin-Heidelberg-New York. pp.274-359.

Siegel C. L.(1965). *Lectures on advanced analytic number theory*, Tata Inst. Bombay.

Siegel C. L.(1954). *A simple proof of* $\eta(-1/\tau) = \eta(\tau)(\tau/i)^{1/2}$, Mathematika **1**. p. 4.

Sivaramakrishnan R.(1988). Classical thery of arithmetic functions, Dekker, New York.

Smith R. A.(1981). *A note on Dirichlet's theorem*, Canad. Math. Bull. **24**. pp.379-380.

Smith A.(1972). *The differentiability of Riemann's function,* Proc. Amer. Math. Soc. **34**. pp.463-468; Correction to "The differentiability of Riemann's function", ibid. **89** (1983), 567-568.

Smith H. J. S.(1881). *On some discontinuous series considered by Riemann*, Mess. Math. (2) **11**. pp. 1-11; Collected Papers Vol. 2, 312-320, 1966.

Srivatava H. M. S. and Choi J. -S.(2001), *Series associated with the zeta and related functions*, Kluwer Academic Publishers, Dordrecht-Boston-London.

Suetuna Z.(1925), *On the product of L-functions*, Japan. J. Math. **2**. pp.19-37.

Suetuna Z.(1931), *Über die Anzahl der Idealtheiler*, J. Frac. Sci. Imp. Univ. Tokyo **2**. pp. 155-177; Collected Papers, Vol. III, Nansosha, (1989) pp. 253-275.

Suetuna Z.(1950), *Analytic theory of numbers*, Iwanami Shoten,(in Japanese).

Suetuna Z.(1950). *Analytic number theory*, Iwanami-Shoten, Tokyo.

Tennenbaum G.(1995), *Introduction to analytic and probabilistic number theory*, Cambridge Univ. Press, London 1995 (Translation from the 1990 French edition).

Terras A.(1985). *Harmonic analysis on symmetric spaces and applications I*, Springer 1985.

Titchmarsh E.C.(1986). *The Theory of the Riemann zeta-function*, 2nd ed., Clarendon Press, Oxford.

Titchmarsh E. C.(1938), *On a series of Lambert's type*, J. London Math. Soc. **12**. pp. 248-253.

Tull J. P.(1958). *The multiplication problem for Dirichlet series*, Proc. Amer. Math. Soc. **9** pp. 332-334.

Tull J. P.(1959). *Dirichlet multiplication for lattice point problems,* Duke Math. J. **26** pp. 73-80.

Tull J. P.(1959). *Dirichlet multiplication for lattice point problems II*, Pacific J. Math. **9**. pp. 609-615.

Tull J. P.(1961). *Average order of arithmetic functions*, Illinois J. Math. **5**. pp. 175-181.

Ulrich P.(1997). *Anmerkungen zum "Riemannschen Beispiel" einer stetigen, nicht differenzierbaren Funktion*, Res. Math. **31**. pp. 245-265.

Venkov A. B.(1981), *Spectral Theory of automorphic functions*, Trudy Mat. Inst.

Steklov **153**. pp. 3-171; English transi. in Proc. Steklov Inst. Math. (1982), issue 4, 1-163.

Venkov A. B.(1990). *Spectral Theory of automorphic function and its applications*, Kluwer Academic Publishers, Dordrecht/Boston.London.

Vidyasagar M. and Kimura 'H.(1983). *Robust controllers for uncertain linear multivariable systems*, Automatica, **22**. pp. 585-601.

Walfisz A. Z.(1923). *Über die summatorischen Funktionen einiger Dirichletscher Reihen*, Inaugural Diss. Göttingnen.

Walfisz A.(1963). *Weylsche Exponentialsummen in der neueren Zahlentheorie*, VEB Deutscher Verl. Wiss., Berlin.

Walker P. L.(1978). *On the functional equation sdtisfied by modular functions*, Mathematika **25**. pp. 185-190.

Walum H.(1991). *Multiplication formulae for periodic functions*. Pacific J. Math. **149**, no. 2. pp. 383-396.

Wang N. L.(2011), *On Riemann's posthumous fragment II on the limit values of the elliptic modular functions*, Ramanujan J., **24**. pp. 129-145.

Wang N., *Arithmetical Fourier series and the limit values of elliptic modular functions*, Proc. indian Acad. Sci. Ser. A, to appear.

Washington L.(1982). *Introduction to cyclotomic fields*, Springer Verl. Wiss., New York.

Watson T.(2001), *Central value of Rankin triple L-function for unramified Maass cusp forms*, thesis, Princeton.

Weaver H. J.(1983), *Applications of discrete and continuous Fourier transforms*, Wiley, New York etc.

Weil A.(1968). *Sur une formule classique*, J. Math. Soc. Japar **20** (1968), pp. 400-402 (= Coll. Papers. Vol. III, Springer Verl. 1980,

Weil A.(1976), *Elliptic functions according to Eisenstein and Kronecker*, Springer Verl., Heidelberg etc..

Weil A.(1976). *Basic number theory*, Springer Verl., Heidelberg etc..

Widder D. W.(1929), *A generalization of Dirichlet's series and of Laplace's integrals by means of a Stieltjes integral*, Trans. Amer. Math. Soc. **31**. pp. 694-743.

Widder D. W.(1989). *Advanced calculus*, 2nd ed. Dover Publ., New York.

Wigert S.(1918). *Sur la série de Lambert et son application à la théorie des nombres*, Acta Math. **41**. pp. 197-218.

Wintner A.(1937). *On a trigonometrical series of Riemann*, Amer. J. Math. **59**. pp. 629-634.

Wintner A.(1944). *On Riemann's fragment concerning elliptic modular functions*, Amer. J. Math. **63** pp. 628-634.

Wintner A.(1944). *Diophantine approximation and Hilbert space*, Amer. J. Math. **66**. pp. 564-578.

Xin X. and Kimura H.(1994), *Singular (J, J')-lossless factorization for strictly proper functions,* Intern. J. Control, **59**, No. 6, pp., 1383-1400.

Yifan Y.(2004), *Transformation formulas for generalized Dedekind eta fucntion*, Bull. London Math. Soc. **36** pp. 671-682.

Yao A. C.(1992), *Theory and applications of the trapdoor functions*, 23rd Annual

Sympos. on Found. Comput. Sci. IEEE, New York pp. 80-91.

Yangbo Y.(2001), *Automorphic forms and trace formulae*, Peking University Press.

Yoshida K.(1981), *Modern analysis*, Kyoritsu-shuppan, Tokyo.

Zagier D.(1981), *Eisenstein series and the Selberg trace formula I*, in: Automorphic forms, representation theory and arithmetic, Tata Inst. of Fundamental Research Studies in Math. No. **10**, Springer, pp. 303-355.

Zames G. and Francis B. A.(1992), *Feedback minimax sensitivity, and optimal robustness* IEEE Trans. Automatic Control, **28**. pp., 759-769.

Zhai W. G.(2005), *Square-free integers as sums of two squares, Number Theory: Tradition and modernization*, Proc. of the 3rd Chinese-Japan Seminar, ed. by W.-P. Zhang and Y. Tanigawa, Springer. New York etc. pp. 223–231.

Index

Printed in the United States
By Bookmasters